How To Prove It

HOW TO PROVE IT

A Structured Approach
Second Edition

Daniel J. Velleman
Department of Mathematics and Computer Science
Amherst College

CAMBRIDGE UNIVERSITY PRESS
Cambridge, New York, Melbourne, Madrid, Cape Town, Singapore, São Paulo, Delhi

Cambridge University Press
32 Avenue of the Americas, New York, NY 10013-2473, USA

www.cambridge.org
Information on this title: www.cambridge.org/9780521675994

First edition first published 1994
Reprinted 1995 (twice), 1996 (twice), 1998, 2006
Second edition first published 2006
Reprinted 2006, 2008 (twice), 2009

Printed in the United States of America

A catalog record for this publication is available from the British Library.

Library of Congress Cataloging in Publication Data

Velleman, Daniel J.
How to prove it : a structured approach / Daniel J. Velleman. – 2nd ed.
p. cm.
Includes bibliographical references and index.
ISBN-13: 978-0-521-86124-3 (hardback)
ISBN-10: 0-521-86124-1 (hardback)
ISBN-13: 978-0-521-67599-4 (pbk.)
ISBN-10: 0-521-67599-5 (pbk.)
1. Logic, Symbolic and mathematical. 2. Mathematics. I. Title.
QA9.V38 2006
511.3 – dc22 2005029447

ISBN 978-0-521-86124-3 hardback
ISBN 978-0-521-67599-4 paperback

To Shelley

Contents

Preface to the Second Edition

I would like to thank all of those who have sent me comments about the first edition. Those comments have resulted in a number of small changes throughout the text. However, the biggest difference between the first edition and the second is the addition of more than 200 new exercises. There is also an appendix containing solutions to selected exercises. Exercises for which solutions are supplied are marked with an asterisk. In most cases, the solution supplied is a complete solution; in some cases, it is a sketch of a solution, or a hint.

Some exercises in Chapters 3 and 4 are also marked with the symbol ℔. This indicates that these exercises can be solved using Proof Designer. Proof Designer is computer software that helps the user write outlines of proofs in elementary set theory, using the methods discussed in this book. Further information about Proof Designer can be found in an appendix, and at the Proof Designer Web site: `http://www.cs.amherst.edu/~djv/pd/pd.html`.

Preface

Students of mathematics and computer science often have trouble the first time they're asked to work seriously with mathematical proofs, because they don't know the "rules of the game." What is expected of you if you are asked to prove something? What distinguishes a correct proof from an incorrect one? This book is intended to help students learn the answers to these questions by spelling out the underlying principles involved in the construction of proofs.

Many students get their first exposure to mathematical proofs in a high school course on geometry. Unfortunately, students in high school geometry are usually taught to think of a proof as a numbered list of statements and reasons, a view of proofs that is too restrictive to be very useful. There is a parallel with computer science here that can be instructive. Early programming languages encouraged a similar restrictive view of computer programs as numbered lists of instructions. Now computer scientists have moved away from such languages and teach programming by using languages that encourage an approach called "structured programming." The discussion of proofs in this book is inspired by the belief that many of the considerations that have led computer scientists to embrace the structured approach to programming apply to proof-writing as well. You might say that this book teaches "structured proving."

In structured programming, a computer program is constructed, not by listing instructions one after another, but by combining certain basic structures such as the if-else construct and do-while loop of the Java programming language. These structures are combined, not only by listing them one after another, but also by *nesting* one within another. For example, a program constructed by nesting an if-else construct within a do-while loop would look like this:

```
do
   if [condition]
      [List of instructions goes here.]
   else
      [Alternate list of instructions goes here.]
while [condition]
```

The indenting in this program outline is not absolutely necessary, but it is a convenient method often used in computer science to display the underlying structure of a program.

Mathematical proofs are also constructed by combining certain basic proof structures. For example, a proof of a statement of the form "if P then Q" often uses what might be called the "suppose-until" structure: We *suppose* that P is true *until* we are able to reach the conclusion that Q is true, at which point we retract this supposition and conclude that the statement "if P then Q" is true. Another example is the "for arbitrary x prove" structure: To prove a statement of the form "for all x, $P(x)$," we *declare x to be an arbitrary object* and then *prove P(x)*. Once we reach the conclusion that $P(x)$ is true we retract the declaration of x as arbitrary and conclude that the statement "for all x, $P(x)$" is true. Furthermore, to prove more complex statements these structures are often combined, not only by listing one after another, but also by nesting one within another. For example, to prove a statement of the form "for all x, if $P(x)$ then $Q(x)$" we would probably nest a "suppose-until" structure within a "for arbitrary x prove" structure, getting a proof of this form:

Let x be arbitrary.
    Suppose $P(x)$ is true.
        [Proof of $Q(x)$ goes here.]
    Thus, if $P(x)$ then $Q(x)$.
Thus, for all x, if $P(x)$ then $Q(x)$.

As before, we have used indenting to make the underlying structure of the proof clear.

Of course, mathematicians don't ordinarily write their proofs in this indented form. Our aim in this book is to teach students to write proofs in ordinary English paragraphs, just as mathematicians do, and not in the indented form. Nevertheless, our approach is based on the belief that if students are to succeed at writing such proofs, they must understand the underlying structure that proofs have. They must learn, for example, that sentences like "Let x be arbitrary" and "Suppose P" are not isolated steps in proofs, but are used to introduce the "for arbitrary x prove" and "suppose-until" proof structures. It is not uncommon for beginning students to use these sentences inappropriately in other ways.

Such mistakes are analogous to the programming error of using a "do" with no matching "while."

Note that in our examples, the choice of proof structure is guided by the logical form of the statement being proven. For this reason, the book begins with elementary logic to familiarize students with the various forms that mathematical statements take. Chapter 1 discusses logical connectives, and quantifiers are introduced in Chapter 2. These chapters also present the basics of set theory, because it is an important subject that is used in the rest of the book (and throughout mathematics), and also because it serves to illustrate many of the points of logic discussed in these chapters.

Chapter 3 covers structured proving techniques in a systematic way, running through the various forms that mathematical statements can take and discussing the proof structures appropriate for each form. The examples of proofs in this chapter are for the most part chosen, not for their mathematical content, but for the proof structures they illustrate. This is especially true early in the chapter, when only a few proof techniques have been discussed, and as a result many of the proofs in this part of the chapter are rather trivial. As the chapter progresses the proofs get more sophisticated and more interesting, mathematically.

Chapters 4 and 5, on relations and functions, serve two purposes. First, they provide subject matter on which students can practice the proof-writing techniques from Chapter 3. And second, they introduce students to some fundamental concepts used in all branches of mathematics.

Chapter 6 is devoted to a method of proof that is very important in both mathematics and computer science: mathematical induction. The presentation builds on the techniques from Chapter 3, which students should have mastered by this point in the book.

Finally, in Chapter 7 many ideas from throughout the rest of the book are brought together to prove some of the most difficult and most interesting theorems in the book.

I would like to thank all those who read earlier drafts of the manuscript and made many helpful suggestions for improvements, in particular Lauren Cowles at Cambridge University Press, my colleague Professor Duane Bailey and his Discrete Mathematics class, who tried out earlier versions of some chapters, and finally my wife, Shelley, without whose constant encouragement this book would never have been written.

Introduction

What is mathematics? High school mathematics is concerned mostly with solving equations and computing answers to numerical questions. College mathematics deals with a wider variety of questions, involving not only numbers, but also sets, functions, and other mathematical objects. What ties them together is the use of *deductive reasoning* to find the answers to questions. When you solve an equation for x you are using the information given by the equation to *deduce* what the value of x must be. Similarly, when mathematicians solve other kinds of mathematical problems, they always justify their conclusions with deductive reasoning.

Deductive reasoning in mathematics is usually presented in the form of a *proof*. One of the main purposes of this book is to help you develop your mathematical reasoning ability in general, and in particular your ability to read and write proofs. In later chapters we'll study how proofs are constructed in detail, but first let's take a look at a few examples of proofs.

Don't worry if you have trouble understanding these proofs. They're just intended to give you a taste of what mathematical proofs are like. In some cases you may be able to follow many of the steps of the proof, but you may be puzzled about why the steps are combined in the way they are, or how anyone could have thought of the proof. If so, we ask you to be patient. Many of these questions will be answered later in this book, particularly in Chapter 3.

All of our examples of proofs in this introduction will involve prime numbers. Recall that an integer larger than 1 is said to be *prime* if it cannot be written as a product of two smaller positive integers. For example, 6 is not a prime number, since $6 = 2 \cdot 3$, but 7 is a prime number.

Before we can give an example of a proof involving prime numbers, we need to find something to prove – some fact about prime numbers whose correctness can be verified with a proof. Sometimes you can find interesting patterns in mathematics just by trying out a calculation on a few numbers. For example, consider the table in Figure 1. For each integer n from 2 to 10, the table shows whether or not both n and $2^n - 1$ are prime, and a surprising pattern emerges. It appears that $2^n - 1$ is prime in precisely those cases in which n is prime!

n	Is n prime?	$2^n - 1$	Is $2^n - 1$ prime?
2	yes	3	yes
3	yes	7	yes
4	no: $4 = 2 \cdot 2$	15	no: $15 = 3 \cdot 5$
5	yes	31	yes
6	no: $6 = 2 \cdot 3$	63	no: $63 = 7 \cdot 9$
7	yes	127	yes
8	no: $8 = 2 \cdot 4$	255	no: $255 = 15 \cdot 17$
9	no: $9 = 3 \cdot 3$	511	no: $511 = 7 \cdot 73$
10	no: $10 = 2 \cdot 5$	1023	no: $1023 = 31 \cdot 33$

Figure 1

Will this pattern continue? It is tempting to guess that it will, but this is only a guess. Mathematicians call such guesses *conjectures*. Thus, we have the following two conjectures:

Conjecture 1. *Suppose n is an integer larger than 1 and n is prime. Then $2^n - 1$ is prime.*

Conjecture 2. *Suppose n is an integer larger than 1 and n is not prime. Then $2^n - 1$ is not prime.*

Unfortunately, if we continue the table in Figure 1, we immediately find that Conjecture 1 is incorrect. It is easy to check that 11 is prime, but $2^{11} - 1 = 2047 = 23 \cdot 89$, so $2^{11} - 1$ is not prime. Thus, 11 is a *counterexample* to Conjecture 1. The existence of even one counterexample establishes that the conjecture is incorrect, but it is interesting to note that in this case there are many counterexamples. If we continue checking numbers up to 30, we find two more counterexamples to Conjecture 1: Both 23 and 29 are prime, but $2^{23} - 1 = 8{,}388{,}607 = 47 \cdot 178{,}481$ and $2^{29} - 1 = 536{,}870{,}911 = 2,089 \cdot 256{,}999$. However, no number up to 30 is a counterexample to Conjecture 2.

Do you think that Conjecture 2 is correct? Having found counterexamples to Conjecture 1, we know that this conjecture is incorrect, but our failure to find a counterexample to Conjecture 2 does not show that it is correct. Perhaps there are counterexamples, but the smallest one is larger than 30. Continuing to check examples might uncover a counterexample, or, if it doesn't, it might increase our confidence in the conjecture. But we can never be sure that the conjecture is correct if we only check examples. No matter how many examples we check, there is always the possibility that the next one will be the first counterexample. The only way we can be sure that Conjecture 2 is correct is to *prove* it.

In fact, Conjecture 2 *is* correct. Here is a proof of the conjecture:

Proof of Conjecture 2. Since n is not prime, there are positive integers a and b such that $a < n$, $b < n$, and $n = ab$. Let $x = 2^b - 1$ and $y = 1 + 2^b + 2^{2b} + \cdots + 2^{(a-1)b}$. Then

$$\begin{aligned} xy &= (2^b - 1) \cdot (1 + 2^b + 2^{2b} + \cdots + 2^{(a-1)b}) \\ &= 2^b \cdot (1 + 2^b + 2^{2b} + \cdots + 2^{(a-1)b}) - (1 + 2^b + 2^{2b} + \cdots + 2^{(a-1)b}) \\ &= (2^b + 2^{2b} + 2^{3b} + \cdots + 2^{ab}) - (1 + 2^b + 2^{2b} + \cdots + 2^{(a-1)b}) \\ &= 2^{ab} - 1 \\ &= 2^n - 1. \end{aligned}$$

Since $b < n$, we can conclude that $x = 2^b - 1 < 2^n - 1$. Also, since $ab = n > a$, it follows that $b > 1$. Therefore, $x = 2^b - 1 > 2^1 - 1 = 1$, so $y < xy = 2^n - 1$. Thus, we have shown that $2^n - 1$ can be written as the product of two positive integers x and y, both of which are smaller than $2^n - 1$, so $2^n - 1$ is not prime. □

Now that the conjecture has been proven, we can call it a *theorem*. Don't worry if you find the proof somewhat mysterious. We'll return to it again at the end of Chapter 3 to analyze how it was constructed. For the moment, the most important point to understand is that if n is any integer larger than 1 that can be written as a product of two smaller positive integers a and b, then the proof gives a method (admittedly, a somewhat mysterious one) of writing $2^n - 1$ as a product of two smaller positive integers x and y. Thus, if n is not prime, then $2^n - 1$ must also not be prime. For example, suppose $n = 12$, so $2^n - 1 = 4095$. Since $12 = 3 \cdot 4$, we could take $a = 3$ and $b = 4$ in the proof. Then according to the formulas for x and y given in the proof, we would have $x = 2^b - 1 = 2^4 - 1 = 15$, and $y = 1 + 2^b + 2^{2b} + \cdots + 2^{(a-1)b} = 1 + 2^4 + 2^8 = 273$. And, just as the formulas in the proof predict, we have $xy = 15 \cdot 273 = 4095 = 2^n - 1$. Of course, there are other ways of factoring 12 into a product of two smaller integers, and these might lead to other ways of factoring 4095. For example, since $12 = 2 \cdot 6$, we could use the values $a = 2$ and $b = 6$. Try computing the corresponding values of x and y and make sure their product is 4095.

Although we already know that Conjecture 1 is incorrect, there are still interesting questions we can ask about it. If we continue checking prime numbers n to see if $2^n - 1$ is prime, will we continue to find counterexamples to the conjecture – examples for which $2^n - 1$ is not prime? Will we continue to find examples for which $2^n - 1$ is prime? If there were only finitely many prime numbers, then we might be able to investigate these questions by simply checking $2^n - 1$ for every prime number n. But in fact there are infinitely many prime numbers. Euclid (circa 350 B.C.) gave a proof of this fact in Book IX of his *Elements*. His proof is one of the most famous in all of mathematics:

Theorem 3. *There are infinitely many prime numbers.*

Proof. Suppose there are only finitely many prime numbers. Let $p_1, p_2, \ldots, p_n$ be a list of all prime numbers. Let $m = p_1 p_2 \cdots p_n + 1$. Note that m is not divisible by p_1, since dividing m by p_1 gives a quotient of $p_2 p_3 \cdots p_n$ and a remainder of 1. Similarly, m is not divisible by any of $p_2, p_3, \ldots, p_n$.

We now use the fact that every integer larger than 1 is either prime or can be written as a product of primes. (We'll see a proof of this fact in Chapter 6.) Clearly m is larger than 1, so m is either prime or a product of primes. Suppose first that m is prime. Note that m is larger than all of the numbers in the list $p_1, p_2, \ldots, p_n$, so we've found a prime number not in this list. But this contradicts our assumption that this was a list of *all* prime numbers.

Now suppose m is a product of primes. Let q be one of the primes in this product. Then m is divisible by q. But we've already seen that m is not divisible by any of the numbers in the list $p_1, p_2, \ldots, p_n$, so once again we have a contradiction with the assumption that this list included all prime numbers.

Since the assumption that there are finitely many prime numbers has led to a contradiction, there must be infinitely many prime numbers. □

Once again, you should not be concerned if some aspects of this proof seem mysterious. After you've read Chapter 3 you'll be better prepared to understand the proof in detail. We'll return to this proof then and analyze its structure.

We have seen that if n is not prime then $2^n - 1$ cannot be prime, but if n is prime then $2^n - 1$ can be either prime or not prime. Because there are infinitely many prime numbers, there are infinitely many numbers of the form $2^n - 1$ that, based on what we know so far, *might* be prime. But how many of them *are* prime?

Prime numbers of the form $2^n - 1$ are called *Mersenne primes*, after Father Marin Mersenne (1588–1647), a French monk and scholar who studied these numbers. Although many Mersenne primes have been found, it is still not known if there are infinitely many of them. Many of the largest known prime numbers are Mersenne primes. As of this writing (April 2005), the largest known prime number is the Mersenne prime $2^{25,964,951} - 1$, a number with 7,816,230 digits.

Mersenne primes are related to perfect numbers, the subject of another famous unsolved problem of mathematics. A positive integer n is said to be *perfect* if n is equal to the sum of all positive integers smaller than n that divide n. (For any two integers m and n, we say that m *divides* n if n is divisible by m; in other words, if there is an integer q such that $n = qm$.) For example, the only positive integers smaller than 6 that divide 6 are 1, 2, and 3, and $1 + 2 + 3 = 6$. Thus, 6 is a perfect number. The next smallest perfect number is 28. (You should check for yourself that 28 is perfect by finding all the positive integers smaller than 28 that divide 28 and adding them up.)

Euclid proved that if $2^n - 1$ is prime, then $2^{n-1}(2^n - 1)$ is perfect. Thus, every Mersenne prime gives rise to a perfect number. Furthermore, about 2000 years after Euclid's proof, the Swiss mathematician Leonhard Euler (1707–1783), the most prolific mathematician in history, proved that every even perfect number arises in this way. (For example, note that $6 = 2^1(2^2 - 1)$ and $28 = 2^2(2^3 - 1)$.) Because it is not known if there are infinitely many Mersenne primes, it is also not known if there are infinitely many even perfect numbers. It is also not known if there are any odd perfect numbers.

Although there are infinitely many prime numbers, the primes thin out as we look at larger and larger numbers. For example, there are 25 primes between 1 and 100, 16 primes between 1000 and 1100, and only six primes between 1,000,000 and 1,000,100. As our last introductory example of a proof, we show that there are long stretches of consecutive positive integers containing no primes at all. In this proof, we'll use the following terminology: For any positive integer n, the product of all integers from 1 to n is called *n factorial* and is denoted $n!$. Thus, $n! = 1 \cdot 2 \cdot 3 \cdots n$. As with our previous two proofs, we'll return to this proof at the end of Chapter 3 to analyze its structure.

Theorem 4. *For every positive integer n, there is a sequence of n consecutive positive integers containing no primes.*

Proof. Suppose n is a positive integer. Let $x = (n + 1)! + 2$. We will show that none of the numbers $x, x + 1, x + 2, \ldots, x + (n - 1)$ is prime. Since this is a sequence of n consecutive positive integers, this will prove the theorem.

To see that x is not prime, note that

$$\begin{aligned} x &= 1 \cdot 2 \cdot 3 \cdot 4 \cdots (n+1) + 2 \\ &= 2 \cdot (1 \cdot 3 \cdot 4 \cdots (n+1) + 1). \end{aligned}$$

Thus, x can be written as a product of two smaller positive integers, so x is not prime.

Similarly, we have

$$\begin{aligned} x + 1 &= 1 \cdot 2 \cdot 3 \cdot 4 \cdots (n+1) + 3 \\ &= 3 \cdot (1 \cdot 2 \cdot 4 \cdots (n+1) + 1), \end{aligned}$$

so $x+1$ is also not prime. In general, consider any number $x+i$, where $0 \leq i \leq n-1$. Then we have

$$\begin{aligned} x + i &= 1 \cdot 2 \cdot 3 \cdot 4 \cdots (n+1) + (i+2) \\ &= (i+2) \cdot (1 \cdot 2 \cdot 3 \cdots (i+1) \cdot (i+3) \cdots (n+1) + 1), \end{aligned}$$

so $x+i$ is not prime. □

Theorem 4 shows that there are sometimes long stretches between one prime and the next prime. But primes also sometimes occur close together. Since 2 is the only even prime number, the only pair of consecutive integers that are both prime is 2 and 3. But there are lots of pairs of primes that differ by only two, for example, 5 and 7, 29 and 31, and 7949 and 7951. Such pairs of primes are called *twin primes*. It is not known whether there are infinitely many twin primes.

Exercises

*1. (a) Factor $2^{15} - 1 = 32{,}767$ into a product of two smaller positive integers.
   (b) Find an integer x such that $1 < x < 2^{32767} - 1$ and $2^{32767} - 1$ is divisible by x.
2. Make some conjectures about the values of n for which $3^n - 1$ is prime or the values of n for which $3^n - 2^n$ is prime. (You might start by making a table similar to Figure 1.)
*3. The proof of Theorem 3 gives a method for finding a prime number different from any in a given list of prime numbers.
   (a) Use this method to find a prime different from 2, 3, 5, and 7.
   (b) Use this method to find a prime different from 2, 5, and 11.
4. Find five consecutive integers that are not prime.

5. Use the table in Figure 1 and the discussion on p. 5 to find two more perfect numbers.
6. The sequence 3, 5, 7 is a list of three prime numbers such that each pair of adjacent numbers in the list differ by two. Are there any more such "triplet primes"?

1

Sentential Logic

1.1. Deductive Reasoning and Logical Connectives

As we saw in the introduction, proofs play a central role in mathematics, and deductive reasoning is the foundation on which proofs are based. Therefore, we begin our study of mathematical reasoning and proofs by examining how deductive reasoning works.

Example 1.1.1. Here are three examples of deductive reasoning:

1. It will either rain or snow tomorrow.
   It's too warm for snow.
   Therefore, it will rain.
2. If today is Sunday, then I don't have to go to work today.
   Today is Sunday.
   Therefore, I don't have to go to work today.
3. I will go to work either tomorrow or today.
   I'm going to stay home today.
   Therefore, I will go to work tomorrow.

In each case, we have arrived at a *conclusion* from the assumption that some other statements, called *premises*, are true. For example, the premises in argument 3 are the statements "I will go to work either tomorrow or today" and "I'm going to stay home today." The conclusion is "I will go to work tomorrow," and it seems to be forced on us somehow by the premises.

But is this conclusion really correct? After all, isn't it possible that I'll stay home today, and then wake up sick tomorrow and end up staying home again? If that happened, the conclusion would turn out to be false. But notice that in that case the first premise, which said that I would go to work either tomorrow or today, would be false as well! Although we have no guarantee that the conclusion is true, it can only be false if at least one of the premises is also false. *If* both premises are true, we can be sure that the conclusion is also true. This is the sense in which the conclusion is forced on us by the premises, and this is the standard we will use to judge the correctness of deductive reasoning. We will say that an argument is *valid* if the premises cannot all be true without the conclusion being true as well. All three of the arguments in our example are valid arguments.

Here's an example of an invalid deductive argument:

Either the butler is guilty or the maid is guilty.
Either the maid is guilty or the cook is guilty.
Therefore, either the butler is guilty or the cook is guilty.

The argument is invalid because the conclusion could be false even if both premises are true. For example, if the maid were guilty, but the butler and the cook were both innocent, then both premises would be true and the conclusion would be false.

We can learn something about what makes an argument valid by comparing the three arguments in Example 1.1.1. On the surface it might seem that arguments 2 and 3 have the most in common, because they're both about the same subject: attendance at work. But in terms of the reasoning used, arguments 1 and 3 are the most similar. They both introduce two possibilities in the first premise, rule out the second one with the second premise, and then conclude that the first possibility must be the case. In other words, both arguments have the form:

P or Q.
not Q.
Therefore, P.

It is this form, and not the subject matter, that makes these arguments valid. You can see that argument 1 has this form by thinking of the letter P as standing for the statement "It will rain tomorrow," and Q as standing for "It will snow tomorrow." For argument 3, P would be "I will go to work tomorrow," and Q would be "I will go to work today."

Replacing certain statements in each argument with letters, as we have in stating the form of arguments 1 and 3, has two advantages. First, it keeps us from being distracted by aspects of the arguments that don't affect their validity. You don't need to know anything about weather forecasting or work habits to recognize that arguments 1 and 3 are valid. That's because both arguments have the form shown earlier, and you can tell that this argument form is valid without even knowing what P and Q stand for. If you don't believe this, consider the following argument:

> Either the framger widget is misfiring, or the wrompal mechanism is out of alignment.
> I've checked the alignment of the wrompal mechanism, and it's fine.
> Therefore, the framger widget is misfiring.

If a mechanic gave this explanation after examining your car, you might still be mystified about why the car won't start, but you'd have no trouble following his logic!

Perhaps more important, our analysis of the forms of arguments 1 and 3 makes clear what *is* important in determining their validity: the words *or* and *not*. In most deductive reasoning, and in particular in mathematical reasoning, the meanings of just a few words give us the key to understanding what makes a piece of reasoning valid or invalid. (Which are the important words in argument 2 in Example 1.1.1?) The first few chapters of this book are devoted to studying those words and how they are used in mathematical writing and reasoning.

In this chapter, we'll concentrate on words used to combine statements to form more complex statements. We'll continue to use letters to stand for statements, but only for unambiguous statements that are either true or false. Questions, exclamations, and vague statements will not be allowed. It will also be useful to use symbols, sometimes called *connective symbols*, to stand for some of the words used to combine statements. Here are our first three connective symbols and the words they stand for:

Symbol	Meaning
$\vee$	or
$\wedge$	and
$\neg$	not

Thus, if P and Q stand for two statements, then we'll write $P \vee Q$ to stand for the statement "P or Q," $P \wedge Q$ for "P and Q," and $\neg P$ for "not P" or "P is false." The statement $P \vee Q$ is sometimes called the *disjunction* of P and Q, $P \wedge Q$ is called the *conjunction* of P and Q, and $\neg P$ is called the *negation* of P.

Example 1.1.2. Analyze the logical forms of the following statements:

1. Either John went to the store, or we're out of eggs.
2. Joe is going to leave home and not come back.
3. Either Bill is at work and Jane isn't, or Jane is at work and Bill isn't.

Solutions

1. If we let P stand for the statement "John went to the store" and Q stand for "We're out of eggs," then this statement could be represented symbolically as $P \lor Q$.
2. If we let P stand for the statement "Joe is going to leave home" and Q stand for "Joe is not going to come back," then we could represent this statement symbolically as $P \land Q$. But this analysis misses an important feature of the statement, because it doesn't indicate that Q is a negative statement. We could get a better analysis by letting R stand for the statement "Joe is going to come back" and then writing the statement Q as $\neg R$. Plugging this into our first analysis of the original statement, we get the improved analysis $P \land \neg R$.
3. Let B stand for the statement "Bill is at work" and J for the statement "Jane is at work." Then the first half of the statement, "Bill is at work and Jane isn't," can be represented as $B \land \neg J$. Similarly, the second half is $J \land \neg B$. To represent the entire statement, we must combine these two with *or*, forming their disjunction, so the solution is $(B \land \neg J) \lor (J \land \neg B)$.

Notice that in analyzing the third statement in the preceding example, we added parentheses when we formed the disjunction of $B \land \neg J$ and $J \land \neg B$ to indicate unambiguously which statements were being combined. This is like the use of parentheses in algebra, in which, for example, the product of $a + b$ and $a - b$ would be written $(a + b) \cdot (a - b)$, with the parentheses serving to indicate unambiguously which quantities are to be multiplied. As in algebra, it is convenient in logic to omit some parentheses to make our expressions shorter and easier to read. However, we must agree on some conventions about how to read such expressions so that they are still unambiguous. One convention is that the symbol $\neg$ always applies only to the statement that comes immediately after it. For example, $\neg P \land Q$ means $(\neg P) \land Q$ rather than $\neg(P \land Q)$. We'll see some other conventions about parentheses later.

Example 1.1.3. What English sentences are represented by the following expressions?

1. $(\neg S \land L) \lor S$, where S stands for "John is stupid" and L stands for "John is lazy."
2. $\neg S \land (L \lor S)$, where S and L have the same meanings as before.
3. $\neg(S \land L) \lor S$, with S and L still as before.

Solutions

1. Either John isn't stupid and he is lazy, or he's stupid.
2. John isn't stupid, and either he's lazy or he's stupid. Notice how the placement of the word *either* in English changes according to where the parentheses are.
3. Either John isn't both stupid and lazy, or John is stupid. The word *both* in English also helps distinguish the different possible positions of parentheses.

It is important to keep in mind that the symbols $\wedge$, $\vee$, and $\neg$ don't really correspond to all uses of the words *and*, *or*, and *not* in English. For example, the symbol $\wedge$ could not be used to represent the use of the word *and* in the sentence "John and Bill are friends," because in this sentence the word *and* is not being used to combine two statements. The symbols $\wedge$ and $\vee$ can only be used *between two statements*, to form their conjunction or disjunction, and the symbol $\neg$ can only be used *before a statement*, to negate it. This means that certain strings of letters and symbols are simply meaningless. For example, $P\neg \wedge Q$, $P \wedge/\vee Q$, and $P\neg Q$ are all "ungrammatical" expressions in the language of logic. "Grammatical" expressions, such as those in Examples 1.1.2 and 1.1.3, are sometimes called *well-formed formulas* or just *formulas*. Once again, it may be helpful to think of an analogy with algebra, in which the symbols $+$, $-$, $\cdot$, and $\div$ can be used *between two numbers*, as operators, and the symbol $-$ can also be used *before a number*, to negate it. These are the only ways that these symbols can be used in algebra, so expressions such as $x - \div y$ are meaningless.

Sometimes, words other than *and*, *or*, and *not* are used to express the meanings represented by $\wedge$, $\vee$, and $\neg$. For example, consider the first statement in Example 1.1.3. Although we gave the English translation "Either John isn't stupid and he is lazy, or he's stupid," an alternative way of conveying the same information would be to say "Either John isn't stupid *but* he is lazy, or he's stupid." Often, the word *but* is used in English to mean *and*, especially when there is some contrast or conflict between the statements being combined. For a more striking example, imagine a weather forecaster ending his forecast with the statement "Rain and snow are the only two possibilities for tomorrow's weather." This is just a roundabout way of saying that it will either rain or snow tomorrow. Thus, even though the forecaster has used the word *and*, the meaning expressed by his statement is a disjunction. The lesson of these examples is that to determine the logical form of a statement you must think about what the statement means, rather than just translating word by word into symbols.

Sometimes logical words are hidden within mathematical notation. For example, consider the statement $3 \leq \pi$. Although it appears to be a simple statement that contains no words of logic, if you read it out loud you will hear the word *or*. If we let P stand for the statement $3 < \pi$ and Q for the statement $3 = \pi$, then the statement $3 \leq \pi$ would be written $P \vee Q$. In this example the statements represented by the letters P and Q are so short that it hardly seems worthwhile to abbreviate them with single letters. In cases like this we will sometimes not bother to replace the statements with letters, so we might also write this statement as $(3 < \pi) \vee (3 = \pi)$.

For a slightly more complicated example, consider the statement $3 \leq \pi < 4$. This statement means $3 \leq \pi$ *and* $\pi < 4$, so once again a word of logic has been hidden in mathematical notation. Filling in the meaning that we just worked out for $3 \leq \pi$, we can write the whole statement as $[(3 < \pi) \vee (3 = \pi)] \wedge (\pi < 4)$. Knowing that the statement has this logical form might be important in understanding a piece of mathematical reasoning involving this statement.

Exercises

*1. Analyze the logical forms of the following statements:
   (a) We'll have either a reading assignment or homework problems, but we won't have both homework problems and a test.
   (b) You won't go skiing, or you will and there won't be any snow.
   (c) $\sqrt{7} \not\leq 2$.

2. Analyze the logical forms of the following statements:
   (a) Either John and Bill are both telling the truth, or neither of them is.
   (b) I'll have either fish or chicken, but I won't have both fish and mashed potatoes.
   (c) 3 is a common divisor of 6, 9, and 15.

3. Analyze the logical forms of the following statements:
   (a) Alice and Bob are not both in the room.
   (b) Alice and Bob are both not in the room.
   (c) Either Alice or Bob is not in the room.
   (d) Neither Alice nor Bob is in the room.

4. Which of the following expressions are well-formed formulas?
   (a) $\neg(\neg P \vee \neg\neg R)$.
   (b) $\neg(P, Q, \wedge R)$.
   (c) $P \wedge \neg P$.
   (d) $(P \wedge Q)(P \vee R)$.

*5. Let P stand for the statement "I will buy the pants" and S for the statement "I will buy the shirt." What English sentences are represented by the following expressions?
 (a) $\neg(P \wedge \neg S)$.
 (b) $\neg P \wedge \neg S$.
 (c) $\neg P \vee \neg S$.

6. Let S stand for the statement "Steve is happy" and G for "George is happy." What English sentences are represented by the following expressions?
 (a) $(S \vee G) \wedge (\neg S \vee \neg G)$.
 (b) $[S \vee (G \wedge \neg S)] \vee \neg G$.
 (c) $S \vee [G \wedge (\neg S \vee \neg G)]$.

7. Identify the premises and conclusions of the following deductive arguments and analyze their logical forms. Do you think the reasoning is valid? (Although you will have only your intuition to guide you in answering this last question, in the next section we will develop some techniques for determining the validity of arguments.)
 (a) Jane and Pete won't both win the math prize. Pete will win either the math prize or the chemistry prize. Jane will win the math prize. Therefore, Pete will win the chemistry prize.
 (b) The main course will be either beef or fish. The vegetable will be either peas or corn. We will not have both fish as a main course and corn as a vegetable. Therefore, we will not have both beef as a main course and peas as a vegetable.
 (c) Either John or Bill is telling the truth. Either Sam or Bill is lying. Therefore, either John is telling the truth or Sam is lying.
 (d) Either sales will go up and the boss will be happy, or expenses will go up and the boss won't be happy. Therefore, sales and expenses will not both go up.

1.2. Truth Tables

We saw in Section 1.1 that an argument is valid if the premises cannot all be true without the conclusion being true as well. Thus, to understand how words such as *and*, *or*, and *not* affect the validity of arguments, we must see how they contribute to the truth or falsity of statements containing them.

When we evaluate the truth or falsity of a statement, we assign to it one of the labels *true* or *false*, and this label is called its *truth value*. It is clear how the word *and* contributes to the truth value of a statement containing it. A statement of the form $P \wedge Q$ can only be true if both P and Q are true; if either P or Q is false, then $P \wedge Q$ will be false too. Because we have assumed that P and

P	Q	$P \wedge Q$
F	F	F
F	T	F
T	F	F
T	T	T

Figure 1

Q both stand for statements that are either true or false, we can summarize all the possibilities with the table shown in Figure 1. This is called a *truth table* for the formula $P \wedge Q$. Each row in the truth table represents one of the four possible combinations of truth values for the statements P and Q. Although these four possibilities can appear in the table in any order, it is best to list them systematically so we can be sure that no possibilities have been skipped. The truth table for $\neg P$ is also quite easy to construct because for $\neg P$ to be true, P must be false. The table is shown in Figure 2.

P	$\neg P$
F	T
T	F

Figure 2

The truth table for $P \vee Q$ is a little trickier. The first three lines should certainly be filled in as shown in Figure 3, but there may be some question about the last line. Should $P \vee Q$ be true or false in the case in which P and Q are both true? In other words, does $P \vee Q$ mean "P or Q, or both" or does it mean "P or Q but not both"? The first way of interpreting the word *or* is called the *inclusive or* (because it *includes* the possibility of both statements being true), and the second is called the *exclusive or*. In mathematics, *or* always means inclusive or, unless specified otherwise, so we will interpret $\vee$ as inclusive or. We therefore complete the truth table for $P \vee Q$ as shown in Figure 4. See exercise 3 for more about the exclusive or.

P	Q	$P \vee Q$
F	F	F
F	T	T
T	F	T
T	T	?

Figure 3

P	Q	$P \vee Q$
F	F	F
F	T	T
T	F	T
T	T	T

Figure 4

Using the rules summarized in these truth tables, we can now work out truth tables for more complex formulas. All we have to do is work out the truth values of the component parts of a formula, starting with the individual letters and working up to more complex formulas a step at a time.

Example 1.2.1. Make a truth table for the formula $\neg(P \vee \neg Q)$.

Solution

P	Q	$\neg Q$	$P \vee \neg Q$	$\neg(P \vee \neg Q)$
F	F	T	T	F
F	T	F	F	T
T	F	T	T	F
T	T	F	T	F

The first two columns of this table list the four possible combinations of truth values of P and Q. The third column, listing truth values for the formula $\neg Q$, is found by simply negating the truth values for Q in the second column. The fourth column, for the formula $P \vee \neg Q$, is found by combining the truth values for P and $\neg Q$ listed in the first and third columns, according to the truth value rule for $\vee$ summarized in Figure 4. According to this rule, $P \vee \neg Q$ will be false only if both P and $\neg Q$ are false. Looking in the first and third columns, we see that this happens only in row two of the table, so the fourth column contains an F in the second row and T's in all other rows. Finally, the truth values for the formula $\neg(P \vee \neg Q)$ are listed in the fifth column, which is found by negating the truth values in the fourth column. (Note that these columns had to be worked out in order, because each was used in computing the next.)

Example 1.2.2. Make a truth table for the formula $\neg(P \wedge Q) \vee \neg R$.

Solution

P	Q	R	$P \wedge Q$	$\neg(P \wedge Q)$	$\neg R$	$\neg(P \wedge Q) \vee \neg R$
F	F	F	F	T	T	T
F	F	T	F	T	F	T
F	T	F	F	T	T	T
F	T	T	F	T	F	T
T	F	F	F	T	T	T
T	F	T	F	T	F	T
T	T	F	T	F	T	T
T	T	T	T	F	F	F

Note that because this formula contains three letters, it takes eight lines to list all possible combinations of truth values for these letters. (If a formula contains n different letters, how many lines will its truth table have?)

Here's a way of making truth tables more compactly. Instead of using separate columns to list the truth values for the component parts of a formula, just list those truth values below the corresponding connective symbol in the original formula. This is illustrated in Figure 5, for the formula from Example 1.2.1.

In the first step, we have listed the truth values for P and Q below these letters where they appear in the formula. In step two, the truth values for $\neg Q$ have been added under the $\neg$ symbol for $\neg Q$. In the third step, we have combined the truth values for P and $\neg Q$ to get the truth values for $P \vee \neg Q$, which are listed under the $\vee$ symbol. Finally, in the last step, these truth values are negated and listed under the initial $\neg$ symbol. The truth values added in the last step give the truth value for the entire formula, so we will call the symbol under which they are listed (the first $\neg$ symbol in this case) the *main connective* of the formula. Notice that the truth values listed under the main connective in this case agree with the values we found in Example 1.2.1.

Step 1

P	Q	$\neg(P \vee \neg Q)$
F	F	**F** **F**
F	T	**F** **T**
T	F	**T** **F**
T	T	**T** **T**

Step 2

P	Q	$\neg(P \vee \neg Q)$
F	F	F **TF**
F	T	F **FT**
T	F	T **TF**
T	T	T **FT**

Step 3

P	Q	$\neg(P \vee \neg Q)$
F	F	F **T** TF
F	T	F **F** FT
T	F	T **T** TF
T	T	T **T** FT

Step 4

P	Q	$\neg(P \vee \neg Q)$
F	F	**F** F T TF
F	T	**T** F F FT
T	F	**F** T T TF
T	T	**F** T T FT

Figure 5

Now that we know how to make truth tables for complex formulas, we're ready to return to the analysis of the validity of arguments. Consider again our first example of a deductive argument:

It will either rain or snow tomorrow.
It's too warm for snow.
Therefore, it will rain.

As we have seen, if we let P stand for the statement "It will rain tomorrow" and Q for the statement "It will snow tomorrow," then we can represent the argument symbolically as follows:

$P \vee Q$

$\neg Q$

$\therefore P$ (The symbol $\therefore$ means *therefore*.)

We can now see how truth tables can be used to verify the validity of this argument. Figure 6 shows a truth table for both premises and the conclusion of the argument. Recall that we decided to call an argument valid if the premises cannot all be true without the conclusion being true as well. Looking at Figure 6 we see that the only row of the table in which both premises come out true is row three, and in this row the conclusion is also true. Thus, the truth table confirms that if the premises are all true, the conclusion must also be true, so the argument is valid.

		Premises		Conclusion
P	Q	$P \lor Q$	$\neg Q$	P
F	F	F	T	F
F	T	T	F	F
T	F	T	T	T
T	T	T	F	T

Figure 6

Example 1.2.3. Determine whether the following arguments are valid.

1. Either John isn't stupid and he is lazy, or he's stupid.
   John is stupid.
   Therefore, John isn't lazy.
2. The butler and the cook are not both innocent.
   Either the butler is lying or the cook is innocent.
   Therefore, the butler is either lying or guilty.

Solutions

1. As in Example 1.1.3, we let S stand for the statement "John is stupid" and L stand for "John is lazy." Then the argument has the form:

$$\begin{array}{l} (\neg S \land L) \lor S \\ \underline{S \qquad\qquad\quad} \\ \therefore \neg L \end{array}$$

Now we make a truth table for both premises and the conclusion. (You should work out the intermediate steps in deriving column three of this table to confirm that it is correct.)

		Premises		Conclusion
S	L	$(\neg S \land L) \lor S$	S	$\neg L$
F	F	F	F	T
F	T	T	F	F
T	F	T	T	T
T	T	T	T	F

Both premises are true in lines three and four of this table. The conclusion is also true in line three, but it is false in line four. Thus, it is possible for both premises to be true and the conclusion false, so the argument is invalid. In fact, the table shows us exactly why the argument is invalid. The problem occurs in the fourth line of the table, in which S and L are both true – in other words, John is both stupid and lazy. Thus, if John is both stupid and lazy, then both premises will be true but the conclusion will be false, so it would be a mistake to infer that the conclusion must be true from the assumption that the premises are true.

2. Let B stand for the statement "The butler is innocent," C for the statement "The cook is innocent," and L for the statement "The butler is lying." Then the argument has the form:

$$\neg(B \wedge C)$$
$$L \vee C$$
$$\therefore L \vee \neg B$$

Here is the truth table for the premises and conclusion:

			Premises		Conclusion
B	C	L	$\neg(B \wedge C)$	$L \vee C$	$L \vee \neg B$
F	F	F	T	F	T
F	F	T	T	T	T
F	T	F	T	T	T
F	T	T	T	T	T
T	F	F	T	F	F
T	F	T	T	T	T
T	T	F	F	T	F
T	T	T	F	T	T

The premises are both true only in lines two, three, four, and six, and in each of these cases the conclusion is true as well. Therefore, the argument is valid.

If you expected the first argument in Example 1.2.3 to turn out to be valid, it's probably because the first premise confused you. It's a rather complicated statement, which we represented symbolically with the formula $(\neg S \wedge L) \vee S$. According to our truth table, this formula is false if S and L are both false, and true otherwise. But notice that this is exactly the same as the truth table for the simpler formula $L \vee S$! Because of this, we say that the formulas $(\neg S \wedge L) \vee S$ and $L \vee S$ are *equivalent*. Equivalent formulas always have the same truth value no matter what statements the letters in them stand for and no matter what the truth values of those statements are. The equivalence of the premise $(\neg S \wedge L) \vee S$ and the simpler formula $L \vee S$ may help you understand why the argument is invalid. Translating the formula $L \vee S$ back into English, we see that the first premise could have been stated more simply as "John is either lazy or stupid (or both)." But from this premise and the second premise (that John is stupid), it clearly doesn't follow that he's not lazy, because he might be both stupid and lazy.

Example 1.2.4. Which of these formulas are equivalent?

$$\neg(P \wedge Q), \qquad \neg P \wedge \neg Q, \qquad \neg P \vee \neg Q.$$

Solution

Here's a truth table for all three statements. (You should check it yourself!)

P	Q	$\neg(P \wedge Q)$	$\neg P \wedge \neg Q$	$\neg P \vee \neg Q$
F	F	T	T	T
F	T	T	F	T
T	F	T	F	T
T	T	F	F	F

The third and fifth columns in this table are identical, but they are different from the fourth column. Therefore, the formulas $\neg(P \wedge Q)$ and $\neg P \vee \neg Q$ are equivalent, but neither is equivalent to the formula $\neg P \wedge \neg Q$. This should make sense if you think about what all the symbols mean. For example, suppose P stands for the statement "The Yankees won last night" and Q stands for "The Red Sox won last night." Then $\neg(P \wedge Q)$ would mean "The Yankees and the Red Sox did not both win last night," and $\neg P \vee \neg Q$ would mean "Either the Yankees or the Red Sox lost last night"; these statements clearly convey the same information. On the other hand, $\neg P \wedge \neg Q$ would mean "The Yankees and the Red Sox both lost last night," which is an entirely different statement.

You can check for yourself by making a truth table that the formula $\neg P \wedge \neg Q$ from Example 1.2.4 is equivalent to the formula $\neg(P \vee Q)$. (To see that this equivalence makes sense, notice that the statements "Both the Yankees and the Red Sox lost last night" and "Neither the Yankees nor the Red Sox won last night" mean the same thing.) This equivalence and the one discovered in Example 1.2.4 are called *DeMorgan's laws*.

In analyzing deductive arguments and the statements that occur in them it is helpful to be familiar with a number of equivalences that come up often. Verify the equivalences in the following list yourself by making truth tables, and check that they make sense by translating the formulas into English, as we did in Example 1.2.4.

DeMorgan's laws

$$\neg(P \wedge Q) \text{ is equivalent to } \neg P \vee \neg Q.$$
$$\neg(P \vee Q) \text{ is equivalent to } \neg P \wedge \neg Q.$$

Commutative laws

$$P \wedge Q \text{ is equivalent to } Q \wedge P.$$
$$P \vee Q \text{ is equivalent to } Q \vee P.$$

Associative laws

$$P \wedge (Q \wedge R) \text{ is equivalent to } (P \wedge Q) \wedge R.$$
$$P \vee (Q \vee R) \text{ is equivalent to } (P \vee Q) \vee R.$$

Idempotent laws

$$P \wedge P \text{ is equivalent to } P.$$
$$P \vee P \text{ is equivalent to } P.$$

Distributive laws

$$P \wedge (Q \vee R) \text{ is equivalent to } (P \wedge Q) \vee (P \wedge R).$$
$$P \vee (Q \wedge R) \text{ is equivalent to } (P \vee Q) \wedge (P \vee R).$$

Absorption laws

$$P \vee (P \wedge Q) \text{ is equivalent to } P.$$
$$P \wedge (P \vee Q) \text{ is equivalent to } P.$$

Double Negation law

$$\neg\neg P \text{ is equivalent to } P.$$

Notice that because of the associative laws we can leave out parentheses in formulas of the forms $P \wedge Q \wedge R$ and $P \vee Q \vee R$ without worrying that the resulting formula will be ambiguous, because the two possible ways of filling in the parentheses lead to equivalent formulas.

Many of the equivalences in the list should remind you of similar rules involving $+$, $\cdot$, and $-$ in algebra. As in algebra, these rules can be applied to more complex formulas, and they can be combined to work out more complicated equivalences. Any of the letters in these equivalences can be replaced by more complicated formulas, and the resulting equivalence will still be true. For example, by replacing P in the double negation law with the formula $Q \vee \neg R$, you can see that $\neg\neg(Q \vee \neg R)$ is equivalent to $Q \vee \neg R$. Also, if two formulas are equivalent, you can always substitute one for the other in any expression and the results will be equivalent. For example, since $\neg\neg P$ is equivalent to

P, if $\neg\neg P$ occurs in any formula, you can always replace it with P and the resulting formula will be equivalent to the original.

Example 1.2.5. Find simpler formulas equivalent to these formulas:

1. $\neg(P \vee \neg Q)$.
2. $\neg(Q \wedge \neg P) \vee P$.

Solutions

1. $\neg(P \vee \neg Q)$

is equivalent to	$\neg P \wedge \neg\neg Q$	(DeMorgan's law),
which is equivalent to	$\neg P \wedge Q$	(double negation law).

You can check that this equivalence is right by making a truth table for $\neg P \wedge Q$ and seeing that it is the same as the truth table for $\neg(P \vee \neg Q)$ found in Example 1.2.1.

2. $\neg(Q \wedge \neg P) \vee P$

is equivalent to	$(\neg Q \vee \neg\neg P) \vee P$	(DeMorgan's law),
which is equivalent to	$(\neg Q \vee P) \vee P$	(double negation law),
which is equivalent to	$\neg Q \vee (P \vee P)$	(associative law),
which is equivalent to	$\neg Q \vee P$	(idempotent law).

Some equivalences are based on the fact that certain formulas are either always true or always false. For example, you can verify by making a truth table that the formula $Q \wedge (P \vee \neg P)$ is equivalent to just Q. But even before you make the truth table, you can probably see why they are equivalent. In every line of the truth table, $P \vee \neg P$ will come out true, and therefore $Q \wedge (P \vee \neg P)$ will come out true when Q is also true, and false when Q is false. Formulas that are always true, such as $P \vee \neg P$, are called *tautologies*. Similarly, formulas that are always false are called *contradictions*. For example, $P \wedge \neg P$ is a contradiction.

Example 1.2.6. Are these statements tautologies, contradictions, or neither?

$$P \vee (Q \vee \neg P), \quad P \wedge \neg(Q \vee \neg Q), \quad P \vee \neg(Q \vee \neg Q).$$

Solution

First we make a truth table for all three statements.

P	Q	$P \vee (Q \vee \neg P)$	$P \wedge \neg(Q \vee \neg Q)$	$P \vee \neg(Q \vee \neg Q)$
F	F	T	F	F
F	T	T	F	F
T	F	T	F	T
T	T	T	F	T

From the truth table it is clear that the first formula is a tautology, the second a contradiction, and the third neither. In fact, since the last column is identical to the first, the third formula is equivalent to P.

We can now state a few more useful laws involving tautologies and contradictions. You should be able to convince yourself that all of these laws are correct by thinking about what the truth tables for the statements involved would look like.

Tautology laws

$P \wedge$ (a tautology) is equivalent to P.

$P \vee$ (a tautology) is a tautology.

$\neg$(a tautology) is a contradiction.

Contradiction laws

$P \wedge$ (a contradiction) is a contradiction.

$P \vee$ (a contradiction) is equivalent to P.

$\neg$(a contradiction) is a tautology.

Example 1.2.7. Find simpler formulas equivalent to these formulas:

1. $P \vee (Q \wedge \neg P)$.
2. $\neg(P \vee (Q \wedge \neg R)) \wedge Q$.

Solutions

1. $P \vee (Q \wedge \neg P)$
   is equivalent to $(P \vee Q) \wedge (P \vee \neg P)$ (distributive law),
   which is equivalent to $P \vee Q$ (tautology law).
   The last step uses the fact that $P \vee \neg P$ is a tautology.
2. $\neg(P \vee (Q \wedge \neg R)) \wedge Q$
   is equivalent to $(\neg P \wedge \neg(Q \wedge \neg R)) \wedge Q$ (DeMorgan's law),
   which is equivalent to $(\neg P \wedge (\neg Q \vee \neg\neg R)) \wedge Q$ (DeMorgan's law),
   which is equivalent to $(\neg P \wedge (\neg Q \vee R)) \wedge Q$ (double negation law),
   which is equivalent to $\neg P \wedge ((\neg Q \vee R) \wedge Q)$ (associative law),
   which is equivalent to $\neg P \wedge (Q \wedge (\neg Q \vee R))$ (commutative law),
   which is equivalent to $\neg P \wedge ((Q \wedge \neg Q) \vee (Q \wedge R))$
   (distributive law),
   which is equivalent to $\neg P \wedge (Q \wedge R)$ (contradiction law).

   The last step uses the fact that $Q \wedge \neg Q$ is a contradiction. Finally, by the associative law for $\wedge$ we can remove the parentheses without making the formula ambiguous, so the original formula is equivalent to the formula $\neg P \wedge Q \wedge R$.

Exercises

*1. Make truth tables for the following formulas:
   (a) $\neg P \vee Q$.
   (b) $(S \vee G) \wedge (\neg S \vee \neg G)$.
2. Make truth tables for the following formulas:
   (a) $\neg[P \wedge (Q \vee \neg P)]$.
   (b) $(P \vee Q) \wedge (\neg P \vee R)$.
3. In this exercise we will use the symbol $+$ to mean *exclusive or*. In other words, $P + Q$ means "P or Q, but not both."
   (a) Make a truth table for $P + Q$.
   (b) Find a formula using only the connectives $\wedge$, $\vee$, and $\neg$ that is equivalent to $P + Q$. Justify your answer with a truth table.
4. Find a formula using only the connectives $\wedge$ and $\neg$ that is equivalent to $P \vee Q$. Justify your answer with a truth table.
*5. Some mathematicians use the symbol $\downarrow$ to mean *nor*. In other words, $P \downarrow Q$ means "neither P nor Q."
   (a) Make a truth table for $P \downarrow Q$.
   (b) Find a formula using only the connectives $\wedge$, $\vee$, and $\neg$ that is equivalent to $P \downarrow Q$.
   (c) Find formulas using only the connective $\downarrow$ that are equivalent to $\neg P$, $P \vee Q$, and $P \wedge Q$.
6. Some mathematicians write $P \mid Q$ to mean "P and Q are not both true." (This connective is called *nand*, and is used in the study of circuits in computer science.)
   (a) Make a truth table for $P \mid Q$.
   (b) Find a formula using only the connectives $\wedge$, $\vee$, and $\neg$ that is equivalent to $P \mid Q$.
   (c) Find formulas using only the connective $\mid$ that are equivalent to $\neg P$, $P \vee Q$, and $P \wedge Q$.
*7. Use truth tables to determine whether or not the arguments in exercise 7 of Section 1.1 are valid.
8. Use truth tables to determine which of the following formulas are equivalent to each other:
   (a) $(P \wedge Q) \vee (\neg P \wedge \neg Q)$.
   (b) $\neg P \vee Q$.
   (c) $(P \vee \neg Q) \wedge (Q \vee \neg P)$.
   (d) $\neg(P \vee Q)$.
   (e) $(Q \wedge P) \vee \neg P$.
*9. Use truth tables to determine which of these statements are tautologies, which are contradictions, and which are neither:

(a) $(P \lor Q) \land (\neg P \lor \neg Q)$.
(b) $(P \lor Q) \land (\neg P \land \neg Q)$.
(c) $(P \lor Q) \lor (\neg P \lor \neg Q)$.
(d) $[P \land (Q \lor \neg R)] \lor (\neg P \lor R)$.

10. Use truth tables to check these laws:
    (a) The second DeMorgan's law. (The first was checked in the text.)
    (b) The distributive laws.

*11. Use the laws stated in the text to find simpler formulas equivalent to these formulas. (See Examples 1.2.5 and 1.2.7.)
    (a) $\neg(\neg P \land \neg Q)$.
    (b) $(P \land Q) \lor (P \land \neg Q)$.
    (c) $\neg(P \land \neg Q) \lor (\neg P \land Q)$.

12. Use the laws stated in the text to find simpler formulas equivalent to these formulas. (See Examples 1.2.5 and 1.2.7.)
    (a) $\neg(\neg P \lor Q) \lor (P \land \neg R)$.
    (b) $\neg(\neg P \land Q) \lor (P \land \neg R)$.
    (c) $(P \land R) \lor [\neg R \land (P \lor Q)]$.

13. Use the first DeMorgan's law and the double negation law to derive the second DeMorgan's law.

*14. Note that the associative laws say only that parentheses are unnecessary when combining *three* statements with $\land$ or $\lor$. In fact, these laws can be used to justify leaving parentheses out when more than three statements are combined. Use associative laws to show that $[P \land (Q \land R)] \land S$ is equivalent to $(P \land Q) \land (R \land S)$.

15. How many lines will there be in the truth table for a statement containing n letters?

*16. Find a formula involving the connectives $\land$, $\lor$, and $\neg$ that has the following truth table:

P	Q	???
F	F	T
F	T	F
T	F	T
T	T	T

17. Find a formula involving the connectives $\land$, $\lor$, and $\neg$ that has the following truth table:

P	Q	???
F	F	F
F	T	T
T	F	T
T	T	F

18. Suppose the conclusion of an argument is a tautology. What can you conclude about the validity of the argument? What if the conclusion is a contradiction? What if one of the premises is either a tautology or a contradiction?

1.3. Variables and Sets

In mathematical reasoning it is often necessary to make statements about objects that are represented by letters called *variables*. For example, if the variable x is used to stand for a number in some problem, we might be interested in the statement "x is a prime number." Although we may sometimes use a single letter, say P, to stand for this statement, at other times we will revise this notation slightly and write $P(x)$, to stress that this is a statement *about* x. The latter notation makes it easy to talk about substituting some number for x in the statement. For example, $P(7)$ would represent the statement "7 is a prime number," and $P(a+b)$ would mean "$a+b$ is a prime number." If a statement contains more than one variable, our abbreviation for the statement will include a list of all the variables involved. For example, we might represent the statement "p is divisible by q" by $D(p, q)$. In this case, $D(12, 4)$ would mean "12 is divisible by 4."

Although you have probably seen variables used most often to stand for numbers, they can stand for anything at all. For example, we could let $M(x)$ stand for the statement "x is a man," and $W(x)$ for "x is a woman." In this case, we are using the variable x to stand for a person. A statement might even contain several variables that stand for different kinds of objects. For example, in the statement "x has y children," the variable x stands for a person, and y stands for a number.

Statements involving variables can be combined using connectives, just like statements without variables.

Example 1.3.1. Analyze the logical forms of the following statements:

1. x is a prime number, and either y or z is divisible by x.
2. x is a man and y is a woman and x likes y, but y doesn't like x.

Solutions

1. We could let P stand for the statement "x is a prime number," D for "y is divisible by x," and E for "z is divisible by x." The entire statement would then be represented by the formula $P \wedge (D \vee E)$. But this analysis, though not incorrect, fails to capture the relationship between the statements

D and E. A better analysis would be to let $P(x)$ stand for "x is a prime number" and $D(y, x)$ for "y is divisible by x." Then $D(z, x)$ would mean "z is divisible by x," so the entire statement would be $P(x) \wedge (D(y, x) \vee D(z, x))$.

2. Let $M(x)$ stand for "x is a man," $W(y)$ for "y is a woman," and $L(x, y)$ for "x likes y." Then $L(y, x)$ would mean "y likes x." (Notice that the order of the variables after the L makes a difference!) The entire statement would then be represented by the formula $M(x) \wedge W(y) \wedge L(x, y) \wedge \neg L(y, x)$.

When studying statements that do not contain variables, we can easily talk about their truth values, since each statement is either true or false. But if a statement contains variables, we can no longer describe the statement as being simply true or false. Its truth value might depend on the values of the variables involved. For example, if $P(x)$ stands for the statement "x is a prime number," then $P(x)$ would be true if $x = 23$, but false if $x = 22$. To solve this problem, we will define *truth sets* for statements containing variables. Before giving this definition, though, it might be helpful to review some basic definitions from set theory.

A *set* is a collection of objects. The objects in the collection are called the *elements* of the set. The simplest way to specify a particular set is to list its elements between braces. For example, $\{3, 7, 14\}$ is the set whose elements are the three numbers 3, 7, and 14. We use the symbol $\in$ to mean *is an element of*. For example, if we let A stand for the set $\{3, 7, 14\}$, then we could write $7 \in A$ to say that 7 is an element of A. To say that 11 is not an element of A, we write $11 \notin A$.

A set is completely determined once its elements have been specified. Thus, two sets that have exactly the same elements are always equal. Also, when a set is defined by listing its elements, all that matters is *which* objects are in the list of elements, not the order in which they are listed. An element can even appear more than once in the list. Thus, $\{3, 7, 14\}$, $\{14, 3, 7\}$, and $\{3, 7, 14, 7\}$ are three different names for the same set.

It may be impractical to define a set that contains a very large number of elements by listing all of its elements, and it would be impossible to give such a definition for a set that contains infinitely many elements. Often this problem can be overcome by listing a few elements with an ellipsis (...) after them, if it is clear how the list should be continued. For example, suppose we define a set B by saying that $B = \{2, 3, 5, 7, 11, 13, 17, \ldots\}$. Once you recognize that the numbers listed in the definition of B are the prime numbers, then you know that, for example, $23 \in B$, even though it wasn't listed explicitly when we defined B. But this method requires recognition of the pattern in the list of numbers in the definition of B, and this requirement introduces an element of ambiguity and subjectivity into our notation that is best avoided in mathematical writing. It is therefore usually better to define such a set by spelling out the pattern that determines the elements of the set.

In this case we could be explicit by defining B as follows:

$$B = \{x \mid x \text{ is a prime number}\}.$$

This is read "$B =$ the set of all x such that x is a prime number," and it means that the elements of B are the values of x that make the statement "x is a prime number" come out true. You should think of the statement "x is a prime number" as an *elementhood test* for the set. Any value of x that makes this statement come out true passes the test and is an element of the set. Anything else fails the test and is not an element. Of course, in this case the values of x that make the statement true are precisely the prime numbers, so this definition says that B is the set whose elements are the prime numbers, exactly as before.

Example 1.3.2. Rewrite these set definitions using elementhood tests:

1. $E = \{2, 4, 6, 8, \ldots\}$.
2. P = {George Washington, John Adams, Thomas Jefferson, James Madison, ...}.

Solutions

Although there might be other ways of continuing these lists of elements, probably the most natural ones are given by the following definitions:

1. $E = \{n \mid n \text{ is a positive even integer}\}$.
2. $P = \{z \mid z \text{ was a president of the United States}\}$.

If a set has been defined using an elementhood test, then that test can be used to determine whether or not something is an element of the set. For example, consider the set $\{x \mid x^2 < 9\}$. If we want to know if 5 is an element of this set, we simply apply the elementhood test in the definition of the set – in other words, we check whether or not $5^2 < 9$. Since $5^2 = 25 > 9$, it fails the test, so $5 \notin \{x \mid x^2 < 9\}$. On the other hand, $(-2)^2 = 4 < 9$, so $-2 \in \{x \mid x^2 < 9\}$. The same reasoning would apply to any other number. For any number y, to determine whether or not $y \in \{x \mid x^2 < 9\}$, we just check whether or not $y^2 < 9$. In fact, we could think of the statement $y \in \{x \mid x^2 < 9\}$ as just a roundabout way of saying $y^2 < 9$.

Notice that because the statement $y \in \{x \mid x^2 < 9\}$ means the same thing as $y^2 < 9$, it is a statement about y, but not x! To determine whether or not $y \in \{x \mid x^2 < 9\}$ you need to know what y is (so you can compare its square to 9), but not what x is. We say that in the statement $y \in \{x \mid x^2 < 9\}$, y is a *free* variable, whereas x is a *bound* variable (or a *dummy* variable). The free variables in a statement stand for objects that the statement says something about. Plugging in different values for a free variable affects the meaning of a statement and may change its truth value. The fact that you can plug in different values for a free variable means that it is free to stand for anything. Bound variables, on the other hand, are simply letters that are used as a convenience to help express an idea and should not be thought of as standing for any particular object. A bound variable can always be replaced by a new variable without changing the meaning of the statement, and often the statement can be rephrased so that the bound variables are eliminated altogether. For example, the statements $y \in \{x \mid x^2 < 9\}$ and $y \in \{w \mid w^2 < 9\}$ mean the same thing, because they both mean "y is an element of the set of all numbers whose squares are less than 9." In this last statement, all bound variables have been eliminated, and the only variable mentioned is the free variable y.

Note that x is a bound variable in the statement $y \in \{x \mid x^2 < 9\}$ even though it is a free variable in the statement $x^2 < 9$. This last statement is a statement about x that would be true for some values of x and false for others. It is only when this statement is used inside the elementhood test notation that x becomes a bound variable. We could say that the notation $\{x \mid \ldots\}$ *binds* the variable x.

Everything we have said about the set $\{x \mid x^2 < 9\}$ would apply to any set defined by an elementhood test. In general, the statement $y \in \{x \mid P(x)\}$ means the same thing as $P(y)$, which is a statement about y but not x. Similarly, $y \notin \{x \mid P(x)\}$ means the same thing as $\neg P(y)$. Of course, the expression $\{x \mid P(x)\}$ is not a statement at all; it is a name for a set. As you learn more mathematical notation, it will become increasingly important to make sure you are careful to distinguish between expressions that are mathematical statements and expressions that are names for mathematical objects.

Example 1.3.3. What do these statements mean? What are the free variables in each statement?

1. $a + b \notin \{x \mid x \text{ is an even number}\}$.
2. $y \in \{x \mid x \text{ is divisible by } w\}$.
3. $2 \in \{w \mid 6 \notin \{x \mid x \text{ is divisible by } w\}\}$.

Solutions

1. This statement says that $a + b$ is not an element of the set of all even numbers, or in other words, $a + b$ is not an even number. Both a and b are free variables, but x is a bound variable. The statement will be true for some values of a and b and false for others.

2. This statement says that y is divisible by w. Both y and w are free variables, but x is a bound variable. The statement is true for some values of y and w and false for others.
3. This looks quite complicated, but if we go a step at a time, we can decipher it. First, note that the statement $6 \notin \{x \mid x \text{ is divisible by } w\}$, which appears inside the given statement, means the same thing as "6 is not divisible by w." Substituting this into the given statement, we find that the original statement is equivalent to the simpler statement $2 \in \{w \mid 6 \text{ is not divisible by } w\}$. But this just means the same thing as "6 is not divisible by 2." Thus, the statement has no free variables, and both x and w are bound variables. Because there are no free variables, the truth value of the statement doesn't depend on the values of any variables. In fact, since 6 *is* divisible by 2, the statement is false.

Perhaps you have guessed by now how we can use set theory to help us understand truth values of statements containing free variables. As we have seen, a statement, say $P(x)$, containing a free variable x, may be true for some values of x and false for others. To distinguish the values of x that make $P(x)$ true from those that make it false, we could form the set of values of x for which $P(x)$ is true. We will call this set the *truth set* of $P(x)$.

Definition 1.3.4. The *truth set* of a statement $P(x)$ is the set of all values of x that make the statement $P(x)$ true. In other words, it is the set defined by using the statement $P(x)$ as an elementhood test:

$$\text{Truth set of } P(x) = \{x \mid P(x)\}.$$

Note that we have defined truth sets only for statements containing *one* free variable. We will discuss truth sets for statements with more than one free variable in Chapter 4.

Example 1.3.5. What are the truth sets of the following statements?

1. Shakespeare wrote x.
2. n is an even prime number.

Solutions

1. $\{x \mid \text{Shakespeare wrote } x\} = \{\text{Hamlet, Macbeth, Twelfth Night}, \ldots\}$.
2. $\{n \mid n \text{ is an even prime number}\}$. Because the only even prime number is 2, this is the set $\{2\}$. Note that 2 and $\{2\}$ are not the same thing! The first is a number, and the second is a set whose only element is a number. Thus, $2 \in \{2\}$, but $2 \neq \{2\}$.

Suppose A is the truth set of a statement $P(x)$. According to the definition of truth set, this means that $A = \{x \mid P(x)\}$. We've already seen that for any object y, the statement $y \in \{x \mid P(x)\}$ means the same thing as $P(y)$. Substituting in A for $\{x \mid P(x)\}$, it follows that $y \in A$ means the same thing as $P(y)$. Thus, we see that in general, if A is the truth set of $P(x)$, then to say that $y \in A$ means the same thing as saying $P(y)$.

When a statement contains free variables, it is often clear from context that these variables stand for objects of a particular kind. The set of all objects of this kind – in other words, the set of all possible values for the variables – is called the *universe of discourse* for the statement, and we say that the variables *range over* this universe. For example, in most contexts the universe for the statement $x^2 < 9$ would be the set of all real numbers; the universe for the statement "x is a man" might be the set of all people.

Certain sets come up often in mathematics as universes of discourse, and it is convenient to have fixed names for them. Here are a few of the most important ones:

$\mathbb{R} = \{x \mid x \text{ is a real number}\}$.
$\mathbb{Q} = \{x \mid x \text{ is a rational number}\}$.
(Recall that a *real* number is any number on the number line, and a *rational* number is a number that can be written as a fraction p/q, where p and q are integers.)
$\mathbb{Z} = \{x \mid x \text{ is an integer}\} = \{\ldots, -3, -2, -1, 0, 1, 2, 3, \ldots\}$.
$\mathbb{N} = \{x \mid x \text{ is a natural number}\} = \{0, 1, 2, 3, \ldots\}$.
(Some books include 0 as a natural number and some don't. In this book, we consider 0 to be a natural number.)

The letters $\mathbb{R}$, $\mathbb{Q}$, and $\mathbb{Z}$ can be followed by a superscript $+$ or $-$ to indicate that only positive or negative numbers are to be included in the set. For example, $\mathbb{R}^+ = \{x \mid x \text{ is a positive real number}\}$, and $\mathbb{Z}^- = \{x \mid x \text{ is a negative integer}\}$.

Although the universe of discourse can usually be determined from context, it is sometimes useful to identify it explicitly. Consider a statement $P(x)$ with a free variable x that ranges over a universe U. Although we have written the truth set of $P(x)$ as $\{x \mid P(x)\}$, if there were any possibility of confusion about what the universe was, we could specify it explicitly by writing $\{x \in U \mid P(x)\}$; this is read "the set of all x in U such that $P(x)$." This notation indicates that only elements of U are to be considered for elementhood in this truth set, and among elements of U, only those that pass the elementhood test $P(x)$ will actually be in the truth set. For example, consider again the statement $x^2 < 9$. If the universe of discourse for this statement were the set of all real numbers, then its truth set would be $\{x \in \mathbb{R} \mid x^2 < 9\}$, or in other words, the set of all real numbers between -3 and 3. But if the universe were the set of all integers, then the truth set would be $\{x \in \mathbb{Z} \mid x^2 < 9\} = \{-2, -1, 0, 1, 2\}$. Thus, for example, $1.58 \in \{x \in \mathbb{R} \mid x^2 < 9\}$ but $1.58 \notin \{x \in \mathbb{Z} \mid x^2 < 9\}$. Clearly, the choice of universe can sometimes make a difference!

Sometimes this explicit notation is used not to specify the universe of discourse but to restrict attention to just a part of the universe. For example, in the case of the statement $x^2 < 9$, we might want to consider the universe of discourse to be the set of all real numbers, but in the course of some reasoning involving this statement we might want to temporarily restrict our attention to only positive real numbers. We might then be interested in the set $\{x \in \mathbb{R}^+ \mid x^2 < 9\}$. As before, this notation indicates that only positive real numbers will be considered for elementhood in this set, and among positive real numbers, only those whose square is less than 9 will be in the set. Thus, for a number to be an element of this set, it must pass two tests: it must be a positive real number, and its square must be less than 9. In other words, the statement $y \in \{x \in \mathbb{R}^+ \mid x^2 < 9\}$ means the same thing as $y \in \mathbb{R}^+ \wedge y^2 < 9$. In general, $y \in \{x \in A \mid P(x)\}$ means the same thing as $y \in A \wedge P(y)$.

When a new mathematical concept has been defined, mathematicians are usually interested in studying any possible extremes of this concept. For example, when we discussed truth tables, the extremes we studied were statements whose truth tables contained only T's (tautologies) or only F's (contradictions). For the concept of the truth set of a statement containing a free variable, the corresponding extremes would be the truth sets of statements that are always true or always false. Suppose $P(x)$ is a statement containing a free variable x that ranges over a universe U. It should be clear that if $P(x)$ comes out true for every value of $x \in U$, then the truth set of $P(x)$ will be the whole universe U. For example, since the statement $x^2 \geq 0$ is true for every real number x, the truth set of this statement is $\{x \in \mathbb{R} \mid x^2 \geq 0\} = \mathbb{R}$. Of course, this is not unrelated to the concept of a tautology. For example, since $P \vee \neg P$ is a tautology, the statement $P(x) \vee \neg P(x)$ will be true for every $x \in U$, no matter what statement $P(x)$ stands for or what the universe U is, and therefore the truth set of the statement $P(x) \vee \neg P(x)$ will be U.

For a statement $P(x)$ that is false for every possible value of x, nothing in the universe can pass the elementhood test for the truth set of $P(x)$, and so this truth set must have no elements. The idea of a set with no elements may sound strange, but it arises naturally when we consider truth sets for statements that are always false. Because a set is completely determined once its elements have been specified, there is only one set that has no elements. It is called the *empty set*, or the *null set*, and is often denoted $\varnothing$. For example, $\{x \in \mathbb{Z} \mid x \neq x\} = \varnothing$.

Since the empty set has no elements, the statement $x \in \emptyset$ is an example of a statement that is always false, no matter what x is.

Another common notation for the empty set is based on the fact that any set can be named by listing its elements between braces. Since the empty set has no elements, we write nothing between the braces, like this: $\emptyset = \{\ \}$. Note that $\{\emptyset\}$ is not correct notation for the empty set. Just as we saw earlier that 2 and $\{2\}$ are not the same thing, $\emptyset$ is not the same as $\{\emptyset\}$. The first is a set with no elements, whereas the second is a set with one element, that one element being $\emptyset$, the empty set.

Exercises

*1. Analyze the logical forms of the following statements:
   (a) 3 is a common divisor of 6, 9, and 15. (Note: You did this in exercise 2 of Section 1.1, but you should be able to give a better answer now.)
   (b) x is divisible by both 2 and 3 but not 4.
   (c) x and y are natural numbers, and exactly one of them is prime.

2. Analyze the logical forms of the following statements:
   (a) x and y are men, and either x is taller than y or y is taller than x.
   (b) Either x or y has brown eyes, and either x or y has red hair.
   (c) Either x or y has both brown eyes and red hair.

*3. Write definitions using elementhood tests for the following sets:
   (a) {Mercury, Venus, Earth, Mars, Jupiter, Saturn, Uranus, Neptune, Pluto}.
   (b) {Brown, Columbia, Cornell, Dartmouth, Harvard, Princeton, University of Pennsylvania, Yale}.
   (c) {Alabama, Alaska, Arizona, ..., Wisconsin, Wyoming}.
   (d) {Alberta, British Columbia, Manitoba, New Brunswick, Newfoundland and Labrador, Northwest Territories, Nova Scotia, Nunavut, Ontario, Prince Edward Island, Quebec, Saskatchewan, Yukon}.

4. Write definitions using elementhood tests for the following sets:
   (a) $\{1, 4, 9, 16, 25, 36, 49, \ldots\}$.
   (b) $\{1, 2, 4, 8, 16, 32, 64, \ldots\}$.
   (c) $\{10, 11, 12, 13, 14, 15, 16, 17, 18, 19\}$.

*5. Simplify the following statements. Which variables are free and which are bound? If the statement has no free variables, say whether it is true or false.
   (a) $-3 \in \{x \in \mathbb{R} \mid 13 - 2x > 1\}$.
   (b) $4 \in \{x \in \mathbb{R}^- \mid 13 - 2x > 1\}$.
   (c) $5 \notin \{x \in \mathbb{R} \mid 13 - 2x > c\}$.

6. Simplify the following statements. Which variables are free and which are bound? If the statement has no free variables, say whether it is true or false.
   (a) $w \in \{x \in \mathbb{R} \mid 13 - 2x > c\}$.
   (b) $4 \in \{x \in \mathbb{R} \mid 13 - 2x \in \{y \mid y \text{ is a prime number}\}\}$. (It might make this statement easier to read if we let $P = \{y \mid y \text{ is a prime number}\}$; using this notation, we could rewrite the statement as $4 \in \{x \in \mathbb{R} \mid 13 - 2x \in P\}$.)
   (c) $4 \in \{x \in \{y \mid y \text{ is a prime number}\} \mid 13 - 2x > 1\}$. (Using the same notation as in part (b), we could write this as $4 \in \{x \in P \mid 13 - 2x > 1\}$.)

*7. What are the truth sets of the following statements? List a few elements of the truth set if you can.
   (a) Elizabeth Taylor was once married to x.
   (b) x is a logical connective studied in Section 1.1.
   (c) x is the author of this book.

8. What are the truth sets of the following statements? List a few elements of the truth set if you can.
   (a) x is a real number and $x^2 - 4x + 3 = 0$.
   (b) x is a real number and $x^2 - 2x + 3 = 0$.
   (c) x is a real number and $5 \in \{y \in \mathbb{R} \mid x^2 + y^2 < 50\}$.

1.4. Operations on Sets

Suppose A is the truth set of a statement $P(x)$ and B is the truth set of $Q(x)$. What are the truth sets of the statements $P(x) \wedge Q(x)$, $P(x) \vee Q(x)$, and $\neg P(x)$? To answer these questions, we introduce some basic operations on sets.

Definition 1.4.1. The *intersection* of two sets A and B is the set $A \cap B$ defined as follows:

$$A \cap B = \{x \mid x \in A \text{ and } x \in B\}.$$

The *union* of A and B is the set $A \cup B$ defined as follows:

$$A \cup B = \{x \mid x \in A \text{ or } x \in B\}.$$

The *difference* of A and B is the set $A \setminus B$ defined as follows:

$$A \setminus B = \{x \mid x \in A \text{ and } x \notin B\}.$$

Remember that the statements that appear in these definitions are *elementhood tests*. Thus, for example, the definition of $A \cap B$ says that for an object to be an element of $A \cap B$, it must be an element of both A and B. In other words, $A \cap B$ is the set consisting of the elements that A and B have in common.

Because the word *or* is always interpreted as *inclusive or* in mathematics, anything that is an element of either A or B, or both, will be an element of $A \cup B$. Thus, we can think of $A \cup B$ as the set resulting from throwing all the elements of A and B together into one set. $A \setminus B$ is the set you would get if you started with the set A and removed from it any elements that were also in B.

Example 1.4.2. Suppose $A = \{1, 2, 3, 4, 5\}$ and $B = \{2, 4, 6, 8, 10\}$. List the elements of the following sets:

1. $A \cap B$.
2. $A \cup B$.
3. $A \setminus B$.
4. $(A \cup B) \setminus (A \cap B)$.
5. $(A \setminus B) \cup (B \setminus A)$.

Solutions

1. $A \cap B = \{2, 4\}$.
2. $A \cup B = \{1, 2, 3, 4, 5, 6, 8, 10\}$.
3. $A \setminus B = \{1, 3, 5\}$.
4. We have just computed $A \cup B$ and $A \cap B$ in solutions 1 and 2, so all we need to do is start with the set $A \cup B$ from solution 2 and remove from it any elements that are also in $A \cap B$. The answer is $(A \cup B) \setminus (A \cap B) = \{1, 3, 5, 6, 8, 10\}$.
5. We already have the elements of $A \setminus B$ listed in solution 3, and $B \setminus A = \{6, 8, 10\}$. Thus, their union is $(A \setminus B) \cup (B \setminus A) = \{1, 3, 5, 6, 8, 10\}$. Is it just a coincidence that this is the same as the answer to part 4?

Example 1.4.3. Suppose $A = \{x \mid x \text{ is a man}\}$ and $B = \{x \mid x \text{ has brown hair}\}$. What are $A \cap B$, $A \cup B$, and $A \setminus B$?

Solution

By definition, $A \cap B = \{x \mid x \in A \text{ and } x \in B\}$. As we saw in the last section, the definitions of A and B tell us that $x \in A$ means the same thing as "x is a man," and $x \in B$ means the same thing as "x has brown hair." Plugging this into the definition of $A \cap B$, we find that

$$A \cap B = \{x \mid x \text{ is a man and } x \text{ has brown hair}\}.$$

Similar reasoning shows that

$$A \cup B = \{x \mid \text{either } x \text{ is a man or } x \text{ has brown hair}\}$$

and

$$A \setminus B = \{x \mid x \text{ is a man and } x \text{ does not have brown hair}\}.$$

Sometimes it is helpful when working with operations on sets to draw pictures of the results of these operations. One way to do this is with diagrams like that in Figure 1. This is called a *Venn diagram*. The interior of the rectangle enclosing the diagram represents the universe of discourse U, and the interiors of the two circles represent the two sets A and B. Other sets formed by combining these sets would be represented by different regions in the diagram. For example, the shaded region in Figure 2 is the region common to the circles representing A and B, and so it represents the set $A \cap B$. Figures 3 and 4 show the regions representing $A \cup B$ and $A \setminus B$, respectively.

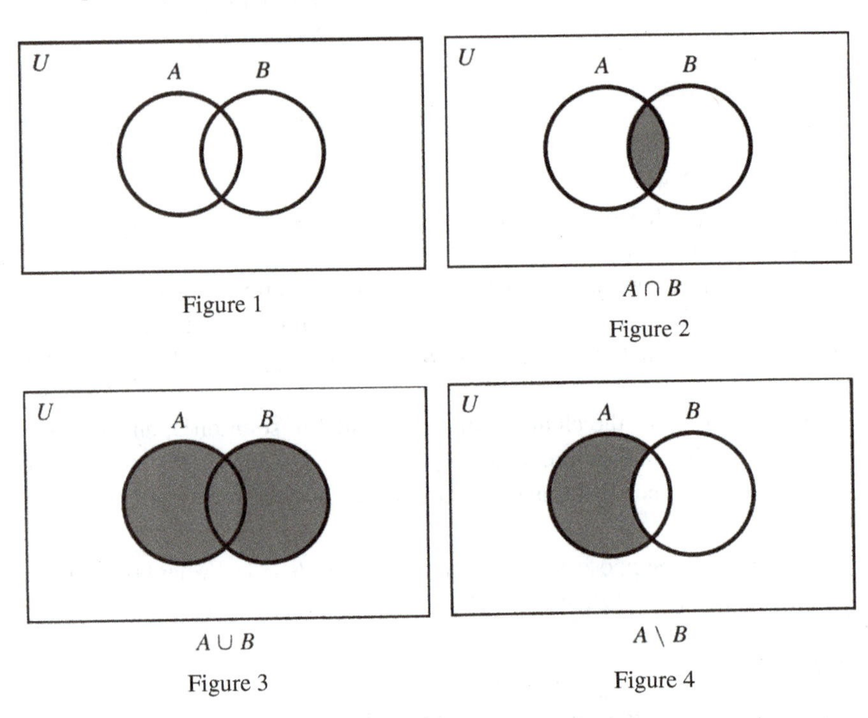

Figure 1

$A \cap B$

Figure 2

$A \cup B$

Figure 3

$A \setminus B$

Figure 4

Here's an example of how Venn diagrams can help us understand operations on sets. In Example 1.4.2 the sets $(A \cup B) \setminus (A \cap B)$ and $(A \setminus B) \cup (B \setminus A)$ turned out to be equal, for a particular choice of A and B. You can see by making Venn diagrams for both sets that this was not a coincidence. You'll find that both Venn diagrams look like Figure 5. Thus, these sets will always be equal, no matter what the sets A and B are, because both sets will always be the set of objects that are elements of either A or B but not both. This set is called the *symmetric difference* of A and B and is written $A \Delta B$. In other words, $A \Delta B = (A \setminus B) \cup (B \setminus A) = (A \cup B) \setminus (A \cap B)$. Later in this section we'll see another explanation of why these sets are always equal.

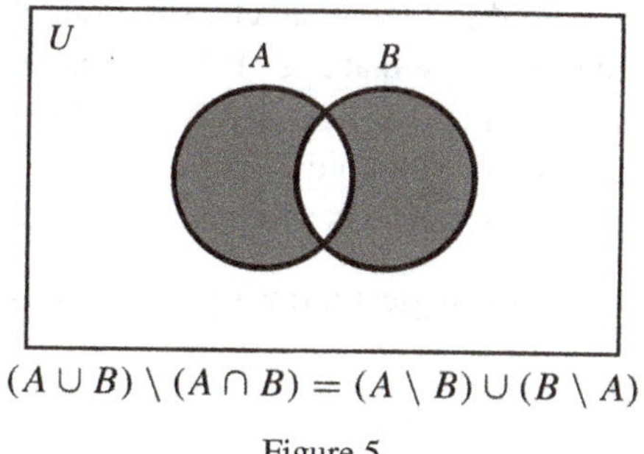

$$(A \cup B) \setminus (A \cap B) = (A \setminus B) \cup (B \setminus A)$$

Figure 5

Let's return now to the question with which we began this section. If A is the truth set of a statement $P(x)$ and B is the truth set of $Q(x)$, then, as we saw in the last section, $x \in A$ means the same thing as $P(x)$ and $x \in B$ means the same thing as $Q(x)$. Thus, the truth set of $P(x) \wedge Q(x)$ is $\{x \mid P(x) \wedge Q(x)\} = \{x \mid x \in A \wedge x \in B\} = A \cap B$. This should make sense. It just says that the truth set of $P(x) \wedge Q(x)$ consists of those elements that the truth sets of $P(x)$ and $Q(x)$ have in common – in other words, the values of x that make both $P(x)$ and $Q(x)$ come out true. We have already seen an example of this. In Example 1.4.3 the sets A and B were the truth sets of the statements "x is a man" and "x has brown hair," and $A \cap B$ turned out to be the truth set of "x is a man and x has brown hair."

Similar reasoning shows that the truth set of $P(x) \vee Q(x)$ is $A \cup B$. To find the truth set of $\neg P(x)$, we need to talk about the universe of discourse U. The truth set of $\neg P(x)$ will consist of those elements of the universe for which $P(x)$ is false, and we can find this set by starting with U and removing from it those elements for which $P(x)$ is true. Thus, the truth set of $\neg P(x)$ is $U \setminus A$.

These observations about truth sets illustrate the fact that the set theory operations $\cap$, $\cup$, and $\setminus$ are related to the logical connectives $\wedge$, $\vee$, and $\neg$. This shouldn't be surprising, since after all the words *and*, *or*, and *not* appear in their definitions. (The word *not* doesn't appear explicitly, but it's there, hidden in the mathematical symbol $\notin$ in the definition of the difference of two sets.) It is important to remember, though, that although the set theory operations and logical connectives are related, they are not interchangeable. The logical connectives can only be used to combine *statements*, whereas the set theory operations must be used to combine *sets*. For example, if A is the truth set of $P(x)$ and B is the truth set of $Q(x)$, then we can say that $A \cap B$ is the truth set of $P(x) \wedge Q(x)$, but expressions such as $A \wedge B$ or $P(x) \cap Q(x)$ are completely meaningless and should never be used.

The relationship between set theory operations and logical connectives also becomes apparent when we analyze the logical forms of statements about intersections, unions, and differences of sets. For example, according to the definition of intersection, to say that $x \in A \cap B$ means that $x \in A \wedge x \in B$. Similarly, to say that $x \in A \cup B$ means that $x \in A \vee x \in B$, and $x \in A \setminus B$ means $x \in A \wedge x \notin B$, or in other words $x \in A \wedge \neg(x \in B)$. We can combine these rules when analyzing statements about more complex sets.

Example 1.4.4. Analyze the logical forms of the following statements:

1. $x \in A \cap (B \cup C)$.
2. $x \in A \setminus (B \cap C)$.
3. $x \in (A \cap B) \cup (A \cap C)$.

Solutions

1. $x \in A \cap (B \cup C)$
   is equivalent to $x \in A \wedge x \in (B \cup C)$ (definition of $\cap$),
   which is equivalent to $x \in A \wedge (x \in B \vee x \in C)$ (definition of $\cup$).
2. $x \in A \setminus (B \cap C)$
   is equivalent to $x \in A \wedge \neg(x \in B \cap C)$ (definition of $\setminus$),
   which is equivalent to $x \in A \wedge \neg(x \in B \wedge x \in C)$ (definition of $\cap$).
3. $x \in (A \cap B) \cup (A \cap C)$
   is equivalent to $x \in (A \cap B) \vee x \in (A \cap C)$ (definition of $\cup$),
   which is equivalent to $(x \in A \wedge x \in B) \vee (x \in A \wedge x \in C)$
   (definition of $\cap$).

Look again at the solutions to parts 1 and 3 of Example 1.4.4. You should recognize that the statements we ended up with in these two parts are equivalent. (If you don't, look back at the distributive laws in Section 1.2.) This equivalence means that the statements $x \in A \cap (B \cup C)$ and $x \in (A \cap B) \cup (A \cap C)$ are equivalent. In other words, the objects that are elements of the set $A \cap (B \cup C)$ will be precisely the same as the objects that are elements of $(A \cap B) \cup (A \cap C)$, no matter what the sets A, B, and C are. But recall that sets with the same elements are equal, so it follows that for any sets A, B, and C, $A \cap (B \cup C) = (A \cap B) \cup (A \cap C)$. Another way to see this is with the Venn diagram in Figure 6. Our earlier Venn diagrams had two circles, because in previous examples only two sets were being combined. This Venn diagram has three circles, which represent the three sets A, B, and C that are being combined in this case. Although it is possible to create Venn diagrams for more than three sets, it is rarely done, because it cannot be done with overlapping circles. For more on Venn diagrams for more than three sets, see exercise 10.

Thus, we see that a distributive law for logical connectives has led to a distributive law for set theory operations. You might guess that because there

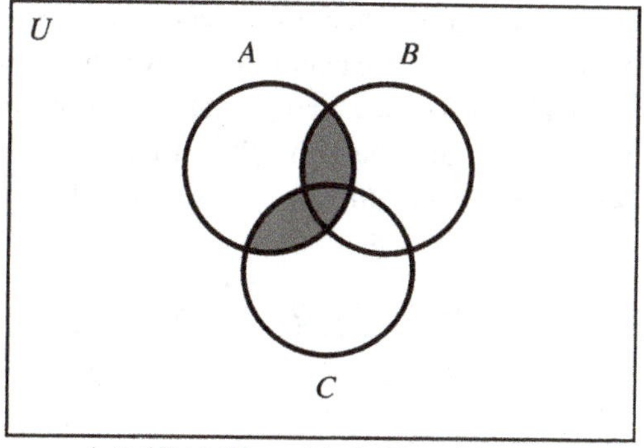

$A \cap (B \cup C) = (A \cap B) \cup (A \cap C)$

Figure 6

were *two* distributive laws for the logical connectives, with $\wedge$ and $\vee$ playing opposite roles in the two laws, there might be two distributive laws for set theory operations too. The second distributive law for sets should say that for any sets A, B, and C, $A \cup (B \cap C) = (A \cup B) \cap (A \cup C)$. You can verify this for yourself by writing out the statements $x \in A \cup (B \cap C)$ and $x \in (A \cup B) \cap (A \cup C)$ using logical connectives and verifying that they are equivalent, using the second distributive law for the logical connectives $\wedge$ and $\vee$. Another way to see it is to make a Venn diagram.

We can derive another set theory identity by finding a statement equivalent to the statement we ended up with in part 2 of Example 1.4.4:

$x \in A \setminus (B \cap C)$
is equivalent to $x \in A \wedge \neg(x \in B \wedge x \in C)$ (Example 1.4.4),
which is equivalent to $x \in A \wedge (x \notin B \vee x \notin C)$ (DeMorgan's law),
which is equivalent to $(x \in A \wedge x \notin B) \vee (x \in A \wedge x \notin C)$
(distributive law),
which is equivalent to $(x \in A \setminus B) \vee (x \in A \setminus C)$ (definition of $\setminus$),
which is equivalent to $x \in (A \setminus B) \cup (A \setminus C)$ (definition of $\cup$).

Thus, we have shown that for any sets A, B, and C, $A \setminus (B \cap C) = (A \setminus B) \cup (A \setminus C)$. Once again, you can verify this with a Venn diagram as well.

Earlier we promised an alternative way to check the identity $(A \cup B) \setminus (A \cap B) = (A \setminus B) \cup (B \setminus A)$. You should see now how this can be done. First, we write out the logical forms of the statements $x \in (A \cup B) \setminus (A \cap B)$ and $x \in (A \setminus B) \cup (B \setminus A)$:

$x \in (A \cup B) \setminus (A \cap B)$ means $(x \in A \vee x \in B) \wedge \neg(x \in A \wedge x \in B)$;

$x \in (A \setminus B) \cup (B \setminus A)$ means $(x \in A \wedge x \notin B) \vee (x \in B \wedge x \notin A)$.

You can now check, using equivalences from Section 1.2, that these statements are equivalent. An alternative way to check the equivalence is with a truth table. To simplify the truth table, let's use P and Q as abbreviations for the statements $x \in A$ and $x \in B$. Then we must check that the formulas $(P \vee Q) \wedge \neg(P \wedge Q)$ and $(P \wedge \neg Q) \vee (Q \wedge \neg P)$ are equivalent. The truth table in Figure 7 shows this.

P	Q	$(P \vee Q) \wedge \neg(P \wedge Q)$	$(P \wedge \neg Q) \vee (Q \wedge \neg P)$
F	F	F	F
F	T	T	T
T	F	T	T
T	T	F	F

Figure 7

Definition 1.4.5. Suppose A and B are sets. We will say that A is a *subset* of B if every element of A is also an element of B. We write $A \subseteq B$ to mean that A is a subset of B. A and B are said to be *disjoint* if they have no elements in common. Note that this is the same as saying that the set of elements they have in common is the empty set, or in other words $A \cap B = \varnothing$.

Example 1.4.6. Suppose $A = \{\text{red, green}\}$, $B = \{\text{red, yellow, green, purple}\}$, and $C = \{\text{blue, purple}\}$. Then the two elements of A, red and green, are both also in B, and therefore $A \subseteq B$. Also, $A \cap C = \varnothing$, so A and C are disjoint.

If we know that $A \subseteq B$, or that A and B are disjoint, then we might draw a Venn diagram for A and B differently to reflect this. Figures 8 and 9 illustrate this.

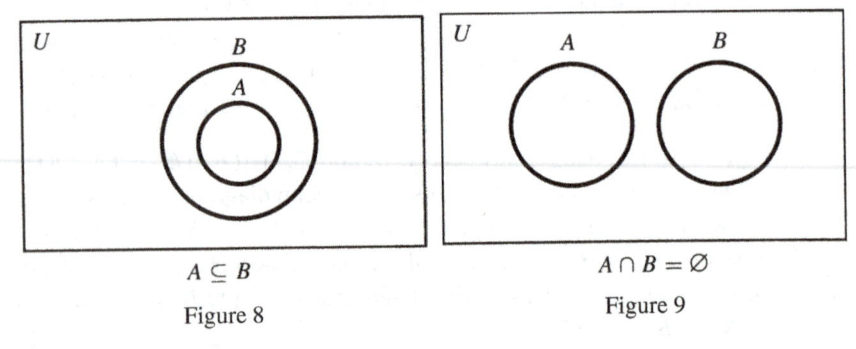

$A \subseteq B$

Figure 8

$A \cap B = \varnothing$

Figure 9

Just as we earlier derived identities showing that certain sets are always equal, it is also sometimes possible to show that certain sets are always disjoint, or that one set is always a subset of another. For example, you can see in a Venn diagram that the sets $A \cap B$ and $A \setminus B$ do not overlap, and therefore they will always be disjoint for any sets A and B. Another way to see this would be to write out what it means to say that $x \in (A \cap B) \cap (A \setminus B)$:

$$x \in (A \cap B) \cap (A \setminus B) \text{ means } (x \in A \wedge x \in B) \wedge (x \in A \wedge x \notin B),$$
$$\text{which is equivalent to } x \in A \wedge (x \in B \wedge x \notin B).$$

But this last statement is clearly a contradiction, so the statement $x \in (A \cap B) \cap (A \setminus B)$ will always be false, no matter what x is. In other words, nothing can be an element of $(A \cap B) \cap (A \setminus B)$, so it must be the case that $(A \cap B) \cap (A \setminus B) = \varnothing$. Therefore, $A \cap B$ and $A \setminus B$ are disjoint.

The next theorem gives another example of a general fact about set operations. The proof of this theorem illustrates that the principles of deductive reasoning we have been studying are actually used in mathematical proofs.

Theorem 1.4.7. *For any sets A and B, $(A \cup B) \setminus B \subseteq A$.*

Proof. We must show that if something is an element of $(A \cup B) \setminus B$, then it must also be an element of A, so suppose that $x \in (A \cup B) \setminus B$. This means that $x \in A \cup B$ and $x \notin B$, or in other words $x \in A \vee x \in B$ and $x \notin B$. But notice that these statements have the logical form $P \vee Q$ and $\neg Q$, and this is precisely the form of the premises of our very first example of a deductive argument in Section 1.1! As we saw in that example, from these premises we can conclude that $x \in A$ must be true. Thus, anything that is an element of $(A \cup B) \setminus B$ must also be an element of A, so $(A \cup B) \setminus B \subseteq A$. □

You might think that such a careful application of logical laws is not needed to understand why Theorem 1.4.7 is correct. The set $(A \cup B) \setminus B$ could be thought of as the result of starting with the set A, adding in the elements of B, and then removing them again. Common sense suggests that the result will just be the original set A; in other words, it appears that $(A \cup B) \setminus B = A$. However, as you are asked to show in exercise 9, this conclusion is incorrect. This illustrates that in mathematics, you must not allow imprecise reasoning to lead you to jump to conclusions. Applying laws of logic carefully, as we did in our proof of Theorem 1.4.7, may help you to avoid jumping to unwarranted conclusions.

Exercises

*1. Let $A = \{1, 3, 12, 35\}$, $B = \{3, 7, 12, 20\}$, and $C = \{x \mid x \text{ is a prime number}\}$. List the elements of the following sets. Are any of the sets below disjoint from any of the others? Are any of the sets below subsets of any others?

(a) $A \cap B$.
(b) $(A \cup B) \setminus C$.
(c) $A \cup (B \setminus C)$.

2. Let $A = \{$United States, Germany, China, Australia$\}$, $B = \{$Germany, France, India, Brazil$\}$, and $C = \{x \mid x$ is a country in Europe$\}$. List the elements of the following sets. Are any of the sets below disjoint from any of the others? Are any of the sets below subsets of any others?
   (a) $A \cup B$.
   (b) $(A \cap B) \setminus C$.
   (c) $(B \cap C) \setminus A$.
3. Verify that the Venn diagrams for $(A \cup B) \setminus (A \cap B)$ and $(A \setminus B) \cup (B \setminus A)$ both look like Figure 5, as stated in this section.

*4. Use Venn diagrams to verify the following identities:
   (a) $A \setminus (A \cap B) = A \setminus B$.
   (b) $A \cup (B \cap C) = (A \cup B) \cap (A \cup C)$.

5. Verify the identities in exercise 4 by writing out (using logical symbols) what it means for an object x to be an element of each set and then using logical equivalences.
6. Use Venn diagrams to verify the following identities:
   (a) $(A \cup B) \setminus C = (A \setminus C) \cup (B \setminus C)$.
   (b) $A \cup (B \setminus C) = (A \cup B) \setminus (C \setminus A)$.
7. Verify the identities in exercise 6 by writing out (using logical symbols) what it means for an object x to be an element of each set and then using logical equivalences.

*8. For each of the following sets, write out (using logical symbols) what it means for an object x to be an element of the set. Then determine which of these sets must be equal to each other by determining which statements are equivalent.
   (a) $(A \setminus B) \setminus C$.
   (b) $A \setminus (B \setminus C)$.
   (c) $(A \setminus B) \cup (A \cap C)$.
   (d) $(A \setminus B) \cap (A \setminus C)$.
   (e) $A \setminus (B \cup C)$.

9. It was shown in this section that for any sets A and B, $(A \cup B) \setminus B \subseteq A$. Give an example of two sets A and B for which $(A \cup B) \setminus B \neq A$.

*10. It is claimed in this section that you cannot make a Venn diagram for four sets using overlapping circles.

(a) What's wrong with the following diagram? (Hint: Where's the set $(A \cap D) \setminus (B \cup C)$?)

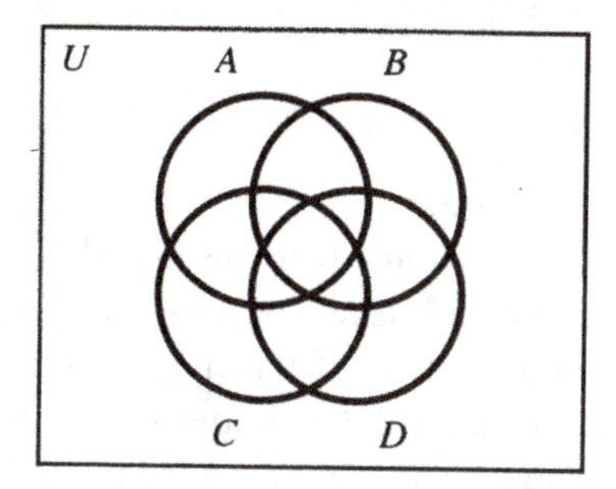

(b) Can you make a Venn diagram for four sets using shapes other than circles?

11. (a) Make Venn diagrams for the sets $(A \cup B) \setminus C$ and $A \cup (B \setminus C)$. What can you conclude about whether one of these sets is necessarily a subset of the other?
    (b) Give an example of sets *A*, *B*, and *C* for which $(A \cup B) \setminus C \neq A \cup (B \setminus C)$.

*12. Use Venn diagrams to show that the associative law holds for symmetric difference; that is, for any sets *A*, *B*, and *C*, $A \triangle (B \triangle C) = (A \triangle B) \triangle C$.

13. Use any method you wish to verify the following identities:
    (a) $(A \triangle B) \cup C = (A \cup C) \triangle (B \setminus C)$.
    (b) $(A \triangle B) \cap C = (A \cap C) \triangle (B \cap C)$.
    (c) $(A \triangle B) \setminus C = (A \setminus C) \triangle (B \setminus C)$.

14. Use any method you wish to verify the following identities:
    (a) $(A \cup B) \triangle C = (A \triangle C) \triangle (B \setminus A)$.
    (b) $(A \cap B) \triangle C = (A \triangle C) \triangle (A \setminus B)$.
    (c) $(A \setminus B) \triangle C = (A \triangle C) \triangle (A \cap B)$.

15. Fill in the blanks to make true identities:
    (a) $(A \triangle B) \cap C = (C \setminus A) \triangle$ ________.
    (b) $C \setminus (A \triangle B) = (A \cap C) \triangle$ ________.
    (c) $(B \setminus A) \triangle C = (A \triangle C) \triangle$ ________.

1.5. The Conditional and Biconditional Connectives

It is time now to return to a question we left unanswered in Section 1.1. We have seen how the reasoning in the first and third arguments in Example 1.1.1 can be understood by analyzing the connectives $\vee$ and $\neg$. But what about the reasoning in the second argument? Recall that the argument went like this:

> If today is Sunday, then I don't have to go to work today.
> Today is Sunday.
> Therefore, I don't have to go to work today.

What makes this reasoning valid?

It appears that the crucial words here are *if* and *then*, which occur in the first premise. We therefore introduce a new logical connective, $\rightarrow$, and write $P \rightarrow Q$ to represent the statement "If P then Q." This statement is sometimes called a *conditional* statement, with P as its *antecedent* and Q as its *consequent*. If we let P stand for the statement "Today is Sunday" and Q for the statement "I don't have to go to work today," then the logical form of the argument would be

$$\begin{array}{l} P \rightarrow Q \\ \underline{P \qquad\quad} \\ \therefore Q \end{array}$$

Our analysis of the new connective $\rightarrow$ should lead to the conclusion that this argument is valid.

Example 1.5.1. Analyze the logical forms of the following statements:

1. If it's raining and I don't have my umbrella, then I'll get wet.
2. If Mary did her homework, then the teacher won't collect it, and if she didn't, then he'll ask her to do it on the board.

Solutions

1. Let R stand for the statement "It's raining," U for "I have my umbrella," and W for "I'll get wet." Then statement 1 would be represented by the formula $(R \wedge \neg U) \rightarrow W$.
2. Let H stand for "Mary did her homework," C for "The teacher will collect it," and B for "The teacher will ask Mary to do the homework on the board." Then the given statement means $(H \rightarrow \neg C) \wedge (\neg H \rightarrow B)$.

To analyze arguments containing the connective $\rightarrow$ we must work out the truth table for the formula $P \rightarrow Q$. Because $P \rightarrow Q$ is supposed to mean that if P is true then Q is also true, we certainly want to say that if P is true and Q is false then $P \rightarrow Q$ is false. If P is true and Q is also true, then it seems reasonable to say that $P \rightarrow Q$ is true. This gives us the last two lines of the truth table in Figure 1. The remaining two lines of the truth table are harder to fill in, although some people might say that if P and Q are both false then

$P \to Q$ should be considered true. Thus, we can sum up our conclusions so far with the table in Figure 1.

P	Q	$P \to Q$
F	F	T?
F	T	?
T	F	F
T	T	T

Figure 1

To help us fill in the undetermined lines in this truth table, let's look at an example. Consider the statement "If $x > 2$ then $x^2 > 4$," which we could represent with the formula $P(x) \to Q(x)$, where $P(x)$ stands for the statement $x > 2$ and $Q(x)$ stands for $x^2 > 4$. Of course, the statements $P(x)$ and $Q(x)$ contain x as a free variable, and each will be true for some values of x and false for others. But surely, no matter what the value of x is, we would say it is true that *if* $x > 2$ then $x^2 > 4$, so the conditional statement $P(x) \to Q(x)$ should be true. Thus, the truth table should be completed in such a way that no matter what value we plug in for x, this conditional statement comes out true.

For example, suppose $x = 3$. In this case $x > 2$ and $x^2 = 9 > 4$, so $P(x)$ and $Q(x)$ are both true. This corresponds to line four of the truth table in Figure 1, and we've already decided that the statement $P(x) \to Q(x)$ should come out true in this case. But now consider the case $x = 1$. Then $x < 2$ and $x^2 = 1 < 4$, so $P(x)$ and $Q(x)$ are both false, corresponding to line one in the truth table. We have tentatively placed a T in this line of the truth table, and now we see that this tentative choice must be right. If we put an F there, then the statement $P(x) \to Q(x)$ would come out false in the case $x = 1$, and we've already decided that it should be true for all values of x.

Finally, consider the case $x = -5$. Then $x < 2$, so $P(x)$ is false, but $x^2 = 25 > 4$, so $Q(x)$ is true. Thus, in this case we find ourselves in the second line of the truth table, and once again, if the conditional statement $P(x) \to Q(x)$ is to be true in this case, we must put a T in this line. So it appears that all the questionable lines in the truth table in Figure 1 must be filled in with T's, and the completed truth table for the connective $\to$ must be as shown in Figure 2.

P	Q	$P \to Q$
F	F	T
F	T	T
T	F	F
T	T	T

Figure 2

Of course, there are many other values of x that could be plugged into our statement "If $x > 2$ then $x^2 > 4$"; but if you try them, you'll find that they all lead to line one, two, or four of the truth table, as our examples $x = 1, -5$, and 3 did. No value of x will lead to line three, because you could never have $x > 2$ but $x^2 \leq 4$. After all, that's why we said that the statement "If $x > 2$ then $x^2 > 4$" was always true, no matter what x was! The point of saying that this conditional statement is always true is simply to say that you will never find a value of x such that $x > 2$ and $x^2 \leq 4$ – in other words, there is no value of x for which $P(x)$ is true but $Q(x)$ is false. Thus, it should make sense that in the truth table for $P \rightarrow Q$, the only line that is false is the line in which P is true and Q is false.

As the truth table in Figure 3 shows, the formula $\neg P \vee Q$ is also true in every case except when P is true and Q is false. Thus, if we accept the truth table in Figure 2 as the correct truth table for the formula $P \rightarrow Q$, then we will be forced to accept the conclusion that the formulas $P \rightarrow Q$ and $\neg P \vee Q$ are equivalent. Is this consistent with the way the words *if* and *then* are used in ordinary language? It may not seem to be at first, but, at least for some uses of the words *if* and *then*, it is.

P	Q	$\neg P \vee Q$
F	F	T
F	T	T
T	F	F
T	T	T

Figure 3

For example, imagine a teacher saying to a class, in a threatening tone of voice, "You won't neglect your homework, or you'll fail the course." Grammatically, this statement has the form $\neg P \vee Q$, where P is the statement "You will neglect your homework" and Q is "You'll fail the course." But what message is the teacher trying to convey with this statement? Clearly the intended message is "If you neglect your homework, then you'll fail the course," or in other words $P \rightarrow Q$. Thus, in this example, the statements $\neg P \vee Q$ and $P \rightarrow Q$ seem to mean the same thing.

There is a similar idea at work in the first statement from Example 1.1.2, "Either John went to the store, or we're out of eggs." In Section 1.1 we represented this statement by the formula $P \vee Q$, with P standing for "John went to the store" and Q for "We're out of eggs." But someone who made this statement would probably be trying to express the idea that if John didn't go to the store, then we're out of eggs, or in other words $\neg P \rightarrow Q$. Thus, this example suggests that $\neg P \rightarrow Q$ means the same thing as $P \vee Q$. In fact, we can derive this equivalence from the previous one by substituting $\neg P$ for P. Because $P \rightarrow Q$

is equivalent to $\neg P \vee Q$, it follows that $\neg P \rightarrow Q$ is equivalent to $\neg\neg P \vee Q$, which is equivalent to $P \vee Q$ by the double negation law.

We can derive another useful equivalence as follows:

$$\begin{aligned} \neg P \vee Q \quad & \text{is equivalent to } \neg P \vee \neg\neg Q \quad \text{(double negation law)}, \\ & \text{which is equivalent to } \neg(P \wedge \neg Q) \quad \text{(DeMorgan's law)}. \end{aligned}$$

Thus, $P \rightarrow Q$ is also equivalent to $\neg(P \wedge \neg Q)$. In fact, this is precisely the conclusion we reached earlier when discussing the statement "If $x > 2$ then $x^2 > 4$." We decided then that the reason this statement is true for every value of x is that there is no value of x for which $x > 2$ and $x^2 \leq 4$. In other words, the statement $P(x) \wedge \neg Q(x)$ is never true, where as before $P(x)$ stands for $x > 2$ and $Q(x)$ for $x^2 > 4$. But that's the same as saying that the statement $\neg(P(x) \wedge \neg Q(x))$ is always true. Thus, to say that $P(x) \rightarrow Q(x)$ is always true means the same thing as saying that $\neg(P(x) \wedge \neg Q(x))$ is always true.

For another example of this equivalence, consider the statement "If it's going to rain, then I'll take my umbrella." Of course, this statement has the form $P \rightarrow Q$, where P stands for the statement "It's going to rain" and Q stands for "I'll take my umbrella." But we could also think of this statement as a declaration that I won't be caught in the rain without my umbrella – in other words, $\neg(P \wedge \neg Q)$.

To summarize, so far we have discovered the following equivalences involving conditional statements:

Conditional laws

$$P \rightarrow Q \text{ is equivalent to } \neg P \vee Q.$$

$$P \rightarrow Q \text{ is equivalent to } \neg(P \wedge \neg Q).$$

In case you're still not convinced that the truth table in Figure 2 is right, we give one more reason. We know that, using this truth table, we can now analyze the validity of deductive arguments involving the words *if* and *then*. We'll find, when we analyze a few simple arguments, that the truth table in Figure 2 leads to reasonable conclusions about the validity of these arguments. But if we were to make any changes in the truth table, we would end up with conclusions that are clearly incorrect. For example, let's return to the argument form with which we started this section:

$$\begin{array}{l} P \rightarrow Q \\ \underline{P \qquad\quad} \\ \therefore Q \end{array}$$

We have already decided that this form of argument should be valid, and the truth table in Figure 4 confirms this. The premises are both true only in line four of the table, and in this line the conclusion is true as well.

		Premises		Conclusion
P	Q	$P \to Q$	P	Q
F	F	T	F	F
F	T	T	F	T
T	F	F	T	F
T	T	T	T	T

Figure 4

You can also see from Figure 4 that both premises are needed to make this argument valid. But if we were to change the truth table for the conditional statement to make $P \to Q$ false in the first line of the table, then the second premise of this argument would no longer be needed. We would end up with the conclusion that, just from the single premise $P \to Q$, we could infer that Q must be true, since in the two lines of the truth table in which the premise $P \to Q$ would still be true, lines two and four, the conclusion Q is true too. But this doesn't seem right. Just knowing that *if* P is true then Q is true, but not knowing that P *is* true, it doesn't seem reasonable that we should be able to conclude that Q is true. For example, suppose we know that the statement "If John didn't go to the store then we're out of eggs" is true. Unless we also know whether or not John has gone to the store, we can't reach any conclusion about whether or not we're out of eggs. Thus, changing the first line of the truth table for $P \to Q$ would lead to an incorrect conclusion about the validity of an argument.

Changing the second line of the truth table would also lead to unacceptable conclusions about the validity of arguments. To see this, consider the argument form:

$$\begin{array}{l} P \to Q \\ \underline{Q} \\ \therefore P \end{array}$$

This should *not* be considered a valid form of reasoning. For example, consider the following argument, which has this form:

> If Jones was convicted of murdering Smith, then he will go to jail.
> Jones will go to jail.
> Therefore, Jones was convicted of murdering Smith.

Even if the premises of this argument are true, the conclusion that Jones was convicted of murdering Smith doesn't follow. Maybe the reason he will go to jail is that he robbed a bank or cheated on his income tax. Thus, the conclusion of this argument could be false even if the premises were true, so the argument isn't valid.

The truth table analysis in Figure 5 agrees with this conclusion. In line two of the table, the conclusion P is false, but both premises are true, so the argument is invalid. But notice that if we were to change the truth table for $P \to Q$ and make it false in line two, then the truth table analysis would say that the argument is valid. Thus, the analysis of this argument seems to support our decision to put a T in the second line of the truth table for $P \to Q$.

		Premises		Conclusion
P	Q	$P \to Q$	Q	P
F	F	T	F	F
F	T	T	T	F
T	F	F	F	T
T	T	T	T	T

Figure 5

The last example shows that from the premises $P \to Q$ and Q it is incorrect to infer P. But it would certainly be correct to infer P from the premises $Q \to P$ and Q. This shows that the formulas $P \to Q$ and $Q \to P$ do *not* mean the same thing. You can check this by making a truth table for both and verifying that they are not equivalent. For example, a person might believe that, in general, the statement "If you are a convicted murderer then you are untrustworthy" is true, without believing that the statement "If you are untrustworthy then you are a convicted murderer" is generally true. The formula $Q \to P$ is called the *converse* of $P \to Q$. It is very important to make sure you never confuse a conditional statement with its converse.

The *contrapositive* of $P \to Q$ is the formula $\neg Q \to \neg P$, and it *is* equivalent to $P \to Q$. This may not be obvious at first, but you can verify it with a truth table. For example, the statements "If John cashed the check I wrote then my bank account is overdrawn" and "If my bank account isn't overdrawn then John hasn't cashed the check I wrote" are equivalent. Both would be true in exactly the same circumstances – namely, if the check I wrote was for more money than I had in my account. The equivalence of conditional statements and their contrapositives is used often in mathematical reasoning. We add it to our list of important equivalences:

Contrapositive law

$$P \to Q \text{ is equivalent to } \neg Q \to \neg P.$$

Example 1.5.2. Which of the following statements are equivalent?

1. If it's either raining or snowing, then the game has been canceled.
2. If the game hasn't been canceled, then it's not raining and it's not snowing.

3. If the game has been canceled, then it's either raining or snowing.
4. If it's raining then the game has been canceled, and if it's snowing then the game has been canceled.
5. If it's neither raining nor snowing, then the game hasn't been canceled.

Solution

We translate all of the statements into the notation of logic, using the following abbreviations: R stands for the statement "It's raining," S stands for "It's snowing," and C stands for "The game has been canceled."

1. $(R \lor S) \to C$.
2. $\neg C \to (\neg R \land \neg S)$. By one of DeMorgan's laws, this is equivalent to $\neg C \to \neg(R \lor S)$. This is the contrapositive of statement 1, so they are equivalent.
3. $C \to (R \lor S)$. This is the converse of statement 1, which is *not* equivalent to it. You can verify this with a truth table, or just think about what the statements mean. Statement 1 says that rain or snow would result in cancelation of the game. Statement 3 says that these are the *only* circumstances in which the game will be canceled.
4. $(R \to C) \land (S \to C)$. This is also equivalent to statement 1, as the following reasoning shows:

   $(R \to C) \land (S \to C)$
   is equivalent to $(\neg R \lor C) \land (\neg S \lor C)$ (conditional law),
   which is equivalent to $(\neg R \land \neg S) \lor C$ (distributive law),
   which is equivalent to $\neg(R \lor S) \lor C$ (DeMorgan's law),
   which is equivalent to $(R \lor S) \to C$ (conditional law).

   You should read statements 1 and 4 again and see if it makes sense to you that they're equivalent.
5. $\neg(R \lor S) \to \neg C$. This is the contrapositive of statement 3, so they are equivalent. It is not equivalent to statements 1, 2, and 4.

Statements that mean $P \to Q$ come up very often in mathematics, but sometimes they are not written in the form "If P then Q." Here are a few other ways of expressing the idea $P \to Q$ that are used often in mathematics:

P implies Q.
Q, if P.
P only if Q.
P is a sufficient condition for Q.
Q is a necessary condition for P.

Some of these may require further explanation. The second expression, "Q, if P," is just a slight rearrangement of the statement "If P then Q," so it should make sense that it means $P \to Q$. As an example of a statement of the form "P only if Q," consider the sentence "You can run for president only if you are a citizen." In this case, P is "You can run for president" and Q is "You are a citizen." What the statement means is that if you're not a citizen, then you can't run for president, or in other words $\neg Q \to \neg P$. But by the contrapositive law, this is equivalent to $P \to Q$.

Think of "P is a sufficient condition for Q" as meaning "The truth of P suffices to guarantee the truth of Q," and it should make sense that this should be represented by $P \to Q$. Finally, "Q is a necessary condition for P" means that in order for P to be true, it is necessary for Q to be true also. This means that if Q isn't true, then P can't be true either, or in other words, $\neg Q \to \neg P$. Once again, by the contrapositive law we get $P \to Q$.

Example 1.5.3. Analyze the logical forms of the following statements:

1. If at least ten people are there, then the lecture will be given.
2. The lecture will be given only if at least ten people are there.
3. The lecture will be given if at least ten people are there.
4. Having at least ten people there is a sufficient condition for the lecture being given.
5. Having at least ten people there is a necessary condition for the lecture being given.

Solutions

Let T stand for the statement "At least ten people are there" and L for "The lecture will be given."

1. $T \to L$.
2. $L \to T$. The given statement means that if there are not at least ten people there, then the lecture will not be given, or in other words $\neg T \to \neg L$. By the contrapositive law, this is equivalent to $L \to T$.
3. $T \to L$. This is just a rephrasing of statement 1.
4. $T \to L$. The statement says that having at least ten people there suffices to guarantee that the lecture will be given, and this means that if there are at least ten people there, then the lecture will be given.
5. $L \to T$. This statement means the same thing as statement 2: If there are not at least ten people there, then the lecture will not be given.

We have already seen that a conditional statement $P \to Q$ and its converse $Q \to P$ are not equivalent. Often in mathematics we want to say that both $P \to Q$ and $Q \to P$ are true, and it is therefore convenient to introduce a new connective symbol, $\leftrightarrow$, to express this. You can think of $P \leftrightarrow Q$ as just an abbreviation for the formula $(P \to Q) \wedge (Q \to P)$. A statement of the form $P \leftrightarrow Q$ is called a *biconditional* statement, because it represents two conditional statements. By making a truth table for $(P \to Q) \wedge (Q \to P)$ you can verify that the truth table for $P \leftrightarrow Q$ is as shown in Figure 6. Note that, by the contrapositive law, $P \leftrightarrow Q$ is also equivalent to $(P \to Q) \wedge (\neg P \to \neg Q)$.

P	Q	$P \leftrightarrow Q$
F	F	T
F	T	F
T	F	F
T	T	T

Figure 6

Because $Q \to P$ can be written "P if Q" and $P \to Q$ can be written "P only if Q," $P \leftrightarrow Q$ means "P if Q and P only if Q," and this is often written "P if and only if Q." The phrase *if and only if* occurs so often in mathematics that there is a common abbreviation for it, *iff*. Thus, $P \leftrightarrow Q$ is usually written "P iff Q." Another statement that means $P \leftrightarrow Q$ is "P is a necessary and sufficient condition for Q."

Example 1.5.4. Analyze the logical forms of the following statements:

1. The game will be canceled iff it's either raining or snowing.
2. Having at least ten people there is a necessary and sufficient condition for the lecture being given.
3. If John went to the store then we have some eggs, and if he didn't then we don't.

Solutions

1. Let C stand for "The game will be canceled," R for "It's raining," and S for "It's snowing." Then the statement would be represented by the formula $C \leftrightarrow (R \vee S)$.
2. Let T stand for "There are at least ten people there" and L for "The lecture will be given." Then the statement means $T \leftrightarrow L$.
3. Let S stand for "John went to the store" and E for "We have some eggs." Then a literal translation of the given statement would be $(S \to E) \wedge (\neg S \to \neg E)$. This is equivalent to $S \leftrightarrow E$.

One of the reasons it's so easy to confuse a conditional statement with its converse is that in everyday speech we sometimes use a conditional statement when what we mean to convey is actually a biconditional. For example, you probably wouldn't say "The lecture will be given if at least ten people are there" unless it was also the case that if there were fewer than ten people, the lecture wouldn't be given. After all, why mention the number ten at all if it's not the minimum number of people required? Thus, the statement actually suggests that the lecture will be given *iff* there are at least ten people there. For another example, suppose a child is told by his parents, "If you don't eat your dinner, you won't get any dessert." The child certainly expects that if he *does* eat his dinner, he *will* get dessert, although that's not literally what his parents said. In other words, the child interprets the statement as meaning "Eating your dinner is a necessary *and sufficient* condition for getting dessert."

Such a blurring of the distinction between *if* and *iff* is never acceptable in mathematics. Mathematicians always use a phrase such as *iff* or *necessary and sufficient condition* when they want to express a biconditional statement. You should never interpret an if-then statement in mathematics as a biconditional statement, the way you might in everyday speech.

Exercises

*1. Analyze the logical forms of the following statements:
   (a) If this gas either has an unpleasant smell or is not explosive, then it isn't hydrogen.
   (b) Having both a fever and a headache is a sufficient condition for George to go to the doctor.
   (c) Both having a fever and having a headache are sufficient conditions for George to go to the doctor.
   (d) If $x \neq 2$, then a necessary condition for x to be prime is that x be odd.

2. Analyze the logical forms of the following statements:
   (a) Mary will sell her house only if she can get a good price and find a nice apartment.
   (b) Having both a good credit history and an adequate down payment is a necessary condition for getting a mortgage.
   (c) John will kill himself, unless someone stops him. (Hint: First try to rephrase this using the words *if* and *then* instead of *unless*.)
   (d) If x is divisible by either 4 or 6, then it isn't prime.

3. Analyze the logical form of the following statement:
   (a) If it is raining, then it is windy and the sun is not shining.

Now analyze the following statements. Also, for each statement determine whether the statement is equivalent to either statement (a) or its converse.

(b) It is windy and not sunny only if it is raining.

(c) Rain is a sufficient condition for wind with no sunshine.

(d) Rain is a necessary condition for wind with no sunshine.

(e) It's not raining, if either the sun is shining or it's not windy.

(f) Wind is a necessary condition for it to be rainy, and so is a lack of sunshine.

(g) Either it is windy only if it is raining, or it is not sunny only if it is raining.

*4. Use truth tables to determine whether or not the following arguments are valid:

(a) Either sales or expenses will go up. If sales go up, then the boss will be happy. If expenses go up, then the boss will be unhappy. Therefore, sales and expenses will not both go up.

(b) If the tax rate and the unemployment rate both go up, then there will be a recession. If the GNP goes up, then there will not be a recession. The GNP and taxes are both going up. Therefore, the unemployment rate is not going up.

(c) The warning light will come on if and only if the pressure is too high and the relief valve is clogged. The relief valve is not clogged. Therefore, the warning light will come on if and only if the pressure is too high.

5. (a) Show that $P \leftrightarrow Q$ is equivalent to $(P \land Q) \lor (\neg P \land \neg Q)$.

(b) Show that $(P \to Q) \lor (P \to R)$ is equivalent to $P \to (Q \lor R)$.

*6. (a) Show that $(P \to R) \land (Q \to R)$ is equivalent to $(P \lor Q) \to R$.

(b) Formulate and verify a similar equivalence involving $(P \to R) \lor (Q \to R)$.

7. (a) Show that $(P \to Q) \land (Q \to R)$ is equivalent to $(P \to R) \land [(P \leftrightarrow Q) \lor (R \leftrightarrow Q)]$.

(b) Show that $(P \to Q) \lor (Q \to R)$ is a tautology.

*8. Find a formula involving only the connectives $\neg$ and $\to$ that is equivalent to $P \land Q$.

9. Find a formula involving only the connectives $\neg$ and $\to$ that is equivalent to $P \leftrightarrow Q$.

10. Which of the following formulas are equivalent?

(a) $P \to (Q \to R)$.

(b) $Q \to (P \to R)$.

(c) $(P \to Q) \land (P \to R)$.

(d) $(P \land Q) \to R$.

(e) $P \to (Q \land R)$.

2

Quantificational Logic

2.1. Quantifiers

We have seen that a statement $P(x)$ containing a free variable x may be true for some values of x and false for others. Sometimes we want to say something about *how many* values of x make $P(x)$ come out true. In particular, we often want to say either that $P(x)$ is true for *every* value of x or that it is true for *at least one* value of x. We therefore introduce two more symbols, called *quantifiers*, to help us express these ideas.

To say that $P(x)$ is true for every value of x in the universe of discourse U, we will write $\forall x P(x)$. This is read "For all x, $P(x)$." Think of the upside down A as standing for the word *all*. The symbol $\forall$ is called the *universal quantifier*, because the statement $\forall x P(x)$ says that $P(x)$ is *universally* true. As we discussed in Section 1.3, to say that $P(x)$ is true for every value of x in the universe means that the truth set of $P(x)$ will be the whole universe U. Thus, you could also think of the statement $\forall x P(x)$ as saying that the truth set of $P(x)$ is equal to U.

We write $\exists x P(x)$ to say that there is at least one value of x in the universe for which $P(x)$ is true. This is read "There exists an x such that $P(x)$." The backward E comes from the word *exists* and is called the *existential quantifier*. Once again, you can interpret this statement as saying something about the truth set of $P(x)$. To say that $P(x)$ is true for at least one value of x means that there is at least one element in the truth set of $P(x)$, or in other words, the truth set is not equal to $\varnothing$.

For example, in Section 1.5 we discussed the statement "If $x > 2$ then $x^2 > 4$," where x ranges over the set of all real numbers, and we claimed that this statement was true for all values of x. We can now write this claim symbolically as $\forall x(x > 2 \rightarrow x^2 > 4)$.

Example 2.1.1. What do the following formulas mean? Are they true or false?

1. $\forall x(x^2 \geq 0)$, where the universe of discourse is $\mathbb{R}$, the set of all real numbers.
2. $\exists x(x^2 - 2x + 3 = 0)$, with universe $\mathbb{R}$ again.
3. $\exists x(M(x) \wedge B(x))$, where the universe of discourse is the set of all people, $M(x)$ stands for the statement "x is a man," and $B(x)$ means "x has brown hair."
4. $\forall x(M(x) \rightarrow B(x))$, with the same universe and the same meanings for $M(x)$ and $B(x)$.
5. $\forall x L(x, y)$, where the universe is the set of all people, and $L(x, y)$ means "x likes y."

Solutions

1. This means that for every real number x, $x^2 \geq 0$. This is true.
2. This means that there is at least one real number x that makes the equation $x^2 - 2x + 3 = 0$ come out true. In other words, the equation has at least one real solution. If you solve the equation, you'll find that this statement is false; the equation has no solutions. (Try either completing the square or using the quadratic formula.)
3. There is at least one person x such that x is a man and x has brown hair. In other words, there is at least one man who has brown hair. Of course, this is true.
4. For every person x, if x is a man then x has brown hair. In other words, all men have brown hair. If you're not convinced that this is what the formula means, it might help to look back at the truth table for the conditional connective. According to this truth table, the statement $M(x) \rightarrow B(x)$ will be false only if $M(x)$ is true and $B(x)$ is false; that is, x is a man and x doesn't have brown hair. Thus, to say that $M(x) \rightarrow B(x)$ is true for every person x means that this situation never occurs, or in other words, that there are no men who don't have brown hair. But that's exactly what it means to say that all men have brown hair. Of course, this statement is false.
5. For every person x, x likes y. In other words, everyone likes y. We can't tell if this is true or false unless we know who y is.

Notice that in the fifth statement in this example, we needed to know who y was to determine if the statement was true or false, but not who x was. The statement says that everyone likes y, and this is a statement about y, but not x. This means that y is a free variable in this statement but x is a bound variable.

Similarly, although all the other statements contain the letter x, we didn't need to know the value of x to determine their truth values, so x is a bound variable in every case. In general, even if x is a free variable in some statement $P(x)$, it is a bound variable in the statements $\forall x P(x)$ and $\exists x P(x)$. For this reason, we say that the quantifiers *bind* a variable. As in Section 1.3, this means that a variable that is bound by a quantifier can always be replaced with a new variable without changing the meaning of the statement, and it is often possible to paraphrase the statement without mentioning the bound variable at all. For example, the statement $\forall x L(x, y)$ from Example 2.1.1 is equivalent to $\forall w L(w, y)$, because both mean the same thing as "Everyone likes y." Words such as *everyone, someone, everything*, or *something* are often used to express the meanings of statements containing quantifiers. If you are translating an English statement into symbols, these words will often tip you off that a quantifier will be needed.

As with the symbol $\neg$, we follow the convention that the expressions $\forall x$ and $\exists x$ apply only to the statements that come immediately after them. For example, $\forall x P(x) \rightarrow Q(x)$ means $(\forall x P(x)) \rightarrow Q(x)$, not $\forall x(P(x) \rightarrow Q(x))$.

Example 2.1.2. Analyze the logical forms of the following statements.

1. Someone didn't do the homework.
2. Everything in that store is either overpriced or poorly made.
3. Nobody's perfect.
4. Susan likes everyone who dislikes Joe.
5. $A \subseteq B$.
6. $A \cap B \subseteq B \setminus C$.

Solutions

1. The word *someone* tips us off that we should use an existential quantifier. As a first step, we write $\exists x$(x didn't do the homework). Now if we let $H(x)$ stand for the statement "x did the homework," then we can rewrite this as $\exists x \neg H(x)$.
2. Think of this statement as saying "If it's in that store, then it's either overpriced or poorly made (no matter what *it* is)." Thus, we start by writing $\forall x$(if x is in that store then x is either overpriced or poorly made). To write the part in parentheses symbolically, we let $S(x)$ stand for "x is in that store," $O(x)$ for "x is overpriced," and $P(x)$ for "x is poorly made." Then our final answer is $\forall x[S(x) \rightarrow (O(x) \vee P(x))]$.

   Note that, like statement 4 in Example 2.1.1, this statement has the form of a universal quantifier applied to a conditional statement. This form occurs quite often, and it is important to learn to recognize what it means and when it should be used. We can check our answer to this problem as we did before, by using the truth table for the conditional connective. The only way that the statement $S(x) \to (O(x) \vee P(x))$ can be false is if x is in that store, but is neither overpriced nor poorly made. Thus, to say that the statement is true for all values of x means that this never happens, which is exactly what it means to say that everything in that store is either overpriced or poorly made.

3. This means $\neg$(somebody is perfect), or in other words $\neg\exists x P(x)$, where $P(x)$ stands for "x is perfect."
4. As in statement 2 in this example, we could think of this as meaning "If a person dislikes Joe then Susan likes that person (no matter who the person is)." Thus, we can start by rewriting the given statement as $\forall x$(if x dislikes Joe then Susan likes x). Let $L(x, y)$ stand for "x likes y." In statements that talk about specific elements of the universe of discourse it is sometimes convenient to introduce letters to stand for those specific elements. In this case we need to talk about Joe and Susan, so let's let j stand for Joe and s for Susan. Thus, we can write $L(s, x)$ to mean "Susan likes x," and $\neg L(x, j)$ for "x dislikes Joe." Filling these in, we end up with the answer $\forall x(\neg L(x, j) \to L(s, x))$. Notice that, once again, we have a universal quantifier applied to a conditional statement. As before, you can check this answer using the truth table for the conditional connective.
5. According to Definition 1.4.5, to say that A is a subset of B means that everything in A is in B. If you've caught on to the pattern of how universal quantifiers and conditionals are combined, you should recognize that this would be written symbolically as $\forall x(x \in A \to x \in B)$.
6. As in the previous statement, we first write this as $\forall x(x \in A \cap B \to x \in B \setminus C)$. Now using the definitions of intersection and difference, we can expand this further to get $\forall x[(x \in A \wedge x \in B) \to (x \in B \wedge x \notin C)]$.

Although all of our examples so far have contained only one quantifier, there's no reason why a statement can't have more than one quantifier. For example, consider the statement "Some students are married." The word *some* indicates that this statement should be written using an existential quantifier, so we can think of it as having the form $\exists x$(x is a student and x is married). Let $S(x)$ stand for "x is a student." We could similarly choose a letter to stand for "x is married," but perhaps a better analysis would be to recognize that to be married means to be married *to someone*. Thus, if we let $M(x, y)$ stand for "x is married to y," then we can write "x is married" as $\exists y M(x, y)$. We can therefore represent the entire statement by the formula $\exists x(S(x) \wedge \exists y M(x, y))$, a formula containing two existential quantifiers.

As another example, let's analyze the statement "All parents are married." We start by writing it as $\forall x$(if x is a parent then x is married). Parenthood, like marriage, is a relationship between two people; to be a parent means to be a parent *of someone*. Thus, it might be best to represent the statement "x is a parent" by the formula $\exists y P(x, y)$, where $P(x, y)$ means "x is a parent of y." If we again represent "x is married" by the formula $\exists y M(x, y)$, then our analysis of the original statement will be $\forall x(\exists y P(x, y) \rightarrow \exists y M(x, y))$. Although this isn't wrong, the double use of the variable y could cause confusion. Perhaps a better solution would be to replace the formula $\exists y M(x, y)$ with the equivalent formula $\exists z M(x, z)$. (Recall that these are equivalent because a bound variable in any statement can be replaced by another without changing the meaning of the statement.) Our improved analysis of the statement would then be $\forall x(\exists y P(x, y) \rightarrow \exists z M(x, z))$.

Example 2.1.3. Analyze the logical forms of the following statements.

1. Everybody in the dorm has a roommate he doesn't like.
2. Nobody likes a sore loser.
3. Anyone who has a friend who has the measles will have to be quarantined.
4. If anyone in the dorm has a friend who has the measles, then everyone in the dorm will have to be quarantined.
5. If $A \subseteq B$, then A and $C \setminus B$ are disjoint.

Solutions

1. This means $\forall x$(if x lives in the dorm then x has a roommate he doesn't like). To say that x has a roommate he doesn't like, we could write $\exists y$(x and y are roommates and x doesn't like y). If we let $R(x, y)$ stand for "x and y are roommates" and $L(x, y)$ for "x likes y," then this becomes $\exists y(R(x, y) \wedge \neg L(x, y))$. Finally, if we let $D(x)$ mean "x lives in the dorm," then the complete analysis of the original statement would be $\forall x[D(x) \rightarrow \exists y(R(x, y) \wedge \neg L(x, y))]$.
2. This is tricky, because the phrase *a sore loser* doesn't refer to a *particular* sore loser, it refers to *all* sore losers. The statement means that all sore losers are disliked, or in other words $\forall x$(if x is a sore loser then nobody likes x). To say nobody likes x we write $\neg$(somebody likes x), which means $\neg\exists y L(y, x)$, where $L(y, x)$ means "y likes x." If we let $S(x)$ mean "x is a sore loser," then the whole statement would be written $\forall x(S(x) \rightarrow \neg\exists y L(y, x))$.
3. You have probably realized by now that it is usually easiest to translate from English into symbols in several steps, translating only a little bit at a time. Here are the steps we might use to translate this statement:
   (i) $\forall x$(if x has a friend who has the measles then x will have to be quarantined).

(ii) $\forall x[\exists y(y$ is a friend of x and y has the measles) $\to x$ will have to be quarantined].

Now, letting $F(y, x)$ stand for "y is a friend of x," $M(y)$ for "y has the measles," and $Q(x)$ for "x will have to be quarantined," we get:

(iii) $\forall x[\exists y(F(y, x) \wedge M(y)) \to Q(x)]$.

4. The word *anyone* is difficult to interpret, because in different statements it means different things. In statement 3 it meant *everyone*, but in this statement it means *someone*. Here are the steps of our analysis:

   (i) (Someone in the dorm has a friend who has the measles) $\to$ (everyone in the dorm will have to be quarantined).

   (ii) $\exists x(x$ lives in the dorm and x has a friend who has the measles) $\to \forall z$(if z lives in the dorm then z will have to be quarantined).

   Using the same abbreviations as in the last statement and letting $D(x)$ stand for "x lives in the dorm," we end up with the following formula:

   (iii) $\exists x[D(x) \wedge \exists y(F(y, x) \wedge M(y))] \to \forall z(D(z) \to Q(z))$.

5. Clearly the answer will have the form of a conditional statement, $(A \subseteq B) \to$ (A and $C \setminus B$ are disjoint). We have already written $A \subseteq B$ symbolically in Example 2.1.2. To say that A and $C \setminus B$ are disjoint means that they have no elements in common, or in other words $\neg\exists x(x \in A \wedge x \in C \setminus B)$. Putting this all together, and filling in the definition of $C \setminus B$, we end up with $\forall x(x \in A \to x \in B) \to \neg\exists x(x \in A \wedge x \in C \wedge x \notin B)$.

When a statement contains more than one quantifier it is sometimes difficult to figure out what it means and whether it is true or false. It may be best in this case to think about the quantifiers one at a time, in order. For example, consider the statement $\forall x \exists y(x + y = 5)$, where the universe of discourse is the set of all real numbers. Thinking first about just the first quantifier expression $\forall x$, we see that the statement means that for every real number x, the statement $\exists y(x + y = 5)$ is true. We can worry later about what $\exists y(x + y = 5)$ means; thinking about two quantifiers at once is too confusing.

If we want to figure out whether or not the statement $\exists y(x + y = 5)$ is true for every value of x, it might help to try out a few values of x. For example, suppose $x = 2$. Then we must determine whether or not the statement $\exists y(2 + y = 5)$ is true. Now it's time to think about the next quantifier, $\exists y$. This statement says that there is at least one value of y for which the equation $2 + y = 5$ holds. In other words, the equation $2 + y = 5$ has at least one solution. Of course, this is true, because the equation has the solution $y = 5 - 2 = 3$. Thus, the statement $\exists y(2 + y = 5)$ is true.

Let's try one more value of x. If $x = 7$, then we are interested in the statement $\exists y(7 + y = 5)$, which says that the equation $7 + y = 5$ has at least one solution.

Once again, this is true, since the solution is $y = 5 - 7 = -2$. In fact, you have probably realized by now that no matter what value we plug in for x, the equation $x + y = 5$ will always have the solution $y = 5 - x$, so the statement $\exists y(x + y = 5)$ will be true. Thus, the original statement $\forall x \exists y(x + y = 5)$ is true.

On the other hand, the statement $\exists y \forall x(x + y = 5)$ means something entirely different. This statement means that there is at least one value of y for which the statement $\forall x(x + y = 5)$ is true. Can we find such a value of y? Suppose, for example, we try $y = 4$. Then we must determine whether or not the statement $\forall x(x + 4 = 5)$ is true. This statement says that no matter what value we plug in for x, the equation $x + 4 = 5$ holds, and this is clearly false. In fact, no value of x other than $x = 1$ works in this equation. Thus, the statement $\forall x(x + 4 = 5)$ is false.

We have seen that when $y = 4$ the statement $\forall x(x + y = 5)$ is false, but maybe some other value of y will work. Remember, we are trying to determine whether or not there is *at least one* value of y that works. Let's try one more, say, $y = 9$. Then we must consider the statement $\forall x(x + 9 = 5)$, which says that no matter what x is, the equation $x + 9 = 5$ holds. Once again this is clearly false, since only $x = -4$ works in this equation. In fact, it should be clear by now that no matter what value we plug in for y, the equation $x + y = 5$ will be true for only one value of x, namely $x = 5 - y$, so the statement $\forall x(x + y = 5)$ will be false. Thus there are *no* values of y for which $\forall x(x + y = 5)$ is true, so the statement $\exists y \forall x(x + y = 5)$ is false.

Notice that we found that the statement $\forall x \exists y(x + y = 5)$ is true, but $\exists y \forall x(x + y = 5)$ is false. Apparently, the order of the quantifiers makes a difference! What is responsible for this difference? The first statement says that for every real number x, there is a real number y such that $x + y = 5$. For example, when we tried $x = 2$ we found that $y = 3$ worked in the equation $x + y = 5$, and with $x = 7$, $y = -2$ worked. Note that for different values of x, we had to use different values of y to make the equation come out true. You might think of this statement as saying that for each real number x there is a *corresponding* real number y such that $x + y = 5$. On the other hand, when we were analyzing the statement $\exists y \forall x(x + y = 5)$ we found ourselves searching for a *single* value of y that made the equation $x + y = 5$ true for all values of x, and this turned out to be impossible. For each value of x there is a corresponding value of y that makes the equation true, but no single value of y works for every x.

For another example, consider the statement $\forall x \exists y L(x, y)$, where the universe of discourse is the set of all people and $L(x, y)$ means "x likes y." This statement says that for every person x, the statement $\exists y L(x, y)$ is true. Now

$\exists y L(x, y)$ could be written as "x likes someone," so the original statement means that for every person x, x likes someone. In other words, everyone likes someone. On the other hand, $\exists y \forall x L(x, y)$ means that there is some person y such that $\forall x L(x, y)$ is true. As we saw in Example 2.1.1, $\forall x L(x, y)$ means "Everyone likes y," so $\exists y \forall x L(x, y)$ means that there is some person y such that everyone likes y. In other words, there is someone who is universally liked. These statements don't mean the same thing. It might be the case that everyone likes someone, but no one is universally liked.

Example 2.1.4. What do the following statements mean? Are they true or false? The universe of discourse in each case is $\mathbb{N}$, the set of all natural numbers.

1. $\forall x \exists y(x < y)$.
2. $\exists y \forall x(x < y)$.
3. $\exists x \forall y(x < y)$.
4. $\forall y \exists x(x < y)$.
5. $\exists x \exists y(x < y)$.
6. $\forall x \forall y(x < y)$.

Solutions

1. This means that for every natural number x, the statement $\exists y(x < y)$ is true. In other words, for every natural number x, there is a natural number bigger than x. This is true. For example, $x + 1$ is always bigger than x.
2. This means that there is some natural number y such that the statement $\forall x(x < y)$ is true. In other words, there is some natural number y such that all natural numbers are smaller than y. This is false. No matter what natural number y we pick, there will always be larger natural numbers.
3. This means that there is a natural number x such that the statement $\forall y(x < y)$ is true. You might be tempted to say that this statement will be true if $x = 0$, but this isn't right. Since 0 is the smallest natural number, the statement $0 < y$ is true for all values of y *except* $y = 0$, but if $y = 0$, then the statement $0 < y$ is false, and therefore $\forall y(0 < y)$ is false. Similar reasoning shows that for every value of x the statement $\forall y(x < y)$ is false, so $\exists x \forall y(x < y)$ is false.
4. This means that for every natural number y, there is a natural number smaller than y. This is true for every natural number y *except* $y = 0$, but there is no natural number smaller than 0. Therefore this statement is false.
5. This means that there is a natural number x such that $\exists y(x < y)$ is true. But as we saw in the first statement, this is actually true for *every* natural number x, so it is certainly true for at least one. Thus, $\exists x \exists y(x < y)$ is true.

6. This means that for every natural number x, the statement $\forall y(x < y)$ is true. But as we saw in the third statement, there isn't even *one* value of x for which this statement is true. Thus, $\forall x \forall y(x < y)$ is false.

Exercises

*1. Analyze the logical forms of the following statements.
   (a) Anyone who has forgiven at least one person is a saint.
   (b) Nobody in the calculus class is smarter than everybody in the discrete math class.
   (c) Everyone likes Mary, except Mary herself.
   (d) Jane saw a police officer, and Roger saw one too.
   (e) Jane saw a police officer, and Roger saw him too.

2. Analyze the logical forms of the following statements.
   (a) Anyone who has bought a Rolls Royce with cash must have a rich uncle.
   (b) If anyone in the dorm has the measles, then everyone who has a friend in the dorm will have to be quarantined.
   (c) If nobody failed the test, then everybody who got an A will tutor someone who got a D.
   (d) If anyone can do it, Jones can.
   (e) If Jones can do it, anyone can.

3. Analyze the logical forms of the following statements. The universe of discourse is $\mathbb{R}$. What are the free variables in each statement?
   (a) Every number that is larger than x is larger than y.
   (b) For every number a, the equation $ax^2 + 4x - 2 = 0$ has at least one solution iff $a \geq -2$.
   (c) All solutions of the inequality $x^3 - 3x < 3$ are smaller than 10.
   (d) If there is a number x such that $x^2 + 5x = w$ and there is a number y such that $4 - y^2 = w$, then w is between -10 and 10.

*4. Translate the following statements into idiomatic English.
   (a) $\forall x[(H(x) \wedge \neg\exists y M(x, y)) \rightarrow U(x)]$, where $H(x)$ means "x is a man," $M(x, y)$ means "x is married to y," and $U(x)$ means "x is unhappy."
   (b) $\exists z(P(z, x) \wedge S(z, y) \wedge W(y))$, where $P(z, x)$ means "z is a parent of x," $S(z, y)$ means "z and y are siblings," and $W(y)$ means "y is a woman."

5. Translate the following statements into idiomatic mathematical English.
   (a) $\forall x[(P(x) \wedge \neg(x = 2)) \rightarrow O(x)]$, where $P(x)$ means "x is a prime number" and $O(x)$ means "x is odd."

(b) $\exists x[P(x) \land \forall y(P(y) \to y \leq x)]$, where $P(x)$ means "x is a perfect number."

6. Are these statements true or false? The universe of discourse is the set of all people, and $P(x, y)$ means "x is a parent of y."
   (a) $\exists x \forall y P(x, y)$.
   (b) $\forall x \exists y P(x, y)$.
   (c) $\neg\exists x \exists y P(x, y)$.
   (d) $\exists x \neg\exists y P(x, y)$.
   (e) $\exists x \exists y \neg P(x, y)$.

*7. Are these statements true or false? The universe of discourse is $\mathbb{N}$.
   (a) $\forall x \exists y(2x - y = 0)$.
   (b) $\exists y \forall x(2x - y = 0)$.
   (c) $\forall x \exists y(x - 2y = 0)$.
   (d) $\forall x(x < 10 \to \forall y(y < x \to y < 9))$.
   (e) $\exists y \exists z(y + z = 100)$.
   (f) $\forall x \exists y(y > x \land \exists z(y + z = 100))$.

8. Same as exercise 7 but with $\mathbb{R}$ as the universe of discourse.
9. Same as exercise 7 but with $\mathbb{Z}$ as the universe of discourse.

2.2. Equivalences Involving Quantifiers

In our study of logical connectives in Chapter 1 we found it useful to examine equivalences between different formulas. In this section, we will see that there are also a number of important equivalences involving quantifiers.

For example, in Example 2.1.2 we represented the statement "Nobody's perfect" by the formula $\neg\exists x P(x)$, where $P(x)$ meant "x is perfect." But another way to express the same idea would be to say that everyone fails to be perfect, or in other words $\forall x \neg P(x)$. This suggests that these two formulas are equivalent, and a little thought should show that they are. No matter what $P(x)$ stands for, the formula $\neg\exists x P(x)$ means that there's no value of x in the universe of discourse for which $P(x)$ is true. But that's the same as saying that for every value of x in the universe, $P(x)$ is false, or in other words $\forall x \neg P(x)$. Thus, $\neg\exists x P(x)$ is equivalent to $\forall x \neg P(x)$.

Similar reasoning shows that $\neg\forall x P(x)$ is equivalent to $\exists x \neg P(x)$. To say that $\neg\forall x P(x)$ means that it is not the case that for all values of x, $P(x)$ is true. That's equivalent to saying there's at least one value of x for which $P(x)$ is false, which is what it means to say $\exists x \neg P(x)$. For example, in Example 2.1.2 we translated "Someone didn't do the homework" as $\exists x \neg H(x)$, where $H(x)$ stands for "x did the homework." An equivalent statement would be "Not everyone did the homework," which would be represented by the formula $\neg\forall x H(x)$.

Thus, we have the following two laws involving negation and quantifiers:

Quantifier Negation laws

$\neg\exists x P(x)$ is equivalent to $\forall x \neg P(x)$.

$\neg\forall x P(x)$ is equivalent to $\exists x \neg P(x)$.

Combining these laws with DeMorgan's laws and other equivalences involving the logical connectives, we can often reexpress a negative statement as an equivalent, but easier to understand, positive statement. This will turn out to be an important skill when we begin to work with negative statements in proofs.

Example 2.2.1. Negate these statements and then reexpress the results as equivalent positive statements.

1. $A \subseteq B$.
2. Everyone has a relative he doesn't like.

Solutions

1. We already know that $A \subseteq B$ means $\forall x(x \in A \rightarrow x \in B)$. To reexpress the negation of this statement as an equivalent positive statement, we reason as follows:

   $\neg\forall x(x \in A \rightarrow x \in B)$
   is equivalent to $\exists x \neg(x \in A \rightarrow x \in B)$ (quantifier negation law),
   which is equivalent to $\exists x \neg(x \notin A \lor x \in B)$ (conditional law),
   which is equivalent to $\exists x(x \in A \land x \notin B)$ (DeMorgan's law).

   Thus, $A \nsubseteq B$ means the same thing as $\exists x(x \in A \land x \notin B)$. If you think about this, it should make sense. To say that A is not a subset of B is the same as saying that there's something in A that is not in B.
2. First of all, let's write the original statement symbolically. You should be able to check that if we let $R(x, y)$ stand for "x is related to y" and $L(x, y)$ for "x likes y," then the original statement would be written $\forall x\exists y(R(x, y) \land \neg L(x, y))$. Now we negate this and try to find a simpler, equivalent positive statement:

   $\neg\forall x\exists y(R(x, y) \land \neg L(x, y))$
   is equivalent to $\exists x\neg\exists y(R(x, y) \land \neg L(x, y))$
   (quantifier negation law),
   which is equivalent to $\exists x\forall y\neg(R(x, y) \land \neg L(x, y))$
   (quantifier negation law),
   which is equivalent to $\exists x\forall y(\neg R(x, y) \lor L(x, y))$
   (DeMorgan's law),
   which is equivalent to $\exists x\forall y(R(x, y) \rightarrow L(x, y))$
   (conditional law).

Let's translate this last formula back into colloquial English. Leaving aside the first quantifier for the moment, the formula $\forall y(R(x, y) \to L(x, y))$ means that for every person y, if x is related to y then x likes y. In other words, x likes all his relatives. Adding $\exists x$ to the beginning of this, we get the statement "There is someone who likes all his relatives." You should take a minute to convince yourself that this really is equivalent to the negation of the original statement "Everyone has a relative he doesn't like."

For another example of how the quantifier negation laws can help us understand statements, consider the statement "Everyone who Patricia likes, Sue doesn't like." If we let $L(x, y)$ stand for "x likes y," and we let p stand for Patricia and s for Sue, then this statement would be represented by the formula $\forall x(L(p, x) \to \neg L(s, x))$. Now we can work out a formula equivalent to this one as follows:

$\forall x(L(p, x) \to \neg L(s, x))$

is equivalent to $\forall x(\neg L(p, x) \vee \neg L(s, x))$ (conditional law),

which is equivalent to $\forall x \neg(L(p, x) \wedge L(s, x))$ (DeMorgan's law),

which is equivalent to $\neg\exists x(L(p, x) \wedge L(s, x))$ (quantifier negation law).

Translating the last formula back into English, we get the statement "There's no one who both Patricia and Sue like," and this does mean the same thing as the statement we started with.

We saw in Section 2.1 that reversing the order of two quantifiers can sometimes change the meaning of a formula. However, if the quantifiers are the same type (both $\forall$ or both $\exists$), it turns out the order can always be switched without affecting the meaning of the formula. For example, consider the statement "Someone has a teacher who is younger than he is." To write this symbolically we first write $\exists x$(x has a teacher who is younger than x). Now to say "x has a teacher who is younger than x" we write $\exists y(T(y, x) \wedge P(y, x))$, where $T(y, x)$ means "y is a teacher of x" and $P(y, x)$ means "y is younger than x." Putting this all together, the original statement would be represented by the formula $\exists x\exists y(T(y, x) \wedge P(y, x))$.

Now what happens if we switch the quantifiers? In other words, what does the formula $\exists y\exists x(T(y, x) \wedge P(y, x))$ mean? You should be able to convince yourself that this formula says that there is a person y such that y is a teacher of someone who is older than y. In other words, someone is a teacher of a person who is older than he is. But this would be true in exactly the same circumstances as the original statement, "Someone has a teacher who is younger than he is"! Both mean that there are people x and y such that y is a teacher of x and y is younger than x. In fact, this suggests that a good way of reading the pair of quantifiers $\exists y \exists x$ or $\exists x \exists y$ would be "there are objects x and y such that. . . ."

Similarly, two universal quantifiers in a row can always be switched without changing the meaning of a formula, because $\forall x \forall y$ and $\forall y \forall x$ can both be thought of as meaning "for all objects x and y, ..." For example, consider the formula $\forall x \forall y(L(x, y) \rightarrow A(x, y))$, where $L(x, y)$ means "x likes y" and $A(x, y)$ means "x admires y." You could think of this formula as saying "For all people x and y, if x likes y then x admires y." In other words, people always admire the people they like. The formula $\forall y \forall x(L(x, y) \rightarrow A(x, y))$ means exactly the same thing.

It is important to realize that when we talk about objects x and y, we are not ruling out the possibility that x and y are the same object. For example, the formula $\forall x \forall y(L(x, y) \rightarrow A(x, y))$ means not just that a person who likes another person always admires that other person, but also that people who like themselves also admire themselves. As another example, suppose we wanted to write a formula that means "x is a bigamist." (Of course, x will be a free variable in this formula.) You might think you could express this with the formula $\exists y \exists z(M(x, y) \wedge M(x, z))$, where $M(x, y)$ means "x is married to y." But to say that x is a bigamist you must say that there are two *different* people to whom x is married, and this formula doesn't say that y and z are different. The right answer is $\exists y \exists z(M(x, y) \wedge M(x, z) \wedge y \neq z)$.

Example 2.2.2. Analyze the logical forms of the following statements.

1. All married couples have fights.
2. Everyone likes at least two people.
3. John likes exactly one person.

Solutions

1. $\forall x \forall y(M(x, y) \rightarrow F(x, y))$, where $M(x, y)$ means "x and y are married to each other" and $F(x, y)$ means "x and y fight with each other."
2. $\forall x \exists y \exists z(L(x, y) \wedge L(x, z) \wedge y \neq z)$, where $L(x, y)$ stands for "x likes y." Note that the statement means that everyone likes at least two *different* people, so it would be incorrect to leave out the "$y \neq z$" at the end.
3. Let $L(x, y)$ mean "x likes y," and let j stand for John. We translate this statement into symbols gradually:
   (i) $\exists x$(John likes x and John doesn't like anyone other than x).
   (ii) $\exists x(L(j, x) \wedge \neg\exists y$(John likes y and $y \neq x$)).
   (iii) $\exists x(L(j, x) \wedge \neg\exists y(L(j, y) \wedge y \neq x))$.

Note that for the third statement in this example we could not have given the simpler answer $\exists x L(j, x)$, because this would mean that John likes *at least* one person, not *exactly* one person. The phrase *exactly one* occurs so often in mathematics that there is a special notation for it. We will write $\exists! x P(x)$ to represent the statement "There is exactly one value of x such that $P(x)$ is true." It is sometimes also read "There is a unique x such that $P(x)$." For example, the third statement in Example 2.2.2 could be written symbolically as $\exists! x L(j, x)$. In fact, we could think of this as just an abbreviation for the formula given in Example 2.2.2 as the answer for statement 3. Similarly, in general we can think of $\exists! x P(x)$ as an abbreviation for the formula $\exists x(P(x) \wedge \neg\exists y(P(y) \wedge y \neq x))$.

Recall that when we were discussing set theory, we sometimes found it useful to write the truth set of $P(x)$ as $\{x \in U \mid P(x)\}$ rather than $\{x \mid P(x)\}$, to make sure it was clear what the universe of discourse was. Similarly, instead of writing $\forall x P(x)$ to indicate that $P(x)$ is true for every value of x in some universe U, we might write $\forall x \in U P(x)$. This is read "For all x in U, $P(x)$." Similarly, we can write $\exists x \in U P(x)$ to say that there is at least one value of x in the universe U such that $P(x)$ is true. For example, the statement $\forall x(x \geq 0)$ would be false if the universe of discourse were the real numbers, but true if it were the natural numbers. We could avoid confusion when discussing this statement by writing either $\forall x \in \mathbb{R}(x \geq 0)$ or $\forall x \in \mathbb{N}(x \geq 0)$, to make it clear which we meant.

As before, we sometimes use this notation not to specify the universe of discourse but to restrict attention to a subset of the universe. For example, if our universe of discourse is the real numbers and we want to say that some real number x has a square root, we could write $\exists y(y^2 = x)$. To say that every *positive* real number has a square root, we would say $\forall x \in \mathbb{R}^+ \exists y(y^2 = x)$. We could say that every positive real number has a negative square root by writing $\forall x \in \mathbb{R}^+ \exists y \in \mathbb{R}^-(y^2 = x)$. In general, for any set A, the formula $\forall x \in A P(x)$ means that for every value of x *in the set* A, $P(x)$ is true, and $\exists x \in A P(x)$ means that there is at least one value of x *in the set* A such that $P(x)$ is true. The quantifiers in these formulas are sometimes called *bounded quantifiers*, because they place *bounds* on which values of x are to be considered. Occasionally we may use variations on this notation to place other kinds of restrictions on quantified variables. For example, the statement that every positive real number has a negative square root could also be written $\forall x > 0 \exists y < 0(y^2 = x)$.

Formulas containing bounded quantifiers can also be thought of as abbreviations for more complicated formulas containing only normal, unbounded quantifiers. To say that $\exists x \in A P(x)$ means that there is some value of x that is in A and that also makes $P(x)$ come out true, and another way to write this would be $\exists x(x \in A \wedge P(x))$. Similarly, you should convince yourself that $\forall x \in A P(x)$ means the same thing as $\forall x(x \in A \to P(x))$. For example, the formula $\forall x \in \mathbb{R}^+ \exists y \in \mathbb{R}^-(y^2 = x)$ discussed earlier means the same thing as $\forall x(x \in \mathbb{R}^+ \to \exists y \in \mathbb{R}^-(y^2 = x))$, which in turn can be expanded as $\forall x(x \in \mathbb{R}^+ \to \exists y(y \in \mathbb{R}^- \wedge y^2 = x))$. By the definitions of $\mathbb{R}^+$ and $\mathbb{R}^-$, an equivalent way to say this would be $\forall x(x > 0 \to \exists y(y < 0 \wedge y^2 = x))$. You should make sure you are convinced that this formula, like the original formula, means that every positive real number has a negative square root. For another example, note that the statement $A \subseteq B$, which by definition means $\forall x(x \in A \to x \in B)$, could also be written as $\forall x \in A(x \in B)$.

It is interesting to note that the quantifier negation laws work for bounded quantifiers as well. In fact, we can derive these bounded quantifier negation laws from the original laws by thinking of the bounded quantifiers as abbreviations, as described earlier. For example,

$\neg\forall x \in A P(x)$
is equivalent to $\neg\forall x(x \in A \to P(x))$ (expanding abbreviation),
which is equivalent to $\exists x \neg(x \in A \to P(x))$ (quantifier negation law),
which is equivalent to $\exists x \neg(x \notin A \vee P(x))$ (conditional law),
which is equivalent to $\exists x(x \in A \wedge \neg P(x))$ (DeMorgan's law),
which is equivalent to $\exists x \in A \neg P(x)$ (abbreviation).

Thus, we have shown that $\neg\forall x \in A P(x)$ is equivalent to $\exists x \in A \neg P(x)$. You are asked in exercise 5 to prove the other bounded quantifier negation law, that $\neg\exists x \in A P(x)$ is equivalent to $\forall x \in A \neg P(x)$.

It should be clear that if $A = \varnothing$ then $\exists x \in A P(x)$ will be false no matter what the statement $P(x)$ is. There can be nothing in A that, when plugged in for x, makes $P(x)$ come out true, because there is nothing in A at all! It may not be so clear whether $\forall x \in A P(x)$ should be considered true or false, but we can find the answer using the quantifier negation laws:

$\forall x \in A P(x)$
is equivalent to $\neg\neg\forall x \in A P(x)$ (double negation law),
which is equivalent to $\neg\exists x \in A \neg P(x)$ (quantifier negation law).

Now if $A = \varnothing$ then this last formula will be true, no matter what the statement $P(x)$ is, because, as we have seen, $\exists x \in A \neg P(x)$ must be false. Thus, $\forall x \in A P(x)$ is always true if $A = \varnothing$. Mathematicians sometimes say that such a statement is *vacuously* true. Another way to see this is to rewrite the statement $\forall x \in A P(x)$ in the equivalent form $\forall x(x \in A \to P(x))$. Now according to the truth table for the conditional connective, the only way this can be false is if there is some value of x such that $x \in A$ is true but $P(x)$ is false. But there is no such value of x, simply because there isn't a value of x for which $x \in A$ is true.

As an application of this principle, we note that the empty set is a subset of every set. To see why, just rewrite the statement $A \subseteq B$ in the equivalent form $\forall x \in A(x \in B)$. Now if $A = \varnothing$ then, as we have just observed, this statement will be vacuously true. Thus, no matter what the set B is, $\varnothing \subseteq B$. Another example of a vacuously true statement is the statement "All unicorns are purple." We could represent this by the formula $\forall x \in A P(x)$, where A is the set of all unicorns and $P(x)$ stands for "x is purple." Since there are no unicorns, A is the empty set, so the statement is vacuously true.

Perhaps you have noticed by now that, although in Chapter 1 we were always able to check equivalences involving logical connectives by making truth tables, we have no such simple way of checking equivalences involving quantifiers. So far, we have justified our equivalences involving quantifiers by just looking at examples and using common sense. As the formulas we work with get more complicated, this method will become unreliable and difficult to use. Fortunately, in Chapter 3 we will develop better methods for reasoning about statements involving quantifiers. To get more practice in thinking about quantifiers, we will work out a few somewhat more complicated equivalences using common sense. If you're not completely convinced that these equivalences are right, you'll be able to check them more carefully when you get to Chapter 3.

Consider the statement "Everyone is bright-eyed and bushy-tailed." If we let $E(x)$ mean "x is bright-eyed" and $T(x)$ mean "x is bushy-tailed," then we could represent this statement by the formula $\forall x(E(x) \wedge T(x))$. Is this equivalent to the formula $\forall x E(x) \wedge \forall x T(x)$? This latter formula means "Everyone is bright-eyed, and also everyone is bushy-tailed," and intuitively this means the same thing as the original statement. Thus, it appears that $\forall x(E(x) \wedge T(x))$ is equivalent to $\forall x E(x) \wedge \forall x T(x)$. In other words, we could say that the universal quantifier *distributes over* conjunction.

However, the corresponding distributive law doesn't work for the existential quantifier. Consider the formulas $\exists x(E(x) \wedge T(x))$ and $\exists x E(x) \wedge \exists x T(x)$. The first means that there is someone who is both bright-eyed and bushy-tailed, and the second means that there is someone who is bright-eyed, and there is also someone who is bushy-tailed. These don't mean the same thing at all. In the second statement the bright-eyed person and the bushy-tailed person don't have to be the same, but in the first statement they do. Another way to see the difference between the two statements is to think about truth sets. Let A be the truth set of $E(x)$ and B the truth set of $T(x)$. In other words, A is the set of bright-eyed people, and B is the set of bushy-tailed people. Then the second statement says that neither A nor B is the empty set, but the first says that $A \cap B$ is not the empty set, or in other words that A and B are not disjoint.

As an application of the distributive law for the universal quantifier and conjunction, suppose A and B are sets and consider the equation $A = B$. We know that two sets are equal when they have exactly the same elements. Thus, the equation $A = B$ means $\forall x(x \in A \leftrightarrow x \in B)$, which is equivalent to $\forall x[(x \in A \rightarrow x \in B) \wedge (x \in B \rightarrow x \in A)]$. Because the universal quantifier distributes over conjuction, this is equivalent to the formula $\forall x(x \in A \rightarrow x \in B) \wedge \forall x(x \in B \rightarrow x \in A)$, and by the definition of subset this means $A \subseteq B \wedge B \subseteq A$. Thus, we have shown that the equation $A = B$ is also equivalent to the formula $A \subseteq B \wedge B \subseteq A$.

We have now introduced seven basic logical symbols: the connectives $\wedge$, $\vee$, $\neg$, $\rightarrow$, and $\leftrightarrow$, and the quantifiers $\forall$ and $\exists$. It is a remarkable fact that the structure of all mathematical statements can be understood using these symbols, and all mathematical reasoning can be analyzed in terms of the proper use of these symbols. To illustrate the power of the symbols we have introduced, we conclude this section by writing out a few more mathematical statements in logical notation.

Example 2.2.3. Analyze the logical forms of the following statements.

1. Statements about the natural numbers. The universe of discourse is $\mathbb{N}$.
   (a) x is a perfect square.
   (b) x is a multiple of y.
   (c) x is prime.
   (d) x is the smallest number that is a multiple of both y and z.
2. Statements about the real numbers. The universe of discourse is $\mathbb{R}$.
   (a) The identity element for addition is 0.
   (b) Every real number has an additive inverse.
   (c) Negative numbers don't have square roots.
   (d) Every positive number has exactly two square roots.

Solutions

1. (a) This means that x is the square of some natural number, or in other words $\exists y(x = y^2)$.
   (b) This means that x is equal to y times some natural number, or in other words $\exists z(x = yz)$.
   (c) This means that $x > 1$, and x cannot be written as a product of two smaller natural numbers. In symbols: $x > 1 \wedge \neg\exists y \exists z(x = yz \wedge y < x \wedge z < x)$.
   (d) We translate this in several steps:
      (i) x is a multiple of both y and z and there is no smaller number that is a multiple of both y and z.

(ii) $\exists a(x = ya) \wedge \exists b(x = zb) \wedge \neg\exists w(w < x \wedge (w$ is a multiple of both y and $z))$.

(iii) $\exists a(x = ya) \wedge \exists b(x = zb) \wedge \neg\exists w(w < x \wedge \exists c(w = yc) \wedge \exists d(w = zd))$.

2. (a) $\forall x(x + 0 = x)$.
   (b) $\forall x \exists y(x + y = 0)$.
   (c) $\forall x(x < 0 \to \neg\exists y(y^2 = x))$.
   (d) We translate this gradually:
      (i) $\forall x(x > 0 \to x$ has exactly two square roots$)$.
      (ii) $\forall x(x > 0 \to \exists y \exists z(y$ and z are square roots of x and $y \neq z$ and nothing else is a square root of $x))$.
      (iii) $\forall x(x > 0 \to \exists y \exists z(y^2 = x \wedge z^2 = x \wedge y \neq z \wedge \neg\exists w(w^2 = x \wedge w \neq y \wedge w \neq z)))$.

Exercises

*1. Negate these statements and then reexpress the results as equivalent positive statements. (See Example 2.2.1.)
   (a) Everyone who is majoring in math has a friend who needs help with his homework.
   (b) Everyone has a roommate who dislikes everyone.
   (c) $A \cup B \subseteq C \setminus D$.
   (d) $\exists x \forall y[y > x \to \exists z(z^2 + 5z = y)]$.

2. Negate these statements and then reexpress the results as equivalent positive statements. (See Example 2.2.1.)
   (a) There is someone in the freshman class who doesn't have a roommate.
   (b) Everyone likes someone, but no one likes everyone.
   (c) $\forall a \in A \exists b \in B(a \in C \leftrightarrow b \in C)$.
   (d) $\forall y > 0 \exists x(ax^2 + bx + c = y)$.

3. Are these statements true or false? The universe of discourse is $\mathbb{N}$.
   (a) $\forall x(x < 7 \to \exists a \exists b \exists c(a^2 + b^2 + c^2 = x))$.
   (b) $\exists! x((x - 4)^2 = 9)$.
   (c) $\exists! x((x - 4)^2 = 25)$.
   (d) $\exists x \exists y((x - 4)^2 = 25 \wedge (y - 4)^2 = 25)$.

*4. Show that the second quantifier negation law, which says that $\neg\forall x P(x)$ is equivalent to $\exists x \neg P(x)$, can be derived from the first, which says that $\neg\exists x P(x)$ is equivalent to $\forall x \neg P(x)$. (Hint: Use the double negation law.)

5. Show that $\neg\exists x \in A P(x)$ is equivalent to $\forall x \in A \neg P(x)$.

*6. Show that the existential quantifier distributes over disjunction. In other words, show that $\exists x(P(x) \vee Q(x))$ is equivalent to $\exists x P(x) \vee \exists x Q(x)$.

(Hint: Use the fact, discussed in this section, that the universal quantifier distributes over conjunction.)

7. Show that $\exists x(P(x) \to Q(x))$ is equivalent to $\forall x P(x) \to \exists x Q(x)$.

*8. Show that $[\forall x \in A P(x)] \wedge [\forall x \in B P(x)]$ is equivalent to $\forall x \in (A \cup B)P(x)$. (Hint: Start by writing out the meanings of the bounded quantifiers in terms of unbounded quantifiers.)

9. Is $\forall x(P(x) \vee Q(x))$ equivalent to $\forall x P(x) \vee \forall x Q(x)$? Explain. (Hint: Try assigning meanings to $P(x)$ and $Q(x)$.)

10. (a) Show that $(\exists x \in A)P(x) \vee (\exists x \in B)P(x)$ is equivalent to $\exists x \in (A \cup B)P(x)$.
    (b) Is $(\exists x \in A)P(x) \wedge (\exists x \in B)P(x)$ equivalent to $\exists x \in (A \cap B)P(x)$? Explain.

*11. Show that the statements $A \subseteq B$ and $A \setminus B = \varnothing$ are equivalent by writing each in logical symbols and then showing that the resulting formulas are equivalent.

12. Let $T(x, y)$ mean "x is a teacher of y." What do the following statements mean? Under what circumstances would each one be true? Are any of them equivalent to each other?
    (a) $\exists! y T(x, y)$.
    (b) $\exists x \exists! y T(x, y)$.
    (c) $\exists! x \exists y T(x, y)$.
    (d) $\exists y \exists! x T(x, y)$.
    (e) $\exists! x \exists! y T(x, y)$.
    (f) $\exists x \exists y[T(x, y) \wedge \neg\exists u \exists v(T(u, v) \wedge (u \neq x \vee v \neq y))]$.

2.3. More Operations on Sets

Now that we know how to work with quantifiers, we are ready to discuss some more advanced topics in set theory.

So far, the only way we have to define sets, other than listing their elements one by one, is to use the elementhood test notation $\{x \mid P(x)\}$. Sometimes this notation is modified by allowing the x before the vertical line to be replaced with a more complex expression. For example, suppose we wanted to define S to be the set of all perfect squares. Perhaps the easiest way to describe this set is to say that it consists of all numbers of the form n^2, where n is a natural number. This is written $S = \{n^2 \mid n \in \mathbb{N}\}$. Note that, using our solution for the first statement from Example 2.2.3, we could also define this set by writing $S = \{x \mid \exists n \in \mathbb{N}(x = n^2)\}$. Thus, $\{n^2 \mid n \in \mathbb{N}\} = \{x \mid \exists n \in \mathbb{N}(x = n^2)\}$, and therefore $x \in \{n^2 \mid n \in \mathbb{N}\}$ means the same thing as $\exists n \in \mathbb{N}(x = n^2)$.

Similar notation is often used if the elements of a set have been numbered. For example, suppose we wanted to form the set whose elements are the first 100 prime numbers. We might start by numbering the prime numbers, calling them $p_1, p_2, p_3, \ldots$. In other words, $p_1 = 2$, $p_2 = 3$, $p_3 = 5$, and so on. Then the set we are looking for would be the set $P = \{p_1, p_2, p_3, \ldots, p_{100}\}$. Another way of describing this set would be to say that it consists of all numbers p_i, for i an element of the set $I = \{1, 2, 3, \ldots, 100\} = \{i \in \mathbb{N} \mid 1 \leq i \leq 100\}$. This could be written $P = \{p_i \mid i \in I\}$. Each element p_i in this set is identified by a number $i \in I$, called the *index* of the element. A set defined in this way is sometimes called an *indexed family*, and I is called the *index set.*

Although the indices for an indexed family are often numbers, they need not be. For example, suppose S is the set of all students at your school. If we wanted to form the set of all mothers of students, we might let m_s stand for the mother of s, for any student s. Then the set of all mothers of students could be written $M = \{m_s \mid s \in S\}$. This is an indexed family in which the index set is S, the set of all students. Each mother in the set is identified by naming the student who is her child. Note that we could also define this set using an elementhood test, by writing $M = \{m \mid m$ is the mother of some student$\} = \{m \mid \exists s \in S(m = m_s)\}$. In general, any indexed family $A = \{x_i \mid i \in I\}$ can also be defined as $A = \{x \mid \exists i \in I(x = x_i)\}$. It follows that the statement $x \in \{x_i \mid i \in I\}$ means the same thing as $\exists i \in I(x = x_i)$.

Example 2.3.1. Analyze the logical forms of the following statements by writing out the definitions of the set theory notation used.

1. $y \in \{\sqrt[3]{x} | x \in \mathbb{Q}\}$.
2. $\{x_i \mid i \in I\} \subseteq A$.
3. $\{n^2 \mid n \in \mathbb{N}\}$ and $\{n^3 \mid n \in \mathbb{N}\}$ are not disjoint.

Solutions

1. $\exists x \in \mathbb{Q}(y = \sqrt[3]{x})$.
2. By the definition of subset we must say that every element of $\{x_i \mid i \in I\}$ is also an element of A, so we could start by writing $\forall x(x \in \{x_i \mid i \in I\} \rightarrow x \in A)$. Filling in the meaning of $x \in \{x_i \mid i \in I\}$, which we worked out earlier, we would end up with $\forall x(\exists i \in I(x = x_i) \rightarrow x \in A)$. But since the elements of $\{x_i \mid i \in I\}$ are just the x_i's, for all $i \in I$, perhaps an easier way of saying that every element of $\{x_i \mid i \in I\}$ is an element of A would be $\forall i \in I(x_i \in A)$. The two answers we have given are equivalent, but showing this would require the methods we will be studying in Chapter 3.

3. We must say that the two sets have a common element, so one solution is to start by writing $\exists x(x \in \{n^2 \mid n \in \mathbb{N}\} \wedge x \in \{n^3 \mid n \in \mathbb{N}\})$. However, as in the last statement, there is an easier way. An element common to the two sets would have to be the square of some natural number and also the cube of some (possibly different) natural number. Thus, we could say that there is such a common element by saying $\exists n \in \mathbb{N} \exists m \in \mathbb{N}(n^2 = m^3)$. Note that it would be wrong to write $\exists n \in \mathbb{N}(n^2 = n^3)$, because this wouldn't allow for the possibility of the two natural numbers being different. By the way, this statement is true, since $64 = 8^2 = 4^3$, so 64 is an element of both sets.

Anything at all can be an element of a set. Some interesting and useful ideas arise when we consider the possibility of a set having *other sets* as elements. For example, suppose $A = \{1, 2, 3\}$, $B = \{4\}$, and $C = \varnothing$. There is no reason why we couldn't form the set $\mathcal{F} = \{A, B, C\}$, whose elements are the three sets A, B, and C. Filling in the definitions of A, B, and C, we could write this in another way: $\mathcal{F} = \{\{1, 2, 3\}, \{4\}, \varnothing\}$. Note that $1 \in A$ and $A \in \mathcal{F}$ but $1 \notin \mathcal{F}$. $\mathcal{F}$ has only three elements, and all three of them are sets, not numbers. Sets such as $\mathcal{F}$, whose elements are all sets, are sometimes called *families* of sets.

It is often convenient to define families of sets as indexed families. For example, suppose we again let S stand for the set of all students, and for each student s we let C_s be the set of courses that s has taken. Then the collection of all of these sets C_s would be an indexed family of sets $\mathcal{F} = \{C_s \mid s \in S\}$. Remember that the elements of this family are not courses but *sets* of courses. If we let t stand for some particular student Tina, and if Tina has taken Calculus, English Composition, and American History, then $C_t =$ {Calculus, English Composition, American History} and $C_t \in \mathcal{F}$, but Calculus $\notin \mathcal{F}$.

An important example of a family of sets is given by the power set of a set.

Definition 2.3.2. Suppose A is a set. The *power set* of A, denoted $\mathscr{P}(A)$, is the set whose elements are all the subsets of A. In other words,

$$\mathscr{P}(A) = \{x \mid x \subseteq A\}.$$

For example, the set $A = \{7, 12\}$ has four subsets: $\varnothing$, $\{7\}$, $\{12\}$, and $\{7, 12\}$. Thus, $\mathscr{P}(A) = \{\varnothing, \{7\}, \{12\}, \{7, 12\}\}$. What about $\mathscr{P}(\varnothing)$? Although $\varnothing$ has no elements, it does have one subset, namely $\varnothing$. Thus, $\mathscr{P}(\varnothing) = \{\varnothing\}$. Note that, as we saw in Section 1.3, $\{\varnothing\}$ is not the same as $\varnothing$.

Any time you are working with some subsets of a set X, it may be helpful to remember that all of these subsets of X are elements of $\mathscr{P}(X)$, by the definition of power set. For example, if we let C be the set of all courses offered at your school, then each of the sets C_s from our previous example is a subset of C. Thus, for each student s, $C_s \in \mathscr{P}(C)$. This means that every element of the family $\mathcal{F} = \{C_s \mid s \in S\}$ is an element of $\mathscr{P}(C)$, so $\mathcal{F} \subseteq \mathscr{P}(C)$.

Example 2.3.3. Analyze the logical forms of the following statements.

1. $x \in \mathscr{P}(A)$.
2. $\mathscr{P}(A) \subseteq \mathscr{P}(B)$.
3. $B \in \{\mathscr{P}(A) \mid A \in \mathcal{F}\}$.
4. $x \in \mathscr{P}(A \cap B)$.
5. $x \in \mathscr{P}(A) \cap \mathscr{P}(B)$.

Solutions

1. By the definition of power set, the elements of $\mathscr{P}(A)$ are the subsets of A. Thus, to say that $x \in \mathscr{P}(A)$ means that $x \subseteq A$, which we already know can be written as $\forall y(y \in x \rightarrow y \in A)$.
2. By the definition of subset, this means $\forall x(x \in \mathscr{P}(A) \rightarrow x \in \mathscr{P}(B))$. Now, writing out $x \in \mathscr{P}(A)$ and $x \in \mathscr{P}(B)$ as before, we get $\forall x[\forall y(y \in x \rightarrow y \in A) \rightarrow \forall y(y \in x \rightarrow y \in B)]$.
3. As before, this means $\exists A \in \mathcal{F}(B = \mathscr{P}(A))$. Now, to say that $B = \mathscr{P}(A)$ means that the elements of B are precisely the subsets of A, or in other words $\forall x(x \in B \leftrightarrow x \subseteq A)$. Filling this in, and writing out the definition of subset, we get our final answer, $\exists A \in \mathcal{F}\, \forall x(x \in B \leftrightarrow \forall y(y \in x \rightarrow y \in A))$.
4. As in the first statement, we start by writing this as $\forall y(y \in x \rightarrow y \in A \cap B)$. Now, filling in the definition of intersection, we get $\forall y(y \in x \rightarrow (y \in A \wedge y \in B))$.
5. By the definition of intersection, this means $(x \in \mathscr{P}(A)) \wedge (x \in \mathscr{P}(B))$. Now, writing out the definition of power set as before, we get $\forall y(y \in x \rightarrow y \in A) \wedge \forall y(y \in x \rightarrow y \in B)$.

Note that for statement 5 in this example we first wrote out the definition of intersection and then used the definition of power set, whereas in statement 4 we started by writing out the definition of power set and then used the definition of intersection. As you learn the definitions of more mathematical terms and symbols, it will become more important to be able to choose which definition to think about first when working out the meaning of a complex mathematical statement. A good rule of thumb is to always start with the "outermost" symbol. In statement 4 in Example 2.3.3, the intersection symbol occurred inside the power set notation, so we wrote out the definition of power set first. In statement 5, the power set notation occurred within both sides of the notation for the intersection of two sets, so we started with the definition of intersection. Similar considerations led us to use the definition of subset first, rather than power set, in statement 2.

It is interesting to note that our answers for statements 4 and 5 in Example 2.3.3 are equivalent. (You are asked to verify this in exercise 10.) As in Section 1.4, it follows that for any sets A and B, $\mathscr{P}(A \cap B) = \mathscr{P}(A) \cap \mathscr{P}(B)$. You are asked in exercise 11 to show that this equation is not true in general if we change $\cap$ to $\cup$.

Consider once again the family of sets $\mathcal{F} = \{C_s \mid s \in S\}$, where S is the set of all students and for each student s, C_s is the set of all courses that s has taken. If we wanted to know which courses had been taken by all students, we would need to find those elements that all the sets in $\mathcal{F}$ have in common. The set of all these common elements is called the intersection of the family $\mathcal{F}$ and is written $\cap\mathcal{F}$. Similarly, the union of the family $\mathcal{F}$, written $\cup\mathcal{F}$, is the set resulting from throwing all the elements of all the sets in $\mathcal{F}$ together into one set. In this case, $\cup\mathcal{F}$ would be the set of all courses that had been taken by any student.

Example 2.3.4. Let $\mathcal{F} = \{\{1, 2, 3, 4\}, \{2, 3, 4, 5\}, \{3, 4, 5, 6\}\}$. Find $\cap\mathcal{F}$ and $\cup\mathcal{F}$.

Solution

$$\cap\mathcal{F} = \{1, 2, 3, 4\} \cap \{2, 3, 4, 5\} \cap \{3, 4, 5, 6\} = \{3, 4\}.$$
$$\cup\mathcal{F} = \{1, 2, 3, 4\} \cup \{2, 3, 4, 5\} \cup \{3, 4, 5, 6\} = \{1, 2, 3, 4, 5, 6\}.$$

Although these examples may make it clear what we mean by $\cap\mathcal{F}$ and $\cup\mathcal{F}$, we still have not given careful definitions for these sets. In general, if $\mathcal{F}$ is any family of sets, then we want $\cap\mathcal{F}$ to contain the elements that all the sets in $\mathcal{F}$ have in common. Thus, to be an element of $\cap\mathcal{F}$, an object will have to be an element of every set in $\mathcal{F}$. On the other hand, anything that is an element of any of the sets in $\mathcal{F}$ should be in $\cup\mathcal{F}$, so to be in $\cup\mathcal{F}$ an object only needs to be an element of at least one set in $\mathcal{F}$. Thus, we are led to the following general definitions.

Definition 2.3.5. Suppose $\mathcal{F}$ is a family of sets. Then the *intersection* and *union* of $\mathcal{F}$ are the sets $\cap\mathcal{F}$ and $\cup\mathcal{F}$ defined as follows:

$$\cap\mathcal{F} = \{x \mid \forall A \in \mathcal{F}(x \in A)\} = \{x \mid \forall A(A \in \mathcal{F} \to x \in A)\}.$$
$$\cup\mathcal{F} = \{x \mid \exists A \in \mathcal{F}(x \in A)\} = \{x \mid \exists A(A \in \mathcal{F} \land x \in A)\}.$$

Some mathematicians consider $\cap\mathcal{F}$ to be undefined if $\mathcal{F} = \varnothing$. For an explanation of the reason for this, see exercise 14. We will use the notation $\cap\mathcal{F}$ only when $\mathcal{F} \neq \varnothing$.

Notice that if A and B are any two sets and $\mathcal{F} = \{A, B\}$, then $\cap\mathcal{F} = A \cap B$ and $\cup\mathcal{F} = A \cup B$. Thus, the definitions of intersection and union of a family of sets are actually generalizations of our old definitions of the intersection and union of two sets.

Example 2.3.6. Analyze the logical forms of the following statements.

1. $x \in \cap\mathcal{F}$.
2. $\cap\mathcal{F} \not\subseteq \cup\mathcal{G}$.
3. $x \in \mathscr{P}(\cup\mathcal{F})$.
4. $x \in \cup\{\mathscr{P}(A) \mid A \in \mathcal{F}\}$.

Solutions

1. By the definition of the intersection of a family of sets, this means $\forall A \in \mathcal{F}(x \in A)$, or equivalently, $\forall A(A \in \mathcal{F} \to x \in A)$.
2. As we saw in Example 2.2.1, to say that one set is not a subset of another means that there is something that is an element of the first but not the second. Thus, we start by writing $\exists x(x \in \cap\mathcal{F} \wedge x \notin \cup\mathcal{G})$. We have already written out what $x \in \cap\mathcal{F}$ means in solution 1. By the definition of the union of a family of sets, $x \in \cup\mathcal{G}$ means $\exists A \in \mathcal{G}(x \in A)$, so $x \notin \cup\mathcal{G}$ means $\neg\exists A \in \mathcal{G}(x \in A)$. By the quantifier negation laws, this is equivalent to $\forall A \in \mathcal{G}(x \notin A)$. Putting this all together, our answer is $\exists x[\forall A \in \mathcal{F}(x \in A) \wedge \forall A \in \mathcal{G}(x \notin A)]$.
3. Because the union symbol occurs within the power set notation, we start by writing out the definition of power set. As in Example 2.3.3, we get $x \subseteq \cup\mathcal{F}$, or in other words $\forall y(y \in x \to y \in \cup\mathcal{F})$. Now we use the definition of union to write out $y \in \cup\mathcal{F}$ as $\exists A \in \mathcal{F}(y \in A)$. The final answer is $\forall y(y \in x \to \exists A \in \mathcal{F}(y \in A))$.
4. This time we start by writing out the definition of union. According to this definition, the statement means that x is an element of at least one of the sets $\mathscr{P}(A)$, for $A \in \mathcal{F}$. In other words, $\exists A \in \mathcal{F}(x \in \mathscr{P}(A))$. Inserting our analysis of the statement $x \in \mathscr{P}(A)$ from Example 2.3.3, we get $\exists A \in \mathcal{F}\forall y(y \in x \to y \in A)$.

Writing complex mathematical statements in logical symbols, as we did in the last example, may sometimes help you understand what the statements mean and whether they are true or false. For example, suppose that we once again let C_s be the set of all courses that have been taken by student s.

Let M be the set of math majors and E the set of English majors, and let $\mathcal{F} = \{C_s \mid s \in M\}$ and $\mathcal{G} = \{C_s \mid s \in E\}$. With these definitions, what does statement 2 of Example 2.3.6 mean, and under what circumstances would it be true? According to our solution for this example, the statement means $\exists x[\forall A \in \mathcal{F}(x \in A) \wedge \forall A \in \mathcal{G}(x \notin A)]$, or in other words, there is something that is an element of each set in $\mathcal{F}$, and that fails to be an element of each set in $\mathcal{G}$. Taking into account the definitions of $\mathcal{F}$ and $\mathcal{G}$ that we are using, this means that there is some course that has been taken by all of the math majors but none of the English majors. If, for example, all of the math majors have taken Calculus but none of the English majors have, then the statement would be true.

As another example, suppose $\mathcal{F} = \{\{1, 2, 3\}, \{2, 3, 4\}, \{3, 4, 5\}\}$, and $x = \{4, 5, 6\}$. With these definitions, would statement 3 of Example 2.3.6 be true? You could determine this by finding $\mathscr{P}(\cup\mathcal{F})$ and then checking to see if x is an element of it, but this would take a very long time, because it turns out that $\mathscr{P}(\cup\mathcal{F})$ has 32 elements. It is easier to use the translation into logical symbols given in our solution for this example. According to that translation, the statement means $\forall y(y \in x \rightarrow \exists A \in \mathcal{F}(y \in A))$; in other words, every element of x is in at least one set in $\mathcal{F}$. Looking back at our definitions of $\mathcal{F}$ and x, it is not hard to see that this is false, because $6 \in x$, but 6 is not in any of the sets in $\mathcal{F}$.

An alternative notation is sometimes used for the union or intersection of an indexed family of sets. Suppose $\mathcal{F} = \{A_i \mid i \in I\}$, where each A_i is a set. Then $\cap\mathcal{F}$ would be the set of all elements common to all the A_i's, for $i \in I$, and this can also be written as $\cap_{i\in I} A_i$. In other words, $\cap\mathcal{F} = \cap_{i\in I} A_i = \{x \mid \forall i \in I(x \in A_i)\}$. Similarly, an alternative notation for $\cup\mathcal{F}$ is $\cup_{i\in I} A_i$, so $\cup\mathcal{F} = \cup_{i\in I} A_i = \{x \mid \exists i \in I(x \in A_i)\}$. Returning to our example of courses taken by students, we could use this notation to write the set of courses taken by all students as $\cap_{s\in S} C_s$.

Example 2.3.7. Let $I = \{1, 2, 3\}$, and for each $i \in I$ let $A_i = \{i, i+1, i+2, i+3\}$. Find $\cap_{i\in I} A_i$ and $\cup_{i\in I} A_i$.

Solution

First we list the elements of the sets A_i, for $i \in I$:

$$A_1 = \{1, 2, 3, 4\}, \qquad A_2 = \{2, 3, 4, 5\}, \qquad A_3 = \{3, 4, 5, 6\}.$$

Then $\cap_{i\in I} A_i = A_1 \cap A_2 \cap A_3 = \{1, 2, 3, 4\} \cap \{2, 3, 4, 5\} \cap \{3, 4, 5, 6\} = \{3, 4\}$, and similarly $\cup_{i\in I} A_i = \{1, 2, 3, 4\} \cup \{2, 3, 4, 5\} \cup \{3, 4, 5, 6\} = \{1, 2, 3, 4, 5, 6\}$. In fact, we can now see that the question asked in this example is exactly the same as the one in Example 2.3.4, but with different notation.

Example 2.3.8. For this example our universe of discourse will be the set S of all students. Let $L(x., y)$ stand for "x likes y" and $A(x, y)$ for "x admires y." For each student s, let L_s be the set of all students that s likes. In other words $L_s = \{t \in S \mid L(s, t)\}$. Similarly, let $A_s = \{t \in S \mid A(s, t)\}$ = the set of all students that s admires. Describe the following sets.

1. $\bigcap_{s \in S} L_s$.
2. $\bigcup_{s \in S} L_s$.
3. $\bigcup_{s \in S} L_s \setminus \bigcup_{s \in S} A_s$.
4. $\bigcup_{s \in S}(L_s \setminus A_s)$.
5. $(\bigcap_{s \in S} L_s) \cap (\bigcap_{s \in S} A_s)$.
6. $\bigcap_{s \in S}(L_s \cap A_s)$.
7. $\bigcup_{b \in B} L_b$, where $B = \bigcap_{s \in S} A_s$.

Solutions

First of all, note that in general $t \in L_s$ means the same thing as $L(s, t)$, and similarly $t \in A_s$ means $A(s, t)$.

1. $\bigcap_{s \in S} L_s = \{t \mid \forall s \in S(t \in L_s)\} = \{t \in S \mid \forall s \in S\, L(s, t)\}$ = the set of all students who are liked by all students.
2. $\bigcup_{s \in S} L_s = \{t \mid \exists s \in S(t \in L_s)\} = \{t \in S \mid \exists s \in S\, L(s, t)\}$ = the set of all students who are liked by at least one student.
3. As we saw in solution 2, $\bigcup_{s \in S} L_s$ = the set of all students who are liked by at least one student. Similarly, $\bigcup_{s \in S} A_s$ = the set of all students who are admired by at least one student. Thus $\bigcup_{s \in S} L_s \setminus \bigcup_{s \in S} A_s = \{t \mid t \in \bigcup_{s \in S} L_s$ and $t \notin \bigcup_{s \in S} A_s\}$ = the set of all students who are liked by at least one student, but are not admired by any students.
4. $\bigcup_{s \in S}(L_s \setminus A_s) = \{t \mid \exists s \in S(t \in L_s \setminus A_s)\} = \{t \in S \mid \exists s \in S(L(s, t) \wedge \neg A(s, t))\}$ = the set of all students t such that some student likes t, but doesn't admire t. Note that this is different from the set in part 3. For a student t to be in this set, there must be a student who likes t but doesn't admire t, but there could be other students who admire t. To be in the set in part 3, t must be admired by *nobody*.
5. $(\bigcap_{s \in S} L_s) \cap (\bigcap_{s \in S} A_s) = \{t \mid t \in \bigcap_{s \in S} L_s$ and $t \in \bigcap_{s \in S} A_s\} = \{t \mid \forall s \in S(t \in L_s) \wedge \forall s \in S(t \in A_s)\} = \{t \in S \mid \forall s \in S\, L(s, t) \wedge \forall s \in S\, A(s, t)\}$ = the set of all students who are liked by all students and also admired by all students.
6. $\bigcap_{s \in S}(L_s \cap A_s) = \{t \mid \forall s \in S\,(t \in L_s \cap A_s)\} = \{t \in S \mid \forall s \in S\,(L(s, t) \wedge A(s, t))\}$ = the set of all students who are both liked and admired by all students. This is the same as the set in part 5. In fact, you can use the

distributive law for universal quantification and conjunction to show that the elementhood tests for the two sets are equivalent.

7. $\bigcup_{b \in B} L_b = \{t \mid \exists b \in B(t \in L_b)\} = \{t \in S \mid \exists b(b \in B \wedge L(b,t))\}$. But B was defined to be the set of all students who are admired by all students, so $b \in B$ means $b \in S \wedge \forall s \in S\, A(s,b)$. Inserting this, we get $\bigcup_{b \in B} L_b = \{t \in S \mid \exists b(b \in S \wedge \forall s \in S\, A(s,b) \wedge L(b,t))\}$ = the set of all students who are liked by some student who is admired by all students.

Exercises

*1. Analyze the logical forms of the following statements. You may use the symbols $\in$, $\notin$, $=$, $\neq$, $\wedge$, $\vee$, $\rightarrow$, $\leftrightarrow$, $\forall$, and $\exists$ in your answers, but not $\subseteq$, $\nsubseteq$, $\mathscr{P}$, $\cap$, $\cup$, $\setminus$, $\{$, $\}$, or $\neg$. (Thus, you must write out the definitions of some set theory notation, and you must use equivalences to get rid of any occurrences of $\neg$.)
   (a) $\mathcal{F} \subseteq \mathscr{P}(A)$.
   (b) $A \subseteq \{2n+1 \mid n \in \mathbb{N}\}$.
   (c) $\{n^2+n+1 \mid n \in \mathbb{N}\} \subseteq \{2n+1 \mid n \in \mathbb{N}\}$.
   (d) $\mathscr{P}(\bigcup_{i \in I} A_i) \nsubseteq \bigcup_{i \in I} \mathscr{P}(A_i)$.

2. Analyze the logical forms of the following statements. You may use the symbols $\in$, $\notin$, $=$, $\neq$, $\wedge$, $\vee$, $\rightarrow$, $\leftrightarrow$, $\forall$, and $\exists$ in your answers, but not $\subseteq$, $\nsubseteq$, $\mathscr{P}$, $\cap$, $\cup$, $\setminus$, $\{$, $\}$, or $\neg$. (Thus, you must write out the definitions of some set theory notation, and you must use equivalences to get rid of any occurrences of $\neg$.)
   (a) $x \in \bigcup\mathcal{F} \setminus \bigcup\mathcal{G}$.
   (b) $\{x \in B \mid x \notin C\} \in \mathscr{P}(A)$.
   (c) $x \in \bigcap_{i \in I}(A_i \cup B_i)$.
   (d) $x \in (\bigcap_{i \in I} A_i) \cup (\bigcap_{i \in I} B_i)$.

3. We've seen that $\mathscr{P}(\varnothing) = \{\varnothing\}$, and $\{\varnothing\} \neq \varnothing$. What is $\mathscr{P}(\{\varnothing\})$?

*4. Suppose $\mathcal{F}$ = {{red, green, blue}, {orange, red, blue}, {purple, red, green, blue}}. Find $\bigcap\mathcal{F}$ and $\bigcup\mathcal{F}$.

5. Suppose $\mathcal{F} = \{\{3,7,12\},\{5,7,16\},\{5,12,23\}\}$. Find $\bigcap\mathcal{F}$ and $\bigcup\mathcal{F}$.

6. Let $I = \{2,3,4,5\}$, and for each $i \in I$ let $A_i = \{i, i+1, i-1, 2i\}$.
   (a) List the elements of all the sets A_i, for $i \in I$.
   (b) Find $\bigcap_{i \in I} A_i$ and $\bigcup_{i \in I} A_i$.

7. Let P = {Johann Sebastian Bach, Napoleon Bonaparte, Johann Wolfgang von Goethe, David Hume, Wolfgang Amadeus Mozart, Isaac Newton, George Washington} and let $Y = \{1750, 1751, 1752, \ldots, 1759\}$.

For each $y \in Y$, let $A_y = \{p \in P \mid \text{the person } p \text{ was alive at some time during the year } y\}$. Find $\bigcup_{y \in Y} A_y$ and $\bigcap_{y \in Y} A_y$.

*8. Let $I = \{2, 3\}$, and for each $i \in I$ let $A_i = \{i, 2i\}$ and $B_i = \{i, i + 1\}$.
   (a) List the elements of the sets A_i and B_i for $i \in I$.
   (b) Find $\bigcap_{i \in I}(A_i \cup B_i)$ and $(\bigcap_{i \in I} A_i) \cup (\bigcap_{i \in I} B_i)$. Are they the same?
   (c) In parts (c) and (d) of exercise 2 you analyzed the statements $x \in \bigcap_{i \in I}(A_i \cup B_i)$ and $x \in (\bigcap_{i \in I} A_i) \cup (\bigcap_{i \in I} B_i)$. What can you conclude from your answer to part (b) about whether or not these statements are equivalent?

9. Give an example of an index set I and indexed families of sets $\{A_i \mid i \in I\}$ and $\{B_i \mid i \in I\}$ such that $\bigcup_{i \in I}(A_i \cap B_i) \neq (\bigcup_{i \in I} A_i) \cap (\bigcup_{i \in I} B_i)$.

10. Show that for any sets A and B, $\mathscr{P}(A \cap B) = \mathscr{P}(A) \cap \mathscr{P}(B)$, by showing that the statements $x \in \mathscr{P}(A \cap B)$ and $x \in \mathscr{P}(A) \cap \mathscr{P}(B)$ are equivalent. (See Example 2.3.3.)

*11. Give examples of sets A and B for which $\mathscr{P}(A \cup B) \neq \mathscr{P}(A) \cup \mathscr{P}(B)$.

12. Verify the following identities by writing out (using logical symbols) what it means for an object x to be an element of each set and then using logical equivalences.
   (a) $\bigcup_{i \in I}(A_i \cup B_i) = (\bigcup_{i \in I} A_i) \cup (\bigcup_{i \in I} B_i)$.
   (b) $(\bigcap \mathcal{F}) \cap (\bigcap \mathcal{G}) = \bigcap(\mathcal{F} \cup \mathcal{G})$.
   (c) $\bigcap_{i \in I}(A_i \setminus B_i) = (\bigcap_{i \in I} A_i) \setminus (\bigcup_{i \in I} B_i)$.

*13. Sometimes each set in an indexed family of sets has *two* indices. For this problem, use the following definitions: $I = \{1, 2\}$, $J = \{3, 4\}$. For each $i \in I$ and $j \in J$, let $A_{i,j} = \{i, j, i + j\}$. Thus, for example, $A_{2,3} = \{2, 3, 5\}$.
   (a) For each $j \in J$ let $B_j = \bigcup_{i \in I} A_{i,j} = A_{1,j} \cup A_{2,j}$. Find B_3 and B_4.
   (b) Find $\bigcap_{j \in J} B_j$. (Note that, replacing B_j with its definition, we could say that $\bigcap_{j \in J} B_j = \bigcap_{j \in J}(\bigcup_{i \in I} A_{i,j})$.)
   (c) Find $\bigcup_{i \in I}(\bigcap_{j \in J} A_{i,j})$. (Hint: You may want to do this in two steps, corresponding to parts (a) and (b).) Are $\bigcap_{j \in J}(\bigcup_{i \in I} A_{i,j})$ and $\bigcup_{i \in I}(\bigcap_{j \in J} A_{i,j})$ equal?
   (d) Analyze the logical forms of the statements $x \in \bigcap_{j \in J}(\bigcup_{i \in I} A_{i,j})$ and $x \in \bigcup_{i \in I}(\bigcap_{j \in J} A_{i,j})$. Are they equivalent?

14. (a) Show that if $\mathcal{F} = \varnothing$, then the statement $x \in \bigcup \mathcal{F}$ will be false no matter what x is. It follows that $\bigcup \varnothing = \varnothing$.
   (b) Show that if $\mathcal{F} = \varnothing$, then the statement $x \in \bigcap \mathcal{F}$ will be true no matter what x is. In a context in which it is clear what the universe of discourse U is, we might therefore want to say that $\bigcap \varnothing = U$. However, this has the unfortunate consequence that the notation $\bigcap \varnothing$ will mean different things in different contexts. Furthermore, when

working with sets whose elements are sets, mathematicians often do not use a universe of discourse at all. (For more on this, see the next exercise.) For these reasons, some mathematicians consider the notation $\cap\varnothing$ to be meaningless. We will avoid this problem in this book by using the notation $\cap\mathcal{F}$ only in contexts in which we can be sure that $\mathcal{F} \neq \varnothing$.

15. In Section 2.3 we saw that a set can have other sets as elements. When discussing sets whose elements are sets, it might seem most natural to consider the universe of discourse to be the collection of all sets. However, as we will see in this problem, assuming that there is such a universe leads to contradictions.

    Suppose U were the collection of all sets. Note that in particular U is a set, so we would have $U \in U$. This is not yet a contradiction; although most sets are not elements of themselves, perhaps some sets are elements of themselves. But it suggests that the sets in the universe U could be split into two categories: the unusual sets that, like U itself, are elements of themselves, and the more typical sets that are not. Let R be the set of sets in the second category. In other words, $R = \{A \in U \mid A \notin A\}$. This means that for any set A in the universe U, A will be an element of R iff $A \notin A$. In other words, we have $\forall A \in U(A \in R \leftrightarrow A \notin A)$.

    (a) Show that applying this last fact to the set R itself (in other words, plugging in R for A) leads to a contradiction. This contradiction was discovered by Bertrand Russell in 1901, and is known as *Russell's Paradox*.

    (b) Think some more about the paradox in part (a). What do you think it tells us about sets?

3

Proofs

3.1. Proof Strategies

Mathematicians are skeptical people. They use many methods, including experimentation with examples, trial and error, and guesswork, to try to find answers to mathematical questions, but they are generally not convinced that an answer is correct unless they can prove it. You have probably seen some mathematical proofs before, but you may not have any experience writing them yourself. In this chapter you'll learn more about how proofs are put together, so you can start writing your own proofs.

Proofs are a lot like jigsaw puzzles. There are no rules about how jigsaw puzzles must be solved. The only rule concerns the final product: All the pieces must fit together, and the picture must look right. The same holds for proofs.

Although there are no rules about how jigsaw puzzles must be solved, some techniques for solving them work better than others. For example, you'd never do a jigsaw puzzle by filling in every *other* piece, and then going back and filling in the holes! But you also don't do it by starting at the top and filling in the pieces in order until you reach the bottom. You probably fill in the border first, and then gradually put other chunks of the puzzle together and figure out where they go. Sometimes you try to put pieces in the wrong places, realize that they don't fit, and feel that you're not making any progress. And every once in a while you see, in a satisfying flash, how two big chunks fit together and feel that you've suddenly made a lot of progress. As the pieces of the puzzle fall into place, a picture emerges. You suddenly realize that the patch of blue you've been putting together is a lake, or part of the sky. But it's only when the puzzle is complete that you can see the whole picture.

Similar things could be said about the process of figuring out a proof. And I think one more similarity should be mentioned. When you finish a jigsaw puzzle, you don't take it apart right away, do you? You probably leave it out for a day or two, so you can admire it. You should do the same thing with a proof. You figured out how to fit it together yourself, and once it's all done, isn't it pretty?

In this chapter we will discuss the proof-writing techniques that mathematicians use most often and explain how to use them to begin writing proofs yourself. Understanding these techniques may also help you read and understand proofs written by other people. Unfortunately, the techniques in this chapter do not give a step-by-step procedure for solving every proof problem. When trying to write a proof you may make a few false starts before finding the right way to proceed, and some proofs may require some cleverness or insight. With practice your proof-writing skills should improve, and you'll be able to tackle more and more challenging proofs.

Mathematicians usually state the answer to a mathematical question in the form of a *theorem* that says that if certain assumptions called the *hypotheses* of the theorem are true, then some conclusion must also be true. Often the hypotheses and conclusion contain free variables, and in this case it is understood that these variables can stand for any elements of the universe of discourse. An assignment of particular values to these variables is called an *instance* of the theorem, and in order for the theorem to be correct it must be the case that for every instance of the theorem that makes the hypotheses come out true, the conclusion is also true. If there is even one instance in which the hypotheses are true but the conclusion is false, then the theorem is incorrect. Such an instance is called a *counterexample* to the theorem.

Example 3.1.1. Consider the following theorem:

Theorem. *Suppose $x > 3$ and $y < 2$. Then $x^2 - 2y > 5$.*

This theorem is correct. (You are asked to prove it in exercise 14.) The hypotheses of the theorem are $x > 3$ and $y < 2$, and the conclusion is $x^2 - 2y > 5$. As an instance of the theorem, we could plug in 5 for x and 1 for y. Clearly with these values of the variables the hypotheses $x > 3$ and $y < 2$ are both true, so the theorem tells us that the conclusion $x^2 - 2y > 5$ must also be true. In fact, plugging in the values of x and y we find that $x^2 - 2y = 25 - 2 = 23$, and certainly $23 > 5$. Note that this calculation does not constitute a proof of the theorem. We have only checked one instance of the theorem, and a proof would have to show that *all* instances are correct.

If we drop the second hypothesis, then we get an incorrect theorem:

Incorrect Theorem. *Suppose $x > 3$. Then $x^2 - 2y > 5$.*

We can see that this theorem is incorrect by finding a counterexample. For example, suppose we let $x = 4$ and $y = 6$. Then the only remaining hypothesis, $x > 3$, is true, but $x^2 - 2y = 16 - 12 = 4$, so the conclusion $x^2 - 2y > 5$ is false.

If you find a counterexample to a theorem, then you can be sure that the theorem is incorrect, but the only way to know for sure that a theorem is correct is to prove it. A proof of a theorem is simply a deductive argument whose premises are the hypotheses of the theorem and whose conclusion is the conclusion of the theorem. Of course the argument should be valid, so we can be sure that if the hypotheses of the theorem are true, then the conclusion must be true as well. How you figure out and write up the proof of a theorem will depend mostly on the logical form of the conclusion. Often it will also depend on the logical forms of the hypotheses. The proof-writing techniques we will discuss in this chapter will tell you which proof strategies are most likely to work for various forms of hypotheses and conclusions.

Proof-writing techniques that are based on the logical forms of the hypotheses usually suggest ways of drawing inferences from the hypotheses. When you draw an inference from the hypotheses, you use the assumption that the hypotheses are true to justify the assertion that some other statement is also true. Once you have shown that a statement is true, you can use it later in the proof exactly as if it were a hypothesis. Perhaps the most important rule to keep in mind when drawing such inferences is this: *Never assert anything until you can justify it completely* using the hypotheses or using conclusions reached from them earlier in the proof. Your motto should be: "I shall make no assertion before its time." Following this rule will prevent you from using circular reasoning or jumping to conclusions and will guarantee that, if the hypotheses are true, then the conclusion must also be true. And this is the primary purpose of any proof: to provide a guarantee that the conclusion is true if the hypotheses are.

To make sure your assertions are adequately justified, you must be skeptical about every inference in your proof. If there is any doubt in your mind about whether the justification you have given for an assertion is adequate, then it isn't. After all, if your own reasoning doesn't even convince *you*, how can you expect it to convince anybody else?

Proof-writing techniques based on the logical form of the conclusion are often somewhat different from techniques based on the forms of the hypotheses. They usually suggest ways of transforming the problem into one that is equivalent but easier to solve. The idea of solving a problem by transforming it into an easier problem should be familiar to you. For example, adding the same number to both sides of an equation transforms the equation into an equivalent equation, and the resulting equation is sometimes easier to solve than the original one. Students who have studied calculus may be familiar with techniques of evaluating integrals, such as substitution or integration by parts, that can be used to transform a difficult integration problem into an easier one.

Proofs that are written using these transformation strategies often include steps in which you assume for the sake of argument that some statement is true without providing any justification for that assumption. It may seem at first that such reasoning would violate the rule that assertions must always be justified, but it doesn't, because *assuming* something is not the same as *asserting* it. To assert a statement is to claim that it is true, and such a claim is never acceptable in a proof unless it can be justified. However, the purpose of making an assumption in a proof is not to make a claim about what *is* true, but rather to enable you to find out what *would be* true *if* the assumption were correct. You must always keep in mind that any conclusion you reach that is based on an assumption might turn out to be false if the assumption is incorrect. Whenever you make a statement in a proof, it's important to be sure you know whether it's an assertion or an assumption.

Perhaps an example will help clarify this. Suppose during the course of a proof you decide to assume that some statement, call it P, is true, and you use this assumption to conclude that another statement Q is true. It would be wrong to call this a proof that Q is true, because you can't be sure that your assumption about the truth of P was correct. All you can conclude at this point is that *if* P is true, then you can be sure that Q is true as well. In other words, you know that the statement $P \rightarrow Q$ is true. If the conclusion of the theorem being proven was Q, then the proof is incomplete at best. But if the conclusion was $P \rightarrow Q$, then the proof is complete. This brings us to our first proof strategy.

To prove a conclusion of the form $P \rightarrow Q$:
Assume P is true and then prove Q.

Here's another way of looking at what this proof technique means. Assuming that P is true amounts to the same thing as adding P to your list of hypotheses. Although P might not originally have been one of your hypotheses, once you have assumed it, you can use it exactly the way you would use any other hypothesis. Proving Q means treating Q as your conclusion and forgetting about the original conclusion. So this technique says that if the conclusion of the theorem you are trying to prove has the form $P \rightarrow Q$, then you can *transform the problem* by adding P to your list of hypotheses and changing your conclusion from $P \to Q$ to Q. This gives you a new, perhaps easier proof problem to work on. If you can solve the new problem, then you will have shown that *if* P is true then Q is also true, thus solving the original problem of proving $P \to Q$. How you solve this new problem will now be guided by the logical form of the new conclusion Q (which might itself be a complex statement), and perhaps also by the logical form of the new hypothesis P.

Note that this technique doesn't tell you how to do the whole proof, it just gives you one step, leaving you with a new problem to solve in order to finish the proof. Proofs are usually not written all at once, but are created gradually by applying several proof techniques one after another. Often the use of these techniques will lead you to transform the problem several times. In discussing this process it will be helpful to have some way to keep track of the results of this sequence of transformations. We therefore introduce the following terminology. We will refer to the statements that are known or assumed to be true at some point in the course of figuring out a proof as *givens*, and the statement that remains to be proven at that point as the *goal*. When you are starting to figure out a proof, the givens will be just the hypotheses of the theorem you are proving, but they may later include other statements that have been inferred from the hypotheses or added as new assumptions as the result of some transformation of the problem. The goal will initially be the conclusion of the theorem, but it may be changed several times in the course of figuring out a proof.

To keep in mind that all of our proof strategies apply not only to the original proof problem but also to the results of any transformation of the problem, we will talk from now on only about givens and goals, rather than hypotheses and conclusions, when discussing proof-writing strategies. For example, the strategy stated earlier should really be called a strategy for proving a *goal* of the form $P \to Q$, rather than a conclusion of this form. Even if the conclusion of the theorem you are proving is not a conditional statement, if you transform the problem in such a way that a conditional statement becomes the goal, then you can apply this strategy as the next step in figuring out the proof.

Example 3.1.2. Suppose a and b are real numbers. Prove that if $0 < a < b$ then $a^2 < b^2$.

Scratch work

We are given as a hypothesis that a and b are real numbers. Our conclusion has the form $P \to Q$, where P is the statement $0 < a < b$ and Q is the statement

$a^2 < b^2$. Thus we start with these statements as given and goal:

Givens	*Goal*
a and b are real numbers	$(0 < a < b) \rightarrow (a^2 < b^2)$

According to our proof technique we should assume that $0 < a < b$ and try to use this assumption to prove that $a^2 < b^2$. In other words, we transform the problem by adding $0 < a < b$ to the list of givens and making $a^2 < b^2$ our goal:

Givens	*Goal*
a and b are real numbers	$a^2 < b^2$
$0 < a < b$	

Comparing the inequalities $a < b$ and $a^2 < b^2$ suggests that multiplying both sides of the given inequality $a < b$ by either a or b might get us closer to our goal. Because we are given that a and b are positive, we won't need to reverse the direction of the inequality if we do this. Multiplying $a < b$ by a gives us $a^2 < ab$, and multiplying it by b gives us $ab < b^2$. Thus $a^2 < ab < b^2$, so $a^2 < b^2$.

Solution

Theorem. *Suppose a and b are real numbers. If* $0 < a < b$ *then* $a^2 < b^2$.

Proof. Suppose $0 < a < b$. Multiplying the inequality $a < b$ by the positive number a we can conclude that $a^2 < ab$, and similarly multiplying by b we get $ab < b^2$. Therefore $a^2 < ab < b^2$, so $a^2 < b^2$, as required. Thus, if $0 < a < b$ then $a^2 < b^2$. □

As you can see from the preceding example, there's a difference between the reasoning you use when you are figuring out a proof and the steps you write down when you write the final version of the proof. In particular, although we will often talk about givens and goals when trying to figure out a proof, the final write-up will rarely refer to them. Throughout this chapter, and sometimes in later chapters as well, we will precede our proofs with the scratch work used to figure out the proof, but this is just to help you understand how proofs are constructed. When mathematicians write proofs, they usually just write the steps needed to justify their conclusions with no explanation of how they thought of them. Some of these steps will be sentences indicating that the problem has been transformed (usually according to some proof strategy based on the logical form of the goal); some steps will be assertions that are justified by inferences from the givens (often using some proof strategy based on the logical form of a given). However, there will usually be no explanation of how the mathematician thought of these transformations and inferences. For example, the proof in Example 3.1.2 starts with the sentence "Suppose $0 < a < b$," indicating that the problem has been transformed according to our strategy, and then proceeds with a sequence of inferences leading to the conclusion that $a^2 < b^2$. No other explanations were necessary to justify the final conclusion, in the last sentence, that if $0 < a < b$ then $a^2 < b^2$.

Although this lack of explanation sometimes makes proofs hard to read, it serves the purpose of keeping two distinct objectives separate: *explaining your thought processes* and *justifying your conclusions*. The first is psychology; the second, mathematics. The primary purpose of a proof is to justify the claim that the conclusion follows from the hypotheses, and no explanation of your thought processes can substitute for adequate justification of this claim. Keeping any discussion of thought processes to a minimum in a proof helps to keep this distinction clear. Occasionally, in a very complicated proof, a mathematician may include some discussion of the strategy behind the proof to make the proof easier to read. Usually, however, it is up to readers to figure this out for themselves. Don't worry if you don't immediately understand the strategy behind a proof you are reading. Just try to follow the justifications of the steps, and the strategy will eventually become clear. If it doesn't, a second reading of the proof might help.

To keep the distinction between the proof and the strategy behind the proof clear, in the future when we state a proof strategy we will often describe both the scratch work you might use to figure out the proof and the form that the final write-up of the proof should take. For example, here's a restatement of the proof strategy we discussed earlier, in the form we will be using to present proof strategies from now on.

To prove a goal of the form $P \rightarrow Q$:
Assume P is true and then prove Q.

Scratch work

Before using strategy:

Givens	*Goal*
—	$P \rightarrow Q$
—	

After using strategy:

Givens	*Goal*
—	Q
—	
P	

Form of final proof:

Suppose P.
 [Proof of Q goes here.]
Therefore $P \rightarrow Q$.

Note that the suggested form for the final proof tells you how the beginning and end of the proof will go, but more steps will have to be added in the middle. The givens and goal list under the heading "After using strategy" tells you what is known or can be assumed and what needs to be proven in order to fill in this gap in the proof. Many of our proof strategies will tell you how to write either the beginning or the end of your proof, leaving a gap to be filled in with further reasoning.

There is a second method that is sometimes used for proving goals of the form $P \rightarrow Q$. Because any conditional statement $P \rightarrow Q$ is equivalent to its contrapositive $\neg Q \rightarrow \neg P$, you can prove $P \rightarrow Q$ by proving $\neg Q \rightarrow \neg P$ instead, using the strategy discussed earlier. In other words:

To prove a goal of the form $P \rightarrow Q$:
Assume Q is false and prove that P is false.

Scratch work

Before using strategy:

Givens	*Goal*
—	$P \rightarrow Q$
—	

After using strategy:

Givens	*Goal*
—	$\neg P$
—	
$\neg Q$	

Form of final proof:

Suppose Q is false.
 [Proof of $\neg P$ goes here.]
Therefore $P \rightarrow Q$.

Example 3.1.3. Suppose a, b, and c are real numbers and $a > b$. Prove that if $ac \leq bc$ then $c \leq 0$.

Scratch work

Givens	*Goal*
$a, b,$ and c are real numbers	$(ac \leq bc) \rightarrow (c \leq 0)$
$a > b$	

The contrapositive of the goal is $\neg(c \leq 0) \rightarrow \neg(ac \leq bc)$, or in other words $(c > 0) \rightarrow (ac > bc)$, so we can prove it by adding $c > 0$ to the list of givens and making $ac > bc$ our new goal:

Givens	*Goal*
$a, b,$ and c are real numbers	$ac > bc$
$a > b$	
$c > 0$	

We can also now write the first and last sentences of the proof. According to the strategy, the final proof should have this form:

Suppose $c > 0$.
[Proof of $ac > bc$ goes here.]
Therefore, if $ac \leq bc$ then $c \leq 0$.

Using the new given $c > 0$, we see that the goal $ac > bc$ follows immediately from the given $a > b$ by multiplying both sides by the positive number c. Inserting this step between the first and last sentences completes the proof.

Solution

Theorem. *Suppose a, b, and c are real numbers and $a > b$. If $ac \leq bc$ then $c \leq 0$.*

Proof. We will prove the contrapositive. Suppose $c > 0$. Then we can multiply both sides of the given inequality $a > b$ by c and conclude that $ac > bc$. Therefore, if $ac \leq bc$ then $c \leq 0$. □

Notice that, although we have used the symbols of logic freely in the scratch work, we have not used them in the final write-up of the proof. Although it would not be incorrect to use logical symbols in a proof, mathematicians usually try to avoid it. Using the notation and rules of logic can be very helpful when you are figuring out the strategy for a proof, but in the final write-up you should try to stick to ordinary English as much as possible.

The reader may be wondering how we knew in Example 3.1.3 that we should use the second method for proving a goal of the form $P \rightarrow Q$

rather than the first. The answer is simple: We tried both methods, and the second worked. When there is more than one strategy for proving a goal of a particular form, you may have to try a few different strategies before you hit on one that works. With practice, you will get better at guessing which strategy is most likely to work for a particular proof.

Notice that in each of the examples we have given our strategy involved making changes in our givens and goal to try to make the problem easier. The beginning and end of the proof, which were supplied for us in the statement of the proof technique, serve to tell a reader of the proof that these changes have been made and how the solution to this revised problem solves the original problem. The rest of the proof contains the solution to this easier, revised problem.

Most of the other proof techniques in this chapter also suggest that you revise your givens and goal in some way. These revisions result in a new proof problem, and in every case the revisions have been designed so that a solution to the new problem, when combined with some beginning or ending sentences explaining these revisions, would also solve the original problem. This means that whenever you use one of these strategies you can write a sentence or two at the beginning or end of the proof and then forget about the original problem and work instead on the new problem, which will usually be easier. Often you will be able to figure out a proof by using the techniques in this chapter to revise your givens and goal repeatedly, making the remaining problem easier and easier until you reach a point at which it is completely obvious that the goal follows from the givens.

Exercises

*1. Consider the following theorem. (This theorem was proven in the introduction.)

Theorem. *Suppose n is an integer larger than 1 and n is not prime. Then $2^n - 1$ is not prime.*

(a) Identify the hypotheses and conclusion of the theorem. Are the hypotheses true when $n = 6$? What does the theorem tell you in this instance? Is it right?
(b) What can you conclude from the theorem in the case $n = 15$? Check directly that this conclusion is correct.
(c) What can you conclude from the theorem in the case $n = 11$?

2. Consider the following theorem. (The theorem is correct, but we will not ask you to prove it here.)

Theorem. *Suppose that $b^2 > 4ac$. Then the quadratic equation $ax^2 + bx + c = 0$ has exactly two real solutions.*

(a) Identify the hypotheses and conclusion of the theorem.
(b) To give an instance of the theorem, you must specify values for a, b, and c, but not x. Why?
(c) What can you conclude from the theorem in the case $a = 2$, $b = -5$, $c = 3$? Check directly that this conclusion is correct.
(d) What can you conclude from the theorem in the case $a = 2$, $b = 4$, $c = 3$?

3. Consider the following incorrect theorem:

Incorrect Theorem. *Suppose n is a natural number larger than 2, and n is not a prime number. Then $2n + 13$ is not a prime number.*

What are the hypotheses and conclusion of this theorem? Show that the theorem is incorrect by finding a counterexample.

*4. Complete the following alternative proof of the theorem in Example 3.1.2.

Proof. Suppose $0 < a < b$. Then $b - a > 0$.
[Fill in a proof of $b^2 - a^2 > 0$ here.]
Since $b^2 - a^2 > 0$, it follows that $a^2 < b^2$. Therefore if $0 < a < b$ then $a^2 < b^2$. □

5. Suppose a and b are real numbers. Prove that if $a < b < 0$ then $a^2 > b^2$.
6. Suppose a and b are real numbers. Prove that if $0 < a < b$ then $1/b < 1/a$.
7. Suppose that a is a real number. Prove that if $a^3 > a$ then $a^5 > a$. (Hint: One approach is to start by completing the following equation: $a^5 - a = (a^3 - a) \cdot \underline{\ ?\ }$.)
8. Suppose $A \setminus B \subseteq C \cap D$ and $x \in A$. Prove that if $x \notin D$ then $x \in B$.
*9. Suppose a and b are real numbers. Prove that if $a < b$ then $\frac{a+b}{2} < b$.
10. Suppose x is a real number and $x \neq 0$. Prove that if $\frac{\sqrt[3]{x}+5}{x^2+6} = \frac{1}{x}$ then $x \neq 8$.
*11. Suppose a, b, c, and d are real numbers, $0 < a < b$, and $d > 0$. Prove that if $ac \geq bd$ then $c > d$.
12. Suppose x and y are real numbers, and $3x + 2y \leq 5$. Prove that if $x > 1$ then $y < 1$.
13. Suppose that x and y are real numbers. Prove that if $x^2 + y = -3$ and $2x - y = 2$ then $x = -1$.

*14. Prove the first theorem in Example 3.1.1. (Hint: You might find it useful to apply the theorem from Example 3.1.2.)

15. Consider the following theorem.

Theorem. *Suppose x is a real number and $x \neq 4$. If $\frac{2x-5}{x-4} = 3$ then $x = 7$.*

(a) What's wrong with the following proof of the theorem?

Proof. Suppose $x = 7$. Then $\frac{2x-5}{x-4} = \frac{2(7)-5}{7-4} = \frac{9}{3} = 3$. Therefore if $\frac{2x-5}{x-4} = 3$ then $x = 7$. □

(b) Give a correct proof of the theorem.

16. Consider the following incorrect theorem:

Incorrect Theorem. *Suppose that x and y are real numbers and $x \neq 3$. If $x^2y = 9y$ then $y = 0$.*

(a) What's wrong with the following proof of the theorem?

Proof. Suppose that $x^2y = 9y$. Then $(x^2 - 9)y = 0$. Since $x \neq 3$, $x^2 \neq 9$, so $x^2 - 9 \neq 0$. Therefore we can divide both sides of the equation $(x^2 - 9)y = 0$ by $x^2 - 9$, which leads to the conclusion that $y = 0$. Thus, if $x^2y = 9y$ then $y = 0$. □

(b) Show that the theorem is incorrect by finding a counterexample.

3.2. Proofs Involving Negations and Conditionals

We turn now to proofs in which the goal has the form $\neg P$. Usually it's easier to prove a positive than a negative statement, so it is often helpful to reexpress a goal of the form $\neg P$ before proving it. Instead of using a goal that says what *shouldn't* be true, see if you can rephrase it as a goal that says what *should* be true. Fortunately, we have already studied several equivalences that will help with this reexpression. Thus, our first strategy for proving negated statements is:

To prove a goal of the form $\neg P$:

If possible, reexpress the goal in some other form and then use one of the proof strategies for this other goal form.

Example 3.2.1. Suppose $A \cap C \subseteq B$ and $a \in C$. Prove that $a \notin A \setminus B$.

Scratch Work

Givens	*Goal*
$A \cap C \subseteq B$	$a \notin A \setminus B$
$a \in C$	

Because the goal is a negated statement, we try to reexpress it:

$a \notin A \setminus B$ is equivalent to $\neg(a \in A \wedge a \notin B)$ (definition of $A \setminus B$),
which is equivalent to $a \notin A \vee a \in B$ (DeMorgan's law),
which is equivalent to $a \in A \rightarrow a \in B$ (conditional law).

Rewriting the goal in this way gives us:

Givens	*Goal*
$A \cap C \subseteq B$	$a \in A \rightarrow a \in B$
$a \in C$	

We now prove the goal in this new form, using the first strategy from Section 3.1. Thus, we add $a \in A$ to our list of givens and make $a \in B$ our goal:

Givens	*Goal*
$A \cap C \subseteq B$	$a \in B$
$a \in C$	
$a \in A$	

The proof is now easy: From the givens $a \in A$ and $a \in C$ we can conclude that $a \in A \cap C$, and then, since $A \cap C \subseteq B$, it follows that $a \in B$.

Solution

Theorem. *Suppose* $A \cap C \subseteq B$ *and* $a \in C$. *Then* $a \notin A \setminus B$.

Proof. Suppose $a \in A$. Then since $a \in C$, $a \in A \cap C$. But then since $A \cap C \subseteq B$ it follows that $a \in B$. Thus, it cannot be the case that a is an element of A but not B, so $a \notin A \setminus B$. □

Sometimes a goal of the form $\neg P$ cannot be reexpressed as a positive statement, and therefore this strategy cannot be used. In this case it is usually best to do a *proof by contradiction*. Start by assuming that P is true, and try to use this assumption to prove something that you know is false. Often this is done by proving a statement that contradicts one of the givens. Because you know that the statement you have proven is false, the assumption that P was true must have been incorrect. The only remaining possibility then is that P is false.

To prove a goal of the form $\neg P$:

Assume P is true and try to reach a contradiction. Once you have reached a contradiction, you can conclude that P must be false.

Scratch work

Before using strategy:

Givens	*Goal*
—	$\neg P$
—	

After using strategy:

Givens	*Goal*
—	Contradiction
—	
P	

Form of final proof:

Suppose P is true.
[Proof of contradiction goes here.]
Thus, P is false.

Example 3.2.2. Prove that if $x^2 + y = 13$ and $y \neq 4$ then $x \neq 3$.

Scratch work

The goal is a conditional statement, so according to the first proof strategy in Section 3.1 we can treat the antecedent as given and make the consequent our new goal:

Givens	*Goal*
$x^2 + y = 13$	$x \neq 3$
$y \neq 4$	

This proof strategy also suggests what form the final proof should take. According to the strategy, the proof should look like this:

Suppose $x^2 + y = 13$ and $y \neq 4$.
[Proof of $x \neq 3$ goes here.]
Thus, if $x^2 + y = 13$ and $y \neq 4$ then $x \neq 3$.

In other words, the first and last sentences of the final proof have already been written, and the problem that remains to be solved is to fill in a proof of $x \neq 3$

between these two sentences. The givens–goal list summarizes what we know and what we have to prove in order to solve this problem.

The goal $x \neq 3$ means $\neg(x = 3)$, but because $x = 3$ has no logical connectives in it, none of the equivalences we know can be used to reexpress this goal in a positive form. We therefore try proof by contradiction and transform the problem as follows:

Givens	*Goal*
$x^2 + y = 13$	Contradiction
$y \neq 4$	
$x = 3$	

Once again, the proof strategy that suggested this transformation also tells us how to fill in a few more sentences of the final proof. As we indicated earlier, these sentences go between the first and last sentences of the proof, which were written before.

Suppose $x^2 + y = 13$ and $y \neq 4$.
  Suppose $x = 3$.
    [Proof of contradiction goes here.]
  Therefore $x \neq 3$.
Thus, if $x^2 + y = 13$ and $y \neq 4$ then $x \neq 3$.

The indenting in this outline of the proof will not be part of the final proof. We have done it here to make the underlying structure of the proof clear. The first and last lines go together and indicate that we are proving a conditional statement by assuming the antecedent and proving the consequent. Between these lines is a proof of the consequent, $x \neq 3$, which we have set off from the first and last lines by indenting it. This inner proof has the form of a proof by contradiction, as indicated by its first and last lines. Between these lines we still need to fill in a proof of a contradiction.

At this point we don't have a particular statement as our goal; any impossible conclusion will do. We must therefore look more closely at the givens to see if some of them contradict others. In this case, the first and third together imply that $y = 4$, which contradicts the second.

Solution

Theorem. *If $x^2 + y = 13$ and $y \neq 4$ then $x \neq 3$.*

Proof. Suppose $x^2 + y = 13$ and $y \neq 4$. Suppose $x = 3$. Substituting this into the equation $x^2 + y = 13$, we get $9 + y = 13$, so $y = 4$. But this contradicts the fact that $y \neq 4$. Therefore $x \neq 3$. Thus, if $x^2 + y = 13$ and $y \neq 4$ then $x \neq 3$. □

You may be wondering at this point why we were justified in concluding, when we reached a contradiction in the proof, that $x \neq 3$. After all, the second list of givens in our scratch work contained three given. How could we be sure, when we reached a contradiction, that the culprit was the third given, $x = 3$? To answer this question, look back at the first givens and goal analysis for this example. According to that analysis, there were two givens, $x^2 + y = 13$ and $y \neq 4$, from which we had to prove the goal $x \neq 3$. Remember that a proof only has to guarantee that the goal is true *if* the givens are. Thus, we didn't have to show that $x \neq 3$, only that *if* $x^2 + y = 13$ and $y \neq 4$ then $x \neq 3$. When we reached a contradiction, we knew that one of the three statements in the second list of givens had to be false. We didn't try to figure out which one it was because we didn't need to. We were certainly justified in concluding that *if* neither of the first two was the culprit, then it had to be the third, and that was all that was required to finish the proof.

Proving a goal by contradiction has the advantage that it allows you to assume that your conclusion is false, providing you with another given to work with. But it has the disadvantage that it leaves you with a rather vague goal: produce a contradiction by proving something that you know is false. Because all the proof strategies we have discussed so far depend on analyzing the logical form of the goal, it appears that none of them will help you to achieve the goal of producing a contradiction. In the preceding proof we were forced to look more closely at our givens to find a contradiction. In this case we did it by proving that $y = 4$, contradicting the given $y \neq 4$. This illustrates a pattern that occurs often in proofs by contradiction: If one of the givens has the form $\neg P$, then you can produce a contradiction by proving P. This is our first strategy based on the logical form of a *given*.

To use a given of the form $\neg P$:

If you're doing a proof by contradiction, try making P your goal. If you can prove P, then the proof will be complete, because P contradicts the given $\neg P$.

Scratch work

Before using strategy:

Givens	*Goal*
$\neg P$	Contradiction
—	
—	

After using strategy:

Givens	*Goal*
$\neg P$	P
—	
—	

Form of final proof:

> [Proof of P goes here.]
> Since we already know $\neg P$, this is a contradiction.

Although we have recommended proof by contradiction for proving goals of the form $\neg P$, it can be used for any goal. Usually it's best to try the other strategies first if any of them apply; but if you're stuck, you can try proof by contradiction in any proof.

The next example illustrates this and also another important rule of proof-writing: In many cases the logical form of a statement can be discovered by *writing out the definition* of some mathematical word or symbol that occurs in the statement. For this reason, knowing the precise statements of the definitions of all mathematical terms is extremely important when you're writing a proof.

Example 3.2.3. Suppose A, B, and C are sets, $A \setminus B \subseteq C$, and x is anything at all. Prove that if $x \in A \setminus C$ then $x \in B$.

Scratch work

We're given that $A \setminus B \subseteq C$, and our goal is $x \in A \setminus C \to x \in B$. Because the goal is a conditional statement, our first step is to transform the problem by adding $x \in A \setminus C$ as a second given and making $x \in B$ our goal:

Givens	*Goal*
$A \setminus B \subseteq C$	$x \in B$
$x \in A \setminus C$	

The form of the final proof will therefore be as follows:

> Suppose $x \in A \setminus C$.
> [Proof of $x \in B$ goes here.]
> Thus, if $x \in A \setminus C$ then $x \in B$.

The goal $x \in B$ contains no logical connectives, so none of the techniques we have studied so far apply, and it is not obvious why the goal follows from the givens. Lacking anything else to do, we try proof by contradiction:

Givens	*Goal*
$A \setminus B \subseteq C$	Contradiction
$x \in A \setminus C$	
$x \notin B$	

As before, this transformation of the problem also enables us to fill in a few more sentences of the proof:

> Suppose $x \in A \setminus C$.
>   Suppose $x \notin B$.
>   [Proof of contradiction goes here.]
>   Therefore $x \in B$.
> Thus, if $x \in A \setminus C$ then $x \in B$.

Because we're doing a proof by contradiction and our last given is now a negated statement, we could try using our strategy for using givens of the form $\neg P$. Unfortunately, this strategy suggests making $x \in B$ our goal, which just gets us back to where we started. We must look at the other givens to try to find the contradiction.

In this case, writing out the definition of the second given is the key to the proof, since this definition also contains a negated statement. By definition, $x \in A \setminus C$ means $x \in A$ and $x \notin C$. Replacing this given by its definition gives us:

Givens	*Goal*
$A \setminus B \subseteq C$	Contradiction
$x \in A$	
$x \notin C$	
$x \notin B$	

Now the third given also has the form $\neg P$, where P is the statement $x \in C$, so we can apply the strategy for using givens of the form $\neg P$ and make $x \in C$ our goal. Showing that $x \in C$ would complete the proof because it would contradict the given $x \notin C$.

Givens	*Goal*
$A \setminus B \subseteq C$	$x \in C$
$x \in A$	
$x \notin C$	
$x \notin B$	

Once again, we can add a little more to the proof we are gradually writing by filling in the fact that we plan to derive our contradiction by proving $x \in C$.

We also add the definition of $x \in A \setminus C$ to the proof, inserting it in what seems like the most logical place, right after we stated that $x \in A \setminus C$:

> Suppose $x \in A \setminus C$. This means that $x \in A$ and $x \notin C$.
>   Suppose $x \notin B$.
>   [Proof of $x \in C$ goes here.]
>   This contradicts the fact that $x \notin C$.
>   Therefore $x \in B$.
> Thus, if $x \in A \setminus C$ then $x \in B$.

We have finally reached a point where the goal follows easily from the givens. From $x \in A$ and $x \notin B$ we conclude that $x \in A \setminus B$. Since $A \setminus B \subseteq C$ it follows that $x \in C$.

Solution

Theorem. *Suppose A, B, and C are sets, $A \setminus B \subseteq C$, and x is anything at all. If $x \in A \setminus C$ then $x \in B$.*

Proof. Suppose $x \in A \setminus C$. This means that $x \in A$ and $x \notin C$. Suppose $x \notin B$. Then $x \in A \setminus B$, so since $A \setminus B \subseteq C$, $x \in C$. But this contradicts the fact that $x \notin C$. Therefore $x \in B$. Thus, if $x \in A \setminus C$ then $x \in B$. □

The strategy we've recommended for using givens of the form $\neg P$ only applies if you are doing a proof by contradiction. For other kinds of proofs, the next strategy can be used. This strategy is based on the fact that givens of the form $\neg P$, like goals of this form, may be easier to work with if they are reexpressed as positive statements.

To use a given of the form $\neg P$:
If possible, reexpress this given in some other form.

We have discussed strategies for working with both givens and goals of the form $\neg P$, but only strategies for goals of the form $P \to Q$. We now fill this gap by giving two strategies for using givens of the form $P \to Q$. We said before that many strategies for using givens suggest ways of drawing inferences from the givens. Such strategies are called *rules of inference*. Both of our strategies for using givens of the form $P \to Q$ are examples of rules of inference.

To use a given of the form $P \to Q$:
If you are also given P, or if you can prove that P is true, then you can use this given to conclude that Q is true. Since it is equivalent to $\neg Q \to \neg P$,

if you can prove that Q is false, you can use this given to conclude that P is false.

The first of these rules of inference says that if you know that both P and $P \to Q$ are true, you can conclude that Q must also be true. Logicians call this rule *modus ponens*. We saw this rule used in one of our first examples of valid deductive reasoning in Chapter 1, argument 2 in Example 1.1.1. The validity of this form of reasoning was verified using the truth table for the conditional connective in Section 1.5.

The second rule, called *modus tollens*, says that if you know that $P \to Q$ is true and Q is false, you can conclude that P must also be false. The validity of this rule can also be checked with truth tables, as you are asked to show in exercise 13. Usually you won't find a given of the form $P \to Q$ to be much use until you are able to prove either P or $\neg Q$. However, if you ever reach a point in your proof where you have determined that P is true, you should probably use this given immediately to conclude that Q is true. Similarly, if you ever establish $\neg Q$, immediately use this given to conclude $\neg P$.

Although most of our examples will involve specific mathematical statements, occasionally we will do examples of proofs containing letters standing for unspecified statements. Later in this chapter we will be able to use this method to verify some of the equivalences from Chapter 2 that could only be justified on intuitive grounds before. Here's an example of this kind, illustrating the use of modus ponens and modus tollens.

Example 3.2.4. Suppose $P \to (Q \to R)$. Prove that $\neg R \to (P \to \neg Q)$.

Scratch work

This could actually be done with a truth table, as you are asked to show in exercise 14, but let's do it using the proof strategies we've been discussing. We start with the following situation:

Givens	*Goal*
$P \to (Q \to R)$	$\neg R \to (P \to \neg Q)$

Our only given is a conditional statement. By the rules of inference just discussed, if we knew P we could use modus ponens to conclude $Q \to R$, and if we knew $\neg(Q \to R)$ we could use modus tollens to conclude $\neg P$. Because we don't, at this point, know either of these, we can't yet do anything with this given. If either P or $\neg(Q \to R)$ ever gets added to the givens list, then we should consider using modus ponens or modus tollens. For now, we need to concentrate on the goal.

The goal is also a conditional statement, so we assume the antecedent and set the consequent as our new goal:

Givens	*Goal*
$P \to (Q \to R)$	$P \to \neg Q$
$\neg R$	

We can also now write a little bit of the proof:

Suppose $\neg R$.
  [Proof of $P \to \neg Q$ goes here.]
Therefore $\neg R \to (P \to \neg Q)$.

We still can't do anything with the givens, but the goal is another conditional, so we use the same strategy again:

Givens	*Goal*
$P \to (Q \to R)$	$\neg Q$
$\neg R$	
P	

Now the proof looks like this:

Suppose $\neg R$.
  Suppose P.
    [Proof of $\neg Q$ goes here.]
  Therefore $P \to \neg Q$.
Therefore $\neg R \to (P \to \neg Q)$.

We've been watching for our chance to use our first given by applying either modus ponens or modus tollens, and now we can do it. Since we know $P \to (Q \to R)$ and P, by modus ponens we can infer $Q \to R$. Any conclusion inferred from the givens can be added to the givens column:

Givens	*Goal*
$P \to (Q \to R)$	$\neg Q$
$\neg R$	
P	
$Q \to R$	

We also add one more line to the proof:

Suppose $\neg R$.
  Suppose P.
    Since P and $P \to (Q \to R)$, it follows that $Q \to R$.
    [Proof of $\neg Q$ goes here.]
  Therefore $P \to \neg Q$.
Therefore $\neg R \to (P \to \neg Q)$.

Finally, our last step is to use modus tollens. We now know $Q \to R$ and $\neg R$, so by modus tollens we can conclude $\neg Q$. This is our goal, so the proof is done.

Solution

Theorem. *Suppose $P \to (Q \to R)$. Then $\neg R \to (P \to \neg Q)$.*
Proof. Suppose $\neg R$. Suppose P. Since P and $P \to (Q \to R)$, it follows that $Q \to R$. But then, since $\neg R$, we can conclude $\neg Q$. Thus, $P \to \neg Q$. Therefore $\neg R \to (P \to \neg Q)$. □

Sometimes if you're stuck you can use rules of inference to work backward. For example, suppose one of your givens has the form $P \to Q$ and your goal is Q. If only you could prove P, you could use modus ponens to reach your goal. This suggests treating P as your goal instead of Q. If you can prove P, then you'll just have to add one more step to the proof to reach your original goal Q.

Example 3.2.5. Suppose that $A \subseteq B$, $a \in A$, and $a \notin B \setminus C$. Prove that $a \in C$.

Scratch work

Givens	*Goal*
$A \subseteq B$	$a \in C$
$a \in A$	
$a \notin B \setminus C$	

Our third given is a negative statement, so we begin by reexpressing it as an equivalent positive statement. According to the definition of the difference of two sets, this given means $\neg(a \in B \land a \notin C)$, and by one of DeMorgan's laws, this is equivalent to $a \notin B \lor a \in C$. Because our goal is $a \in C$, it is probably more useful to rewrite this in the equivalent form $a \in B \to a \in C$:

Givens	*Goal*
$A \subseteq B$	$a \in C$
$a \in A$	
$a \in B \to a \in C$	

Now we can use our strategy for using givens of the form $P \to Q$. Our goal is $a \in C$, and we are given that $a \in B \to a \in C$. If we could prove that $a \in B$, then we could use modus ponens to reach our goal. So let's try treating $a \in B$ as our goal and see if that makes the problem easier:

Givens	*Goal*
$A \subseteq B$	$a \in B$
$a \in A$	
$a \in B \rightarrow a \in C$	

Now it is clear how to reach the goal. Since $a \in A$ and $A \subseteq B$, $a \in B$.

Solution

Theorem. *Suppose that $A \subseteq B$, $a \in A$, and $a \notin B \setminus C$. Then $a \in C$.*

Proof. Since $a \in A$ and $A \subseteq B$, we can conclude that $a \in B$. But $a \notin B \setminus C$, so it follows that $a \in C$. □

Exercises

*1. This problem could be solved by using truth tables, but don't do it that way. Instead, use the methods for writing proofs discussed so far in this chapter. (See Example 3.2.4.)
    (a) Suppose $P \rightarrow Q$ and $Q \rightarrow R$ are both true. Prove that $P \rightarrow R$ is true.
    (b) Suppose $\neg R \rightarrow (P \rightarrow \neg Q)$ is true. Prove that $P \rightarrow (Q \rightarrow R)$ is true.
2. This problem could be solved by using truth tables, but don't do it that way. Instead, use the methods for writing proofs discussed so far in this chapter. (See Example 3.2.4.)
    (a) Suppose $P \rightarrow Q$ and $R \rightarrow \neg Q$ are both true. Prove that $P \rightarrow \neg R$ is true.
    (b) Suppose that P is true. Prove that $Q \rightarrow \neg(Q \rightarrow \neg P)$ is true.
3. Suppose $A \subseteq C$, and B and C are disjoint. Prove that if $x \in A$ then $x \notin B$.
4. Suppose that $A \setminus B$ is disjoint from C and $x \in A$. Prove that if $x \in C$ then $x \in B$.
*5. Use the method of proof by contradiction to prove the theorem in Example 3.2.1.
6. Use the method of proof by contradiction to prove the theorem in Example 3.2.5.
7. Suppose that $y + x = 2y - x$, and x and y are not both zero. Prove that $y \neq 0$.

*8. Suppose that a and b are nonzero real numbers. Prove that if $a < 1/a < b < 1/b$ then $a < -1$.

9. Suppose that x and y are real numbers. Prove that if $x^2y = 2x + y$, then if $y \neq 0$ then $x \neq 0$.

10. Suppose that x and y are real numbers. Prove that if $x \neq 0$, then if $y = \frac{3x^2+2y}{x^2+2}$ then $y = 3$.

*11. Consider the following incorrect theorem:

Incorrect Theorem. *Suppose x and y are real numbers and $x + y = 10$. Then $x \neq 3$ and $y \neq 8$.*

(a) What's wrong with the following proof of the theorem?

Proof. Suppose the conclusion of the theorem is false. Then $x = 3$ and $y = 8$. But then $x + y = 11$, which contradicts the given information that $x + y = 10$. Therefore the conclusion must be true. □

(b) Show that the theorem is incorrect by finding a counterexample.

12. Consider the following incorrect theorem:

Incorrect Theorem. *Suppose that $A \subseteq C$, $B \subseteq C$, and $x \in A$. Then $x \in B$.*

(a) What's wrong with the following proof of the theorem?

Proof. Suppose that $x \notin B$. Since $x \in A$ and $A \subseteq C$, $x \in C$. Since $x \notin B$ and $B \subseteq C$, $x \notin C$. But now we have proven both $x \in C$ and $x \notin C$, so we have reached a contradiction. Therefore $x \in B$. □

(b) Show that the theorem is incorrect by finding a counterexample.

13. Use truth tables to show that modus tollens is a valid rule of inference.

*14. Use truth tables to check the correctness of the theorem in Example 3.2.4.

15. Use truth tables to check the correctness of the statements in exercise 1.

16. Use truth tables to check the correctness of the statements in exercise 2.

17. Can the proof in Example 3.2.2 be modified to prove that if $x^2 + y = 13$ and $x \neq 3$ then $y \neq 4$? Explain.

3.3. Proofs Involving Quantifiers

Look again at Example 3.2.3. In that example we said that x could be anything at all, and we proved the statement $x \in A \setminus C \rightarrow x \in B$. Because the reasoning we used would apply no matter what x was, our proof actually shows that $x \in A \setminus C \rightarrow x \in B$ is true for all x. In other words, we can conclude $\forall x(x \in A \setminus C \rightarrow x \in B)$.

This illustrates the easiest and most straightforward way of proving a goal of the form $\forall x P(x)$. If you can give a proof of the goal $P(x)$ that would work no matter what x was, then you can conclude that $\forall x P(x)$ must be true. To make sure that your proof would work for any value of x, it is important to start your proof with no assumptions about x. Mathematicians express this by saying that x must be *arbitrary*. In particular, you must not assume that x is equal to any other object already under discussion in the proof. Thus, if the letter x is already being used in the proof to stand for some particular object, then you cannot use it to stand for an arbitrary object. In this case you must choose a different variable that is not already being used in the proof, say y, and replace the goal $\forall x P(x)$ with the equivalent statement $\forall y P(y)$. Now you can proceed by letting y stand for an arbitrary object and proving $P(y)$.

To prove a goal of the form $\forall x P(x)$:

Let x stand for an arbitrary object and prove $P(x)$. The letter x must be a new variable in the proof. If x is already being used in the proof to stand for something, then you must choose an unused variable, say y, to stand for the arbitrary object, and prove $P(y)$.

Scratch work

Before using strategy:

Givens	*Goal*
—	$\forall x P(x)$
—	

After using strategy:

Givens	*Goal*
—	$P(x)$
—	

Form of final proof:

Let x be arbitrary.

[Proof of $P(x)$ goes here.]

Since x was arbitrary, we can conclude that $\forall x P(x)$.

Example 3.3.1. Suppose A, B, and C are sets, and $A \setminus B \subseteq C$. Prove that $A \setminus C \subseteq B$.

Scratch work

Givens	*Goal*
$A \setminus B \subseteq C$	$A \setminus C \subseteq B$

As usual, we look first at the logical form of the goal to plan our strategy. In this case we must write out the definition of $\subseteq$ to determine the logical form of the goal.

Givens	*Goal*
$A \setminus B \subseteq C$	$\forall x(x \in A \setminus C \rightarrow x \in B)$

Because the goal has the form $\forall x P(x)$, where $P(x)$ is the statement $x \in A \setminus C \rightarrow x \in B$, we will introduce a new variable x into the proof to stand for an arbitrary object and then try to prove $x \in A \setminus C \rightarrow x \in B$. Note that x *is* a new variable in the proof. It appeared in the logical form of the goal as a bound variable, but remember that bound variables don't stand for anything in particular. We have not yet used x as a free variable in any statement, so it has not been used to stand for any particular object. To make sure x is arbitrary we must be careful not to add any assumptions about x to the givens column. However, we do change our goal:

Givens	*Goal*
$A \setminus B \subseteq C$	$x \in A \setminus C \rightarrow x \in B$

According to our strategy, the final proof should look like this:

> Let x be arbitrary.
> [Proof of $x \in A \setminus C \rightarrow x \in B$ goes here.]
> Since x was arbitrary, we can conclude that $\forall x(x \in A \setminus C \rightarrow x \in B)$, so $A \setminus C \subseteq B$.

The problem is now exactly the same as in Example 3.2.3, so the rest of the solution is the same as well. In other words, we can simply insert the proof we wrote in Example 3.2.3 between the first and last sentences of the proof written here.

Solution

Theorem. *Suppose A, B, and C are sets, and $A \setminus B \subseteq C$. Then $A \setminus C \subseteq B$.*

Proof. Let x be arbitrary. Suppose $x \in A \setminus C$. This means that $x \in A$ and $x \notin C$. Suppose $x \notin B$. Then $x \in A \setminus B$, so since $A \setminus B \subseteq C$, $x \in C$. But this contradicts the fact that $x \notin C$. Therefore $x \in B$. Thus, if $x \in A \setminus C$ then $x \in B$. Since x was arbitrary, we can conclude that $\forall x(x \in A \setminus C \to x \in B)$, so $A \setminus C \subseteq B$. □

Notice that, although this proof shows that every element of $A \setminus C$ is also an element of B, it does not contain phrases such as "every element of $A \setminus C$" or "all elements of $A \setminus C$." For most of the proof we simply reason about x, which is treated as a single, fixed element of $A \setminus C$. We pretend that x stands for some particular element of $A \setminus C$, being careful to make no assumptions about *which* element it stands for. It is only at the end of the proof that we observe that, because x was arbitrary, our conclusions about x would be true no matter what x was. This is the main advantage of using this strategy to prove a goal of the form $\forall x P(x)$. It enables you to prove a goal about *all* objects by reasoning about only *one* object, as long as that object is arbitrary. If you are proving a goal of the form $\forall x P(x)$ and you find yourself saying a lot about "all x's" or "every x," you are probably making your proof unnecessarily complicated by not using this strategy.

As we saw in Chapter 2, statements of the form $\forall x(P(x) \to Q(x))$ are quite common in mathematics. It might be worthwhile, therefore, to consider how the strategies we've discussed can be combined to prove a goal of this form. Because the goal starts with $\forall x$, the first step is to let x be arbitrary and try to prove $P(x) \to Q(x)$. To prove this goal, you will probably want to assume that $P(x)$ is true and prove $Q(x)$. Thus, the proof will probably start like this: "Let x be arbitrary. Suppose $P(x)$." It will then proceed with the steps needed to reach the goal $Q(x)$. Often in this type of proof the statement that x is arbitrary is left out, and the proof simply starts with "Suppose $P(x)$." When a new variable x is introduced into a proof in this way, it is usually understood that x is arbitrary. In other words, no assumptions are being made about x other than the stated one that $P(x)$ is true.

An important example of this type of proof is a proof in which the goal has the form $\forall x \in A\, P(x)$. Recall that $\forall x \in A\, P(x)$ means the same thing as $\forall x(x \in A \to P(x))$, so according to our strategy the proof should start with "Suppose $x \in A$" and then proceed with the steps needed to conclude that $P(x)$ is true. Once again, it is understood that no assumptions are being made about x other than the stated assumption that $x \in A$, so x stands for an arbitrary element of A.

Mathematicians sometimes skip other steps in proofs, if knowledgeable readers could be expected to fill them in themselves. In particular, many of our proof strategies have suggested that the proof end with a sentence that sums up why the reasoning that has been given in the proof leads to the desired conclusion.

In a proof in which several of these strategies have been combined, there might be several of these summing up sentences, one after another, at the end of the proof. Mathematicians often condense this summing up into one sentence, or even skip it entirely. When you are reading a proof written by someone else, you may find it helpful to fill in these skipped steps.

Example 3.3.2. Suppose A and B are sets. Prove that if $A \cap B = A$ then $A \subseteq B$.

Scratch work

Our goal is $A \cap B = A \rightarrow A \subseteq B$. Because the goal is a conditional statement, we add the antecedent to the givens list and make the consequent the goal. We will also write out the definition of $\subseteq$ in the new goal to show what its logical form is.

Givens	*Goal*
$A \cap B = A$	$\forall x(x \in A \rightarrow x \in B)$

Now the goal has the form $\forall x(P(x) \rightarrow Q(x))$, where $P(x)$ is the statement $x \in A$ and $Q(x)$ is the statement $x \in B$. We therefore let x be arbitrary, assume $x \in A$, and prove $x \in B$:

Givens	*Goal*
$A \cap B = A$	$x \in B$
$x \in A$	

Combining the proof strategies we have used, we see that the final proof will have this form:

Suppose $A \cap B = A$.
  Let x be arbitrary.
    Suppose $x \in A$.
      [Proof of $x \in B$ goes here.]
    Therefore $x \in A \rightarrow x \in B$.
  Since x was arbitrary, we can conclude that $\forall x(x \in A \rightarrow x \in B)$, so $A \subseteq B$.
Therefore, if $A \cap B = A$ then $A \subseteq B$.

As discussed earlier, when we write up the final proof we can skip the sentence "Let x be arbitrary," and we can also skip some or all of the last three sentences.

We have now reached the point at which we can analyze the logical form of the goal no further. Fortunately, when we look at the givens, we discover that the goal follows easily. Since $x \in A$ and $A \cap B = A$, it follows that $x \in A \cap B$, so $x \in B$. (In this last step we are using the definition of $\cap$: $x \in A \cap B$ means $x \in A$ and $x \in B$.)

Solution

Theorem. *Suppose A and B are sets. If $A \cap B = A$ then $A \subseteq B$.*
Proof. Suppose $A \cap B = A$, and suppose $x \in A$. Then since $A \cap B = A$, $x \in A \cap B$, so $x \in B$. Since x was an arbitrary element of A, we can conclude that $A \subseteq B$. □

Proving a goal of the form $\exists x P(x)$ also involves introducing a new variable x into the proof and proving $P(x)$, but in this case x will not be arbitrary. Because you only need to prove that $P(x)$ is true for *at least one* x, it suffices to assign a particular value to x and prove $P(x)$ for this one value of x.

To prove a goal of the form $\exists x P(x)$:

Try to find a value of x for which you think $P(x)$ will be true. Then start your proof with "Let $x =$ (the value you decided on)" and proceed to prove $P(x)$ for this value of x. Once again, x should be a new variable. If the letter x is already being used in the proof for some other purpose, then you should choose an unused variable, say y, and rewrite the goal in the equivalent form $\exists y P(y)$. Now proceed as before by starting your proof with "Let $y =$ (the value you decided on)" and prove $P(y)$.

Scratch work

Before using strategy:

Givens	*Goal*
—	$\exists x P(x)$
—	

After using strategy:

Givens	*Goal*
—	$P(x)$
—	
$x =$ (the value you decided on)	

Form of final proof:

Let $x =$ (the value you decided on).
[Proof of $P(x)$ goes here.]
Thus, $\exists x P(x)$.

Finding the right value to use for x may be difficult in some cases. One method that is sometimes helpful is to assume that $P(x)$ is true and then see if you can figure out what x must be, based on this assumption. If $P(x)$ is an equation involving x, this amounts to solving the equation for x. However, if this doesn't work, you may use any other method you please to try to find a value to use for x, including trial-and-error and guessing. The reason you have such freedom with this step is that *the reasoning you use to find a value for x will not appear in the final proof.* This is because of our rule that a proof should only contain the reasoning needed to justify the conclusion of the proof, not an explanation of how you thought of that reasoning. To justify the conclusion that $\exists x P(x)$ is true it is only necessary to verify that $P(x)$ comes out true when x is assigned some particular value. How you thought of that value is your own business, and not part of the justification of the conclusion.

Example 3.3.3. Prove that for every real number x, if $x > 0$ then there is a real number y such that $y(y+1) = x$.

Scratch work

In symbols, our goal is $\forall x(x > 0 \to \exists y[y(y+1) = x])$, where the variables x and y in this statement are understood to range over $\mathbb{R}$. We therefore start by letting x be an arbitrary real number, and we then assume that $x > 0$ and try to prove that $\exists y[y(y+1) = x]$. Thus, we now have the following given and goal:

Givens	*Goal*
$x > 0$	$\exists y[y(y+1) = x]$

Because our goal has the form $\exists y P(y)$, where $P(y)$ is the statement $y(y+1) = x$, according to our strategy we should try to find a value of y for which $P(y)$ is true. In this case we can do it by solving the equation $y(y+1) = x$ for y. It's a quadratic equation and can be solved using the quadratic formula:

$$y(y+1) = x \quad \Rightarrow \quad y^2 + y - x = 0 \quad \Rightarrow \quad y = \frac{-1 \pm \sqrt{1+4x}}{2}.$$

Note that $\sqrt{1+4x}$ is defined, since we have $x > 0$ as a given. We have actually found two solutions for y, but to prove that $\exists y[y(y+1) = x]$ we only need to exhibit one value of y that makes the equation $y(y+1) = x$ true. Either of the two solutions could be used in the proof. We will use the solution $y = (-1 + \sqrt{1+4x})/2$.

The steps we've used to solve for y should not appear in the final proof. In the final proof we will simply say "Let $y = (-1 + \sqrt{1+4x})/2$" and then prove that $y(y+1) = x$. In other words, the final proof will have this form:

> Let x be an arbitrary real number.
> Suppose $x > 0$.
> Let $y = (-1 + \sqrt{1+4x})/2$.
> [Proof of $y(y+1) = x$ goes here.]
> Thus, $\exists y[y(y+1) = x]$.
> Therefore $x > 0 \rightarrow \exists y[y(y+1) = x]$.
> Since x was arbitrary, we can conclude that $\forall x(x > 0 \rightarrow \exists y[y(y+1) = x])$.

To see what must be done to fill in the remaining gap in the proof, we add $y = (-1 + \sqrt{1+4x})/2$ to the givens list and make $y(y+1) = x$ the goal:

Givens	*Goal*
$x > 0$	$y(y+1) = x$
$y = \dfrac{-1+\sqrt{1+4x}}{2}$	

We can now prove that the equation $y(y+1) = x$ is true by simply substituting $(-1 + \sqrt{1+4x})/2$ for y and verifying that the resulting equation is true.

Solution

Theorem. *For every real number x, if $x > 0$ then there is a real number y such that $y(y+1) = x$.*

Proof. Let x be an arbitrary real number, and suppose $x > 0$. Let

$$y = \frac{-1+\sqrt{1+4x}}{2},$$

which is defined since $x > 0$. Then,

$$\begin{aligned}
y(y+1) &= \left(\frac{-1+\sqrt{1+4x}}{2}\right)\cdot\left(\frac{-1+\sqrt{1+4x}}{2}+1\right)\\
&= \left(\frac{\sqrt{1+4x}-1}{2}\right)\cdot\left(\frac{\sqrt{1+4x}+1}{2}\right)\\
&= \frac{1+4x-1}{4} = \frac{4x}{4} = x.
\end{aligned}$$

□

Sometimes when you're proving a goal of the form $\exists y Q(y)$ you won't be able to tell just by looking at the statement $Q(y)$ what value you should plug in for y. In this case you may want to look more closely at the givens to see if they suggest a value to use for y. In particular, a given of the form $\exists x P(x)$ may be helpful in this situation. This given says that an object with a certain property exists. It is probably a good idea to imagine that a particular object with this property has been chosen and to introduce a new variable, say x_0, into the proof to stand for this object. Thus, for the rest of the proof you will be using x_0 to stand for some particular object, and you can assume that with x_0 standing for this object, $P(x_0)$ is true. In other words, you can add $P(x_0)$ to your givens list. This object x_0, or something related to it, might turn out to be the right thing to plug in for y to make $Q(y)$ come out true.

To use a given of the form $\exists x P(x)$:

Introduce a new variable x_0 into the proof to stand for an object for which $P(x_0)$ is true. This means that you can now assume that $P(x_0)$ is true. Logicians call this rule of inference *existential instantiation.*

Note that using a given of the form $\exists x P(x)$ is very different from proving a goal of the form $\exists x P(x)$, because when using a given of the form $\exists x P(x)$, *you don't get to choose a particular value to plug in for* x. You can assume that x_0 stands for some object for which $P(x_0)$ is true, but you can't assume anything else about x_0. On the other hand, a given of the form $\forall x P(x)$ says that $P(x)$ would be true *no matter what* value is assigned to x. You can therefore *choose any value you wish* to plug in for x and use this given to conclude that $P(x)$ is true.

To use a given of the form $\forall x P(x)$:

You can plug in any value, say a, for x and use this given to conclude that $P(a)$ is true. This rule is called *universal instantiation.*

Usually, if you have a given of the form $\exists x P(x)$, you should apply existential instantiation to it immediately. On the other hand, you won't be able to apply universal instantiation to a given of the form $\forall x P(x)$ unless you have a particular value a to plug in for x, so you might want to wait until a likely choice for a pops up in the proof. For example, consider a given of the form $\forall x(P(x) \to Q(x))$. You can use this given to conclude that $P(a) \to Q(a)$ for any a, but according to our rule for using givens that are conditional statements, this conclusion probably won't be very useful unless you know either $P(a)$ or $\neg Q(a)$. You should probably wait until an object a appears in the proof for which you know either $P(a)$ or $\neg Q(a)$, and plug this a in for x when it appears.

We've already used this technique in some of our earlier proofs when dealing with givens of the form $A \subseteq B$. For instance, in Example 3.2.5 we used the givens $A \subseteq B$ and $a \in A$ to conclude that $a \in B$. The justification for this reasoning is that $A \subseteq B$ means $\forall x(x \in A \rightarrow x \in B)$, so by universal instantiation we can plug in a for x and conclude that $a \in A \rightarrow a \in B$. Since we also know $a \in A$, it follows by modus ponens that $a \in B$.

Example 3.3.4. Suppose $\mathcal{F}$ and $\mathcal{G}$ are families of sets and $\mathcal{F} \cap \mathcal{G} \neq \emptyset$. Prove that $\cap\mathcal{F} \subseteq \cup\mathcal{G}$.

Scratch work

Our first step in analyzing the logical form of the goal is to write out the meaning of the subset symbol, which gives us the statement $\forall x(x \in \cap\mathcal{F} \rightarrow x \in \cup\mathcal{G})$. We could go further with this analysis by writing out the definitions of union and intersection, but the part of the analysis that we have already done will be enough to allow us to decide how to get started on the proof. The definitions of union and intersection will be needed later in the proof, but we will wait until they are needed before filling them in. When analyzing the logical forms of givens and goals in order to figure out a proof, it is usually best to do only as much of the analysis as is needed to determine the next step of the proof. Going further with the logical analysis usually just introduces unnecessary complication, without providing any benefit.

Because the goal means $\forall x(x \in \cap\mathcal{F} \rightarrow x \in \cup\mathcal{G})$, we let x be arbitrary, assume $x \in \cap\mathcal{F}$, and try to prove $x \in \cup\mathcal{G}$.

Givens	*Goal*
$\mathcal{F} \cap \mathcal{G} \neq \emptyset$	$x \in \cup\mathcal{G}$
$x \in \cap\mathcal{F}$	

The new goal means $\exists A \in \mathcal{G}(x \in A)$, so to prove it we should try to find a value that will "work" for A. Just looking at the goal doesn't make it clear how to choose A, so we look more closely at the givens. We begin by writing them out in logical symbols:

Givens	*Goal*
$\exists A(A \in \mathcal{F} \cap \mathcal{G})$	$\exists A \in \mathcal{G}(x \in A)$
$\forall A \in \mathcal{F}(x \in A)$	

The second given starts with $\forall A$, so we may not be able to use this given until a likely value to plug in for A pops up during the course of the proof. In particular, we should keep in mind that if we ever come across an element of $\mathcal{F}$ while trying to figure out the proof, we can plug it in for A in the second given and conclude that it contains x as an element. The first given, however, starts with $\exists A$, so we should use it immediately. It says that there is some object that is an element of $\mathcal{F} \cap \mathcal{G}$. By existential instantiation, we can introduce a name, say A_0, for this object. Thus, we can treat $A_0 \in \mathcal{F} \cap \mathcal{G}$ as a given from now on. Because we now have a name, A_0, for a particular element of $\mathcal{F} \cap \mathcal{G}$, it would be redundant to continue to discuss the given statement $\exists A(A \in \mathcal{F} \cap \mathcal{G})$, so we will drop it from our list of givens. Since our new given $A_0 \in \mathcal{F} \cap \mathcal{G}$ means $A_0 \in \mathcal{F}$ and $A_0 \in \mathcal{G}$, we now have the following situation:

Givens	*Goal*
$A_0 \in \mathcal{F}$	$\exists A \in \mathcal{G}(x \in A)$
$A_0 \in \mathcal{G}$	
$\forall A \in \mathcal{F}(x \in A)$	

If you've been paying close attention, you should know what the next step should be. We decided before to keep our eyes open for any elements of $\mathcal{F}$ that might come up during the proof, because we might want to plug them in for A in the last given. An element of $\mathcal{F}$ has come up: A_0! Plugging A_0 in for A in the last given, we can conclude that $x \in A_0$. Any conclusions can be treated in the future as givens, so you can add this statement to the givens column if you like.

Remember that we decided to look at the givens because we didn't know what value to assign to A in the goal. What we need is a value for A that is in $\mathcal{G}$ and that will make the statement $x \in A$ come out true. Has this consideration of the givens suggested a value to use for A? Yes! Use $A = A_0$.

Although we translated the given statements $x \in \cap\mathcal{F}$, $x \in \cup\mathcal{G}$, and $\mathcal{F} \cap \mathcal{G} \neq \varnothing$ into logical symbols in order to figure out how to use them in the proof, these translations are not usually written out when the proof is written up in final form. In the final proof we just write these statements in their original form and leave it to the reader of the proof to work out their logical forms in order to follow our reasoning.

Solution

Theorem. *Suppose $\mathcal{F}$ and $\mathcal{G}$ are families of sets, and $\mathcal{F} \cap \mathcal{G} \neq \varnothing$. Then $\cap\mathcal{F} \subseteq \cup\mathcal{G}$.*

Proof. Suppose $x \in \cap\mathcal{F}$. Since $\mathcal{F} \cap \mathcal{G} \neq \varnothing$, we can let A_0 be an element of $\mathcal{F} \cap \mathcal{G}$. Thus, $A_0 \in \mathcal{F}$ and $A_0 \in \mathcal{G}$. Since $x \in \cap\mathcal{F}$ and $A_0 \in \mathcal{F}$, it follows that $x \in A_0$. But we also know that $A_0 \in \mathcal{G}$, so we can conclude that $x \in \cup\mathcal{G}$. $\square$

Proofs involving the quantifiers *for all* and *there exists* are often difficult for them.

That last sentence confused you, didn't it? You're probably wondering, "Who are *they*?" Readers of your proofs will experience the same sort of confusion if you use variables without explaining what they stand for. Beginning proof-writers are sometimes careless about this, and that's why proofs involving the quantifiers *for all* and *there exists* are often difficult for them. (It made more sense that time, didn't it?) When you use the strategies we've discussed in this section, you'll be introducing new variables into your proof, and when you do this, you must always be careful to make it clear to the reader what they stand for.

For example, if you were proving a goal of the form $\forall x \in A\, P(x)$, you would probably start by introducing a variable x to stand for an arbitrary element of A. Your reader won't know what x means, though, unless you begin your proof with "Let x be an arbitrary element of A," or "Suppose $x \in A$." Of course, you must be clear in your own mind about what x stands for. In particular, because x is to be arbitrary, you must be careful not to assume anything about x other than the fact that $x \in A$. It might help to think of x as being chosen by *someone else*; you have no control over which element of A they'll pick. Using a given of the form $\exists x P(x)$ is similar. This given tells you that you can introduce a new variable x_0 into the proof to stand for some object for which $P(x_0)$ is true, but you cannot assume anything else about x_0. On the other hand, if you are *proving* $\exists x P(x)$, your proof will probably start "Let $x = \ldots$" This time *you* get to choose the value of x, and you must tell the reader explicitly that you are choosing the value of x and what value you have chosen.

It's also important, when you're introducing a new variable x, to be sure you know what *kind* of object x is. Is it a number? a set? a function? a matrix? You'd better not write $a \in X$ unless X is a set, for example. If you aren't careful about this, you might end up writing nonsense. You also sometimes need to know what kind of object a variable stands for to figure out the logical form of a statement involving that variable. For example, $A = B$ means $\forall x(x \in A \leftrightarrow x \in B)$ if A and B are sets, but not if they're numbers.

The most important thing to keep in mind about introducing variables into a proof is simply the fact that variables must always be introduced before they are used. If you make a statement about x (i.e., a statement in which x occurs as a free variable) without first explaining what x stands for, a reader of your proof won't know what you're talking about – and there's a good chance that you won't know what you're talking about either!

Because proofs involving quantifiers may require more practice than the other proofs we have discussed so far, we end this section with two more examples.

Example 3.3.5. Suppose B is a set and $\mathcal{F}$ is a family of sets. Prove that if $\cup\mathcal{F} \subseteq B$ then $\mathcal{F} \subseteq \mathscr{P}(B)$.

Scratch Work

We assume $\cup\mathcal{F} \subseteq B$ and try to prove $\mathcal{F} \subseteq \mathscr{P}(B)$. Because this goal means $\forall x(x \in \mathcal{F} \rightarrow x \in \mathscr{P}(B))$, we let x be arbitrary, assume $x \in \mathcal{F}$, and set $x \in \mathscr{P}(B)$ as our goal. Recall that $\mathcal{F}$ is a family of sets, so since $x \in \mathcal{F}$, x is a set. Thus, we now have the following givens and goal:

Givens	*Goal*
$\cup\mathcal{F} \subseteq B$	$x \in \mathscr{P}(B)$
$x \in \mathcal{F}$	

To figure out how to prove this goal, we must use the definition of power set. The statement $x \in \mathscr{P}(B)$ means $x \subseteq B$, or in other words $\forall y(y \in x \rightarrow y \in B)$. We must therefore introduce another arbitrary object into the proof. We let y be arbitrary, assume $y \in x$, and try to prove $y \in B$.

Givens	*Goal*
$\cup\mathcal{F} \subseteq B$	$y \in B$
$x \in \mathcal{F}$	
$y \in x$	

The goal can be analyzed no further, so we must look more closely at the givens. Our goal is $y \in B$, and the only given that even mentions B is the first. In fact, the first given would enable us to reach this goal, if only we knew that $y \in \cup\mathcal{F}$. This suggests that we might try treating $y \in \cup\mathcal{F}$ as our goal. If we can reach this goal, then we can just add one more step, applying the first given, and the proof will be done.

Givens	*Goal*
$\cup\mathcal{F} \subseteq B$	$y \in \cup\mathcal{F}$
$x \in \mathcal{F}$	
$y \in x$	

Once again, we have a goal whose logical form can be analyzed, so we use the form of the goal to guide our strategy. The goal means $\exists A \in \mathcal{F}(y \in A)$, so to prove it we must find a set A such that $A \in \mathcal{F}$ and $y \in A$. Looking at the givens, we see that x is such a set, so the proof is done.

Solution

Theorem. *Suppose B is a set and $\mathcal{F}$ is a family of sets. If $\cup\mathcal{F} \subseteq B$ then $\mathcal{F} \subseteq \mathscr{P}(B)$.*

Proof. Suppose $\cup\mathcal{F} \subseteq B$. Let x be an arbitrary element of $\mathcal{F}$. Let y be an arbitrary element of x. Since $y \in x$ and $x \in \mathcal{F}$, clearly $y \in \cup\mathcal{F}$. But then since $\cup\mathcal{F} \subseteq B$, $y \in B$. Since y was an arbitrary element of x, we can conclude that $x \subseteq B$, so $x \in \mathscr{P}(B)$. But x was an arbitrary element of $\mathcal{F}$, so this shows that $\mathcal{F} \subseteq \mathscr{P}(B)$, as required. □

This is probably the most complex proof we've done so far. Read it again and make sure you understand its structure and the purpose of every sentence. Isn't it remarkable how much logical complexity has been packed into just a few lines?

It is not uncommon for a short proof to have such a rich logical structure. This efficiency of exposition is one of the most attractive features of proofs, but it also often makes them difficult to read. Although we've been concentrating so far on *writing* proofs, it is also important to learn how to *read* proofs written by other people. To give you some practice with this, we present our last proof in this section without the scratch work. See if you can follow the structure of the proof as you read it. We'll provide a commentary after the proof that should help you to understand it.

For this proof we need the following definition: For any integers x and y, we'll say that *x divides y* (or *y is divisible by x*) if $\exists k \in \mathbb{Z}(kx = y)$. We use the notation $x \mid y$ to mean "x divides y." For example, $4 \mid 20$, since $5 \cdot 4 = 20$.

Theorem 3.3.6. *For all integers a, b, and c, if $a \mid b$ and $b \mid c$ then $a \mid c$.*

Proof. Let a, b, and c be arbitrary integers and suppose $a \mid b$ and $b \mid c$. Since $a \mid b$, we can choose some integer m such that $ma = b$. Similarly, since $b \mid c$, we can choose an integer n such that $nb = c$. Therefore $c = nb = nma$, so since nm is an integer, $a \mid c$. □

Commentary. The theorem says $\forall a \in \mathbb{Z}\forall b \in \mathbb{Z}\forall c \in \mathbb{Z}(a \mid b \wedge b \mid c \rightarrow a \mid c)$, so the most natural way to proceed is to let a, b, and c be arbitrary integers, assume $a \mid b$ and $b \mid c$, and then prove $a \mid c$. The first sentence of the proof indicates that this strategy is being used, so the goal for the rest of the proof must be to prove that $a \mid c$. The fact that this is the goal for the rest of the proof is not explicitly stated. You are expected to figure this out for yourself by using your knowledge of proof strategies. You might even want to make a givens and goal list to help you keep track of what is known and what remains to be proven as you continue to read the proof. At this point in the proof, the list would look like this:

Givens	*Goal*
a, b, and c are integers	$a \mid c$
$a \mid b$	
$b \mid c$	

Because the new goal means $\exists k \in \mathbb{Z}(ka = c)$, the proof will probably proceed by finding an integer k such that $ka = c$. As with many proofs of existential statements, the first step in finding such a k involves looking more closely at the givens. The next sentence of the proof uses the given $a \mid b$ to conclude that we can choose an integer m such that $ma = b$. The proof doesn't say what rule of inference justifies this. It is up to you to figure it out by working out the logical form of the given statement $a \mid b$, using the definition of *divides*. Because this given means $\exists k \in \mathbb{Z}(ka = b)$, you should recognize that the rule of inference being used is existential instantiation. Existential instantiation is also used in the next sentence of the proof to justify choosing an integer n such that $nb = c$. The equations $ma = b$ and $nb = c$ can now be added to the list of givens.

Some steps have also been skipped in the last sentence of the proof. We expected that the goal $a \mid c$ would be proven by finding an integer k such that $ka = c$. From the equation $c = nma$ and the fact that nm is an integer, it follows that $k = nm$ will work, but the proof doesn't explicitly say that this value of k is being used; in fact, the variable k is not mentioned at all in the proof. Of course, the variable k is not mentioned in the statement of the theorem either. It is not uncommon for a proof of an existential statement to be written in this way, especially when, as in this case, the goal is not written out explicitly in the statement of the theorem as an existential statement. In this case, the existential nature of the goal became apparent only when we filled in the definition of *divides*.

Exercises

Note: Exercises marked with the symbol ℔ can be done with Proof Designer. For more information about Proof Designer, see Appendix 2.

*1. In exercise 7 of Section 2.2 you used logical equivalences to show that $\exists x(P(x) \to Q(x))$ is equivalent to $\forall x P(x) \to \exists x Q(x)$. Now use the methods of this section to prove that if $\exists x(P(x) \to Q(x))$ is true, then $\forall x P(x) \to \exists x Q(x)$ is true. (Note: The other direction of the equivalence is quite a bit harder to prove. See exercise 29 of Section 3.5.)

2. Prove that if A and $B \setminus C$ are disjoint, then $A \cap B \subseteq C$.

*3. Prove that if $A \subseteq B \setminus C$ then A and C are disjoint.

℔4. Suppose $A \subseteq \mathscr{P}(A)$. Prove that $\mathscr{P}(A) \subseteq \mathscr{P}(\mathscr{P}(A))$.

5. The hypothesis of the theorem proven in exercise 4 is $A \subseteq \mathscr{P}(A)$.
   (a) Can you think of a set A for which this hypothesis is true?
   (b) Can you think of another?

6. Suppose x is a real number.
   (a) Prove that if $x \neq 1$ then there is a real number y such that $\frac{y+1}{y-2} = x$.
   (b) Prove that if there is a real number y such that $\frac{y+1}{y-2} = x$, then $x \neq 1$.

*7. Prove that for every real number x, if $x > 2$ then there is a real number y such that $y + \frac{1}{y} = x$.

℔8. Prove that if $\mathcal{F}$ is a family of sets and $A \in \mathcal{F}$, then $A \subseteq \cup\mathcal{F}$.

*9. Prove that if $\mathcal{F}$ is a family of sets and $A \in \mathcal{F}$, then $\cap\mathcal{F} \subseteq A$.

10. Suppose that $\mathcal{F}$ is a nonempty family of sets, B is a set, and $\forall A \in \mathcal{F}(B \subseteq A)$. Prove that $B \subseteq \cap\mathcal{F}$.

11. Suppose that $\mathcal{F}$ is a family of sets. Prove that if $\varnothing \in \mathcal{F}$ then $\cap\mathcal{F} = \varnothing$.

℔*12. Suppose $\mathcal{F}$ and $\mathcal{G}$ are families of sets. Prove that if $\mathcal{F} \subseteq \mathcal{G}$ then $\cup\mathcal{F} \subseteq \cup\mathcal{G}$.

13. Suppose $\mathcal{F}$ and $\mathcal{G}$ are nonempty families of sets. Prove that if $\mathcal{F} \subseteq \mathcal{G}$ then $\cap\mathcal{G} \subseteq \cap\mathcal{F}$.

*14. Suppose $\{A_i \mid i \in I\}$ is an indexed family of sets. Prove that $\cup_{i\in I}\mathscr{P}(A_i) \subseteq \mathscr{P}(\cup_{i\in I} A_i)$. (Hint: First make sure you know what all the notation means!)

15. Suppose $\{A_i \mid i \in I\}$ is an indexed family of sets and $I \neq \varnothing$. Prove that $\cap_{i\in I} A_i \in \cap_{i\in I}\mathscr{P}(A_i)$.

℔16. Prove the converse of the statement proven in Example 3.3.5. In other words, prove that if $\mathcal{F} \subseteq \mathscr{P}(B)$ then $\cup\mathcal{F} \subseteq B$.

*17. Suppose $\mathcal{F}$ and $\mathcal{G}$ are nonempty families of sets, and every element of $\mathcal{F}$ is a subset of every element of $\mathcal{G}$. Prove that $\cup\mathcal{F} \subseteq \cap\mathcal{G}$.

18. In this problem all variables range over $\mathbb{Z}$, the set of all integers.
    (a) Prove that if $a \mid b$ and $a \mid c$, then $a \mid (b + c)$.
    (b) Prove that if $ac \mid bc$ and $c \neq 0$, then $a \mid b$.

19. (a) Prove that for all real numbers x and y there is a real number z such that $x + z = y - z$.
    (b) Would the statement in part (a) be correct if "real number" were changed to "integer"? Justify your answer.

*20. Consider the following theorem:

    Theorem. *For every real number x, $x^2 \geq 0$.*

    What's wrong with the following proof of the theorem?

Proof. Suppose not. Then for every real number x, $x^2 < 0$. In particular, plugging in $x = 3$ we would get $9 < 0$, which is clearly false. This contradiction shows that for every number x, $x^2 \geq 0$. □

21. Consider the following incorrect theorem:

    Incorrect Theorem. *If* $\forall x \in A(x \neq 0)$ and $A \subseteq B$ then $\forall x \in B(x \neq 0)$.

    (a) What's wrong with the following proof of the theorem?

    Proof. Let x be an arbitrary element of A. Since $\forall x \in A(x \neq 0)$, we can conclude that $x \neq 0$. Also, since $A \subseteq B$, $x \in B$. Since $x \in B$, $x \neq 0$, and x was arbitrary, we can conclude that $\forall x \in B(x \neq 0)$. □

    (b) Find a counterexample to the theorem. In other words, find an example of sets A and B for which the hypotheses of the theorem are true but the conclusion is false.

*22. Consider the following incorrect theorem:

    Incorrect Theorem. $\exists x \in \mathbb{R} \forall y \in \mathbb{R}(xy^2 = y - x)$.

    What's wrong with the following proof of the theorem?

    Proof. Let $x = y/(y^2 + 1)$. Then

$$y - x = y - \frac{y}{y^2+1} = \frac{y^3}{y^2+1} = \frac{y}{y^2+1} \cdot y^2 = xy^2. \qquad \square$$

23. Consider the following incorrect theorem:

    Incorrect Theorem. *Suppose $\mathcal{F}$ and $\mathcal{G}$ are families of sets. If $\cup\mathcal{F}$ and $\cup\mathcal{G}$ are disjoint, then so are $\mathcal{F}$ and $\mathcal{G}$.*

    (a) What's wrong with the following proof of the theorem?

    Proof. Suppose $\cup\mathcal{F}$ and $\cup\mathcal{G}$ are disjoint. Suppose $\mathcal{F}$ and $\mathcal{G}$ are not disjoint. Then we can choose some set A such that $A \in \mathcal{F}$ and $A \in \mathcal{G}$. Since $A \in \mathcal{F}$, by exercise 8, $A \subseteq \cup\mathcal{F}$, so every element of A is in $\cup\mathcal{F}$. Similarly, since $A \in \mathcal{G}$, every element of A is in $\cup\mathcal{G}$. But then every element of A is in both $\cup\mathcal{F}$ and $\cup\mathcal{G}$, and this is impossible since $\cup\mathcal{F}$ and $\cup\mathcal{G}$ are disjoint. Thus, we have reached a contradiction, so $\mathcal{F}$ and $\mathcal{G}$ must be disjoint. □

    (b) Find a counterexample to the theorem.

24. Consider the following putative theorem:

Theorem? *For all real numbers x and y*, $x^2 + xy - 2y^2 = 0$.

(a) What's wrong with the following proof of the theorem?

Proof. Let x and y be equal to some arbitrary real number r. Then

$$x^2 + xy - 2y^2 = r^2 + r \cdot r - 2r^2 = 0.$$

Since x and y were both arbitrary, this shows that for all real numbers x and y, $x^2 + xy - 2y^2 = 0$. □

(b) Is the theorem correct? Justify your answer with either a proof or a counterexample.

*25. Prove that for every real number x there is a real number y such that for every real number z, $yz = (x + z)^2 - (x^2 + z^2)$.

26. (a) Comparing the various rules for dealing with quantifiers in proofs, you should see a similarity between the rules for goals of the form $\forall x P(x)$ and givens of the form $\exists x P(x)$. What is this similarity? What about the rules for goals of the form $\exists x P(x)$ and givens of the form $\forall x P(x)$?

(b) Can you think of a reason why these similarities might be expected? (Hint: Think about how proof by contradiction works when the goal starts with a quantifier.)

3.4. Proofs Involving Conjunctions and Biconditionals

The method for proving a goal of the form $P \wedge Q$ is so simple it hardly seems worth mentioning:

To prove a goal of the form $P \wedge Q$:
Prove P and Q separately.

In other words, a goal of the form $P \wedge Q$ is treated as two separate goals: P, and Q. The same is true of givens of the form $P \wedge Q$:

To use a given of the form $P \wedge Q$:
Treat this given as two separate givens: P, and Q.

We've already used these ideas, without mention, in some of our previous examples. For example, the definition of the given $x \in A \setminus C$ in Example 3.2.3 was $x \in A \wedge x \notin C$, but we treated it as two separate givens: $x \in A$, and $x \notin C$.

Example 3.4.1. Suppose $A \subseteq B$, and A and C are disjoint. Prove that $A \subseteq B \setminus C$.

Scratch work

Givens	*Goal*
$A \subseteq B$	$A \subseteq B \setminus C$
$A \cap C = \varnothing$	

Analyzing the logical form of the goal, we see that it has the form $\forall x(x \in A \rightarrow x \in B \setminus C)$, so we let x be arbitrary, assume $x \in A$, and try to prove that $x \in B \setminus C$. The new goal $x \in B \setminus C$ means $x \in B \wedge x \notin C$, so according to our strategy we should split this into two goals, $x \in B$ and $x \notin C$, and prove them separately.

Givens	*Goals*
$A \subseteq B$	$x \in B$
$A \cap C = \varnothing$	$x \notin C$
$x \in A$	

The final proof will have this form:

Let x be arbitrary.
  Suppose $x \in A$.
    [Proof of $x \in B$ goes here.]
    [Proof of $x \notin C$ goes here.]
   Thus, $x \in B \wedge x \notin C$, so $x \in B \setminus C$.
  Therefore $x \in A \rightarrow x \in B \setminus C$.
Since x was arbitrary, $\forall x(x \in A \rightarrow x \in B \setminus C)$, so $A \subseteq B \setminus C$.

The first goal, $x \in B$, clearly follows from the fact that $x \in A$ and $A \subseteq B$. The second goal, $x \notin C$, follows from $x \in A$ and $A \cap C = \varnothing$. You can see this by analyzing the logical form of the statement $A \cap C = \varnothing$. It is a negative statement, but it can be reexpressed as an equivalent positive statement:

$A \cap C = \varnothing$ is equivalent to $\neg\exists y(y \in A \wedge y \in C)$ (definitions of $\cap$ and $\varnothing$),
  which is equivalent to $\forall y \neg(y \in A \wedge y \in C)$ (quantifier negation law),
  which is equivalent to $\forall y(y \notin A \vee y \notin C)$ (DeMorgan's law),
  which is equivalent to $\forall y(y \in A \rightarrow y \notin C)$ (conditional law).

Plugging in x for y in this last statement, we see that $x \in A \rightarrow x \notin C$, and since we already know $x \in A$, we can conclude that $x \notin C$.

Solution

Theorem. *Suppose $A \subseteq B$, and A and C are disjoint. Then $A \subseteq B \setminus C$*

Proof. Suppose $x \in A$. Since $A \subseteq B$, it follows that $x \in B$, and since A and C are disjoint, we must have $x \notin C$. Thus, $x \in B \setminus C$. Since x was an arbitrary element of A, we can conclude that $A \subseteq B \setminus C$. □

Using our strategies for working with conjunctions, we can now work out the proper way to deal with statements of the form $P \leftrightarrow Q$ in proofs. Because $P \leftrightarrow Q$ is equivalent to $(P \to Q) \wedge (Q \to P)$, according to our strategies a given or goal of the form $P \leftrightarrow Q$ should be treated as two separate givens or goals: $P \to Q$, and $Q \to P$.

To prove a goal of the form $P \leftrightarrow Q$:
Prove $P \to Q$ and $Q \to P$ separately.

To use a given of the form $P \leftrightarrow Q$:
Treat this as two separate givens: $P \to Q$, and $Q \to P$.

This is illustrated in the next example, in which we use the following definitions: An integer x is *even* if $\exists k \in \mathbb{Z}(x = 2k)$, and x is *odd* if $\exists k \in \mathbb{Z}(x = 2k + 1)$. We also use the fact that every integer is either even or odd, but not both. We'll see a proof of this fact in Chapter 6.

Example 3.4.2. Suppose x is an integer. Prove that x is even iff x^2 is even.

Scratch work

The goal is (x is even) $\leftrightarrow$ (x^2 is even), so we prove the two goals (x is even) $\to$ (x^2 is even) and (x^2 is even) $\to$ (x is even) separately. For the first, we assume that x is even and prove that x^2 is even:

Givens	*Goal*
$x \in \mathbb{Z}$	x^2 is even
x is even	

Writing out the definition of *even* in both the given and the goal will reveal their logical forms:

Givens	*Goal*
$x \in \mathbb{Z}$	$\exists k \in \mathbb{Z}(x^2 = 2k)$
$\exists k \in \mathbb{Z}(x = 2k)$	

Because the second given starts with $\exists k$, we immediately use it and let k stand for some particular integer for which the statement $x = 2k$ is true. Thus, we have two new given statements: $k \in \mathbb{Z}$, and $x = 2k$.

Givens	*Goal*
$x \in \mathbb{Z}$	$\exists k \in \mathbb{Z}(x^2 = 2k)$
$k \in \mathbb{Z}$	
$x = 2k$	

The goal starts with $\exists k$, but since k is already being used to stand for a particular number, we cannot assign a new value to k to prove the goal. We must therefore switch to a different letter, say j. One way to understand this is to think of rewriting the goal in the equivalent form $\exists j \in \mathbb{Z}(x^2 = 2j)$. To prove this goal we must come up with a value to plug in for j. It must be an integer, and it must satisfy the equation $x^2 = 2j$. Using the given equation $x = 2k$, we see that $x^2 = (2k)^2 = 4k^2 = 2(2k^2)$, so it looks like the right value to choose for j is $j = 2k^2$. Clearly $2k^2$ is an integer, so this choice for j will work to complete the proof of our first goal.

To prove the second goal (x^2 is even) $\rightarrow$ (x is even), we'll prove the contrapositive (x is not even) $\rightarrow$ (x^2 is not even) instead. Since any integer is either even or odd but not both, this is equivalent to the statement (x is odd) $\rightarrow$ (x^2 is odd).

Givens	*Goal*
$x \in \mathbb{Z}$	x^2 is odd
x is odd	

The steps are now quite similar to the first part of the proof. As before, we begin by writing out the definition of *odd* in both the second given and the goal. This time, to avoid the conflict of variable names we ran into in the first part of the proof, we use different names for the bound variables in the two statements.

Givens	*Goal*
$x \in \mathbb{Z}$	$\exists j \in \mathbb{Z}(x^2 = 2j + 1)$
$\exists k \in \mathbb{Z}(x = 2k + 1)$	

Next we use the second given and let k stand for a particular integer for which $x = 2k + 1$.

Givens	*Goal*
$x \in \mathbb{Z}$	$\exists j \in \mathbb{Z}(x^2 = 2j + 1)$
$k \in \mathbb{Z}$	
$x = 2k + 1$	

We must now find an integer j such that $x^2 = 2j + 1$. Plugging in $2k + 1$ for x we get $x^2 = (2k + 1)^2 = 4k^2 + 4k + 1 = 2(2k^2 + 2k) + 1$, so $j = 2k^2 + 2k$ looks like the right choice.

Before giving the final write-up of the proof, we should make a few explanatory remarks. The two conditional statements we've proven can be thought of as representing the two directions $\rightarrow$ and $\leftarrow$ of the biconditional symbol $\leftrightarrow$ in the original goal. These two parts of the proof are sometimes labeled with the symbols $\rightarrow$ and $\leftarrow$. In each part, we end up proving a statement that asserts the existence of a number with certain properties. We called this number j in the scratch work, but note that j was not mentioned explicitly in the statement of the problem. As in the proof of Theorem 3.3.6, we have chosen not to mention j explicitly in the final proof either.

Solution

Theorem. *Suppose x is an integer. Then x is even iff x^2 is even.*

Proof. ($\rightarrow$) Suppose x is even. Then for some integer k, $x = 2k$. Therefore, $x^2 = 4k^2 = 2(2k^2)$, so since $2k^2$ is an integer, x^2 is even. Thus, if x is even then x^2 is even.

($\leftarrow$) Suppose x is odd. Then $x = 2k + 1$ for some integer k. Therefore, $x^2 = (2k + 1)^2 = 4k^2 + 4k + 1 = 2(2k^2 + 2k) + 1$, so since $2k^2 + 2k$ is an integer, x^2 is odd. Thus, if x^2 is even then x is even. □

Using the proof techniques we've developed, we can now verify some of the equivalences that we were only able to justify on intuitive grounds in Chapter 2. As an example of this, let's prove that the formulas $\forall x \neg P(x)$ and $\neg \exists x P(x)$ are equivalent. To say that these formulas are equivalent means that they will always have the same truth value. In other words, no matter what statement $P(x)$ stands for, the statement $\forall x \neg P(x) \leftrightarrow \neg \exists x P(x)$ will be true. We can prove this using our technique for proving biconditional statements.

Example 3.4.3. Prove that $\forall x \neg P(x) \leftrightarrow \neg \exists x P(x)$.

Scratch work

($\rightarrow$) We must prove $\forall x \neg P(x) \rightarrow \neg \exists x P(x)$, so we assume $\forall x \neg P(x)$ and try to prove $\neg \exists x P(x)$. Our goal is now a negated statement, and reexpressing it would require the use of the very equivalence that we are trying to prove! We therefore fall back on our only other strategy for dealing with negative goals, proof by contradiction. We now have the following situation:

Givens	*Goal*
$\forall x \neg P(x)$	Contradiction
$\exists x P(x)$	

The second given starts with an existential quantifier, so we use it immediately and let x_0 stand for some object for which the statement $P(x_0)$ is true. But now plugging in x_0 for x in the first given we can conclude that $\neg P(x_0)$, which gives us the contradiction we need.

($\leftarrow$) For this direction of the biconditional we should assume $\neg\exists x P(x)$ and try to prove $\forall x \neg P(x)$. Because this goal starts with a universal quantifier, we let x be arbitrary and try to prove $\neg P(x)$. Once again, we now have a negated goal that can't be reexpressed, so we use proof by contradiction:

Givens	*Goal*
$\neg\exists x P(x)$	Contradiction
$P(x)$	

Our first given is also a negated statement, and this suggests that we could get the contradiction we need by proving $\exists x P(x)$. We therefore set this as our goal.

Givens	*Goal*
$\neg\exists x P(x)$	$\exists x P(x)$
$P(x)$	

To keep from confusing the x that appears as a free variable in the second given (the arbitrary x introduced earlier in the proof) with the x that appears as a bound variable in the goal, you might want to rewrite the goal in the equivalent form $\exists y P(y)$. To prove this goal we have to find a value of y that makes $P(y)$ come out true. But this is easy! Our second given, $P(x)$, tells us that our arbitrary x is the value we need.

Solution

Theorem. $\forall x \neg P(x) \leftrightarrow \neg\exists x P(x)$.

Proof. ($\rightarrow$) Suppose $\forall x \neg P(x)$, and suppose $\exists x P(x)$. Then we can choose some x_0 such that $P(x_0)$ is true. But since $\forall x \neg P(x)$, we can conclude that $\neg P(x_0)$, and this is a contradiction. Therefore $\forall x \neg P(x) \rightarrow \neg\exists x P(x)$.

($\leftarrow$) Suppose $\neg\exists x P(x)$. Let x be arbitrary, and suppose $P(x)$. Since we have a specific x for which $P(x)$ is true, it follows that $\exists x P(x)$, which is a contradiction. Therefore, $\neg P(x)$. Since x was arbitrary, we can conclude that $\forall x \neg P(x)$, so $\neg\exists x P(x) \rightarrow \forall x \neg P(x)$. □

Sometimes in a proof of a goal of the form $P \leftrightarrow Q$ the steps in the proof of $Q \rightarrow P$ are the same as the steps used to prove $P \rightarrow Q$, but in reverse order. In this case you may be able to simplify the proof by writing it as a string of equivalences, starting with P and ending with Q. For example, suppose you found that you could prove $P \rightarrow Q$ by first assuming P, then using P to infer some other statement R, and then using R to deduce Q; and suppose that the same steps could be used, in reverse order, to prove that $Q \to P$. In other words, you could assume Q, use this assumption to conclude that R was true, and then use R to prove P. Since you would be asserting both $P \to R$ and $R \to P$, you could sum up these two steps by saying $P \leftrightarrow R$. Similarly, the other two steps of the proof tell you that $R \leftrightarrow Q$. These two statements imply the goal $P \leftrightarrow Q$. Mathematicians sometimes present this kind of proof by simply writing the string of equivalences

$$P \text{ iff } R \text{ iff } Q.$$

You can think of this as an abbreviation for "P iff R and R iff Q (and therefore P iff Q)." This is illustrated in the next example.

Example 3.4.4. Suppose A, B, and C are sets. Prove that $A \cap (B \setminus C) = (A \cap B) \setminus C$.

Scratch work

As we saw in Chapter 2, the equation $A \cap (B \setminus C) = (A \cap B) \setminus C$ means $\forall x(x \in A \cap (B \setminus C) \leftrightarrow x \in (A \cap B) \setminus C)$, but it is also equivalent to the statement $[A \cap (B \setminus C) \subseteq (A \cap B) \setminus C] \land [(A \cap B) \setminus C \subseteq A \cap (B \setminus C)]$. This suggests two approaches to the proof. We could let x be arbitrary and then prove $x \in A \cap (B \setminus C) \leftrightarrow x \in (A \cap B) \setminus C$, or we could prove the two statements $A \cap (B \setminus C) \subseteq (A \cap B) \setminus C$ and $(A \cap B) \setminus C \subseteq A \cap (B \setminus C)$. In fact, almost every proof that two sets are equal will involve one of these two approaches. In this case we will use the first approach, so once we have introduced our arbitrary x, we will have an iff goal.

For the ($\to$) half of the proof we assume $x \in A \cap (B \setminus C)$ and try to prove $x \in (A \cap B) \setminus C$:

Givens	*Goal*
$x \in A \cap (B \setminus C)$	$x \in (A \cap B) \setminus C$

To see the logical forms of the given and goal, we write out their definitions as follows:

$$x \in A \cap (B \setminus C) \text{ iff } x \in A \land x \in B \setminus C \text{ iff } x \in A \land x \in B \land x \notin C;$$
$$x \in (A \cap B) \setminus C \text{ iff } x \in A \cap B \land x \notin C \text{ iff } x \in A \land x \in B \land x \notin C.$$

At this point it is clear that the given implies the goal, since the last steps in both strings of equivalences turned out to be identical. In fact, it is also clear that the reasoning involved in the ($\leftarrow$) direction of the proof will be exactly the same, but with the given and goal columns reversed. Thus, we might try to shorten the proof by writing it as a string of equivalences, starting with $x \in A \cap (B \setminus C)$ and ending with $x \in (A \cap B) \setminus C$. In this case, if we start with $x \in A \cap (B \setminus C)$ and follow the first string of equivalences displayed above, we come to a statement that is the same as the last statement in the second string of equivalences. We can then continue by following the second string of equivalences *backward*, ending with $x \in (A \cap B) \setminus C$.

Solution

Theorem. *Suppose A, B, and C are sets. Then $A \cap (B \setminus C) = (A \cap B) \setminus C$.*
Proof. Let x be arbitrary. Then

$$\begin{aligned} x \in A \cap (B \setminus C) &\text{ iff } x \in A \wedge x \in B \setminus C \\ &\text{ iff } x \in A \wedge x \in B \wedge x \notin C \\ &\text{ iff } x \in (A \cap B) \wedge x \notin C \\ &\text{ iff } x \in (A \cap B) \setminus C. \end{aligned}$$

Thus, $\forall x(x \in A \cap (B \setminus C) \leftrightarrow x \in (A \cap B) \setminus C)$, so $A \cap (B \setminus C) = (A \cap B) \setminus C$.

□

The technique of figuring out a sequence of equivalences in one order and then writing it in the reverse order is used quite often in proofs. The order in which the steps should be written in the final proof is determined by our rule that an assertion should never be made until it can be justified. In particular, if you are trying to prove $P \leftrightarrow Q$, it is wrong to start your write-up of the proof with the unjustified statement $P \leftrightarrow Q$ and then work out the meanings of the two sides P and Q, showing that they are the same. You should instead start with equivalences you can justify and string them together to produce a justification of the goal $P \leftrightarrow Q$ before you assert this goal. A similar technique can sometimes be used to figure out proofs of equations, as the next example shows.

Example 3.4.5. Prove that for any real numbers a and b,

$$(a + b)^2 - 4(a - b)^2 = (3b - a)(3a - b).$$

Scratch work

The goal has the form $\forall a \forall b((a + b)^2 - 4(a - b)^2 = (3b - a)(3a - b))$, so we start by letting a and b be arbitrary real numbers and try to prove the equation.

Multiplying out both sides gives us:

$$\begin{aligned}(a+b)^2 - 4(a-b)^2 &= a^2 + 2ab + b^2 - 4(a^2 - 2ab + b^2)\\ &= -3a^2 + 10ab - 3b^2;\\ (3b-a)(3a-b) = 9ab - 3a^2 - 3b^2 + ab &= -3a^2 + 10ab - 3b^2.\end{aligned}$$

Clearly the two sides are equal. The simplest way to write the proof of this is to write a string of equalities starting with $(a+b)^2 - 4(a-b)^2$ and ending with $(3b-a)(3a-b)$. We can do this by copying down the first string of equalities displayed above, and then following it with the second line, written backward.

Solution

Theorem. *For any real numbers a and b,*

$$(a+b)^2 - 4(a-b)^2 = (3b-a)(3a-b).$$

Proof. Let a and b be arbitrary real numbers. Then

$$\begin{aligned}(a+b)^2 - 4(a-b)^2 &= a^2 + 2ab + b^2 - 4(a^2 - 2ab + b^2)\\ &= -3a^2 + 10ab - 3b^2\\ &= 9ab - 3a^2 - 3b^2 + ab = (3b-a)(3a-b). \quad \square\end{aligned}$$

We end this section by presenting another proof without preliminary scratch work, but with a commentary to help you read the proof.

Theorem 3.4.6. *For every integer n, $6 \mid n$ iff $2 \mid n$ and $3 \mid n$.*

Proof. Let n be an arbitrary integer.

($\rightarrow$) Suppose $6 \mid n$. Then we can choose an integer k such that $6k = n$. Therefore $n = 6k = 2(3k)$, so $2 \mid n$, and similarly $n = 6k = 3(2k)$, so $3 \mid n$.

($\leftarrow$) Suppose $2 \mid n$ and $3 \mid n$. Then we can choose integers j and k such that $n = 2j$ and $n = 3k$. Therefore $6(j-k) = 6j - 6k = 3(2j) - 2(3k) = 3n - 2n = n$, so $6 \mid n$. $\square$

Commentary. The statement to be proven is $\forall n \in \mathbb{Z}(6 \mid n \leftrightarrow (2 \mid n \wedge 3 \mid n))$, and the most natural strategy for proving a goal of this form is to let n be arbitrary and then prove both directions of the biconditional separately. It should be clear that this is the strategy being used in the proof.

For the left-to-right direction of the biconditional, we assume $6 \mid n$ and then prove $2 \mid n$ and $3 \mid n$, treating this as two separate goals. The introduction of the integer k is justified by existential instantiation, since the assumption $6 \mid n$ means $\exists k \in \mathbb{Z}(6k = n)$. At this point in the proof we have the following givens and goals:

Givens	*Goals*
$n \in \mathbb{Z}$	$2 \mid n$
$k \in \mathbb{Z}$	$3 \mid n$
$6k = n$	

The first goal, $2 \mid n$, means $\exists j \in \mathbb{Z}(2j = n)$, so we must find an integer j such that $2j = n$. Although the proof doesn't say so explicitly, the equation $n = 2(3k)$, which is derived in the proof, suggests that the value being used for j is $j = 3k$. Clearly, $3k$ is an integer (another step skipped in the proof), so this choice for j works. The proof of $3 \mid n$ is similar.

For the right-to-left direction we assume $2 \mid n$ and $3 \mid n$ and prove $6 \mid n$. Once again, the introduction of j and k is justified by existential instantiation. No explanation is given for why we should compute $6(j - k)$, but a proof need not provide such explanations. The reason for the calculation should become clear when, surprisingly, it turns out that $6(j - k) = n$. Such surprises provide part of the pleasure of working with proofs. As in the first half of the proof, since $j - k$ is an integer, this shows that $6 \mid n$.

Exercises

*1. Use the methods of this chapter to prove that $\forall x(P(x) \wedge Q(x))$ is equivalent to $\forall x P(x) \wedge \forall x Q(x)$.

PD 2. Prove that if $A \subseteq B$ and $A \subseteq C$ then $A \subseteq B \cap C$.

PD 3. Suppose $A \subseteq B$. Prove that for every set C, $C \setminus B \subseteq C \setminus A$.

PD *4. Prove that if $A \subseteq B$ and $A \nsubseteq C$ then $B \nsubseteq C$.

PD 5. Prove that if $A \subseteq B \setminus C$ and $A \neq \varnothing$ then $B \nsubseteq C$.

6. Prove that for any sets A, B, and C, $A \setminus (B \cap C) = (A \setminus B) \cup (A \setminus C)$, by finding a string of equivalences starting with $x \in A \setminus (B \cap C)$ and ending with $x \in (A \setminus B) \cup (A \setminus C)$. (See Example 3.4.4.)

PD *7. Use the methods of this chapter to prove that for any sets A and B, $\mathscr{P}(A \cap B) = \mathscr{P}(A) \cap \mathscr{P}(B)$.

PD 8. Prove that $A \subseteq B$ iff $\mathscr{P}(A) \subseteq \mathscr{P}(B)$.

*9. Prove that if x and y are odd integers, then xy is odd.

10. Prove that for every integer n, n^3 is even iff n is even.

11. Consider the following putative theorem:

    Theorem? *Suppose m is an even integer and n is an odd integer. Then $n^2 - m^2 = n + m$.*

    (a) What's wrong with the following proof of the theorem?

    Proof. Since m is even, we can choose some integer k such that $m = 2k$. Similarly, since n is odd we have $n = 2k + 1$. Therefore

    $$\begin{aligned} n^2 - m^2 &= (2k+1)^2 - (2k)^2 = 4k^2 + 4k + 1 - 4k^2 = 4k + 1 \\ &= (2k+1) + (2k) = n + m. \end{aligned}$$ □

    (b) Is the theorem correct? Justify your answer with either a proof or a counterexample.

*12. Prove that $\forall x \in \mathbb{R}[\exists y \in \mathbb{R}(x + y = xy) \leftrightarrow x \neq 1]$.

13. Prove that $\exists z \in \mathbb{R} \forall x \in \mathbb{R}^+[\exists y \in \mathbb{R}(y - x = y/x) \leftrightarrow x \neq z]$.

PD14. Suppose B is a set and $\mathcal{F}$ is a family of sets. Prove that $\cup\{A \setminus B \mid A \in \mathcal{F}\} \subseteq \cup(\mathcal{F} \setminus \mathscr{P}(B))$.

*15. Suppose $\mathcal{F}$ and $\mathcal{G}$ are nonempty families of sets and every element of $\mathcal{F}$ is disjoint from some element of $\mathcal{G}$. Prove that $\cup\mathcal{F}$ and $\cap\mathcal{G}$ are disjoint.

PD16. Prove that for any set A, $A = \cup\mathscr{P}(A)$.

PD*17. Suppose $\mathcal{F}$ and $\mathcal{G}$ are families of sets.

    (a) Prove that $\cup(\mathcal{F} \cap \mathcal{G}) \subseteq (\cup\mathcal{F}) \cap (\cup\mathcal{G})$.

    (b) What's wrong with the following proof that $(\cup\mathcal{F}) \cap (\cup\mathcal{G}) \subseteq \cup(\mathcal{F} \cap \mathcal{G})$?

    Proof. Suppose $x \in (\cup\mathcal{F}) \cap (\cup\mathcal{G})$. This means that $x \in \cup\mathcal{F}$ and $x \in \cup\mathcal{G}$, so $\exists A \in \mathcal{F}(x \in A)$ and $\exists A \in \mathcal{G}(x \in A)$. Thus, we can choose a set A such that $A \in \mathcal{F}$, $A \in \mathcal{G}$, and $x \in A$. Since $A \in \mathcal{F}$ and $A \in \mathcal{G}$, $A \in \mathcal{F} \cap \mathcal{G}$. Therefore $\exists A \in \mathcal{F} \cap \mathcal{G}(x \in A)$, so $x \in \cup(\mathcal{F} \cap \mathcal{G})$. Since x was arbitrary, we can conclude that $(\cup\mathcal{F}) \cap (\cup\mathcal{G}) \subseteq \cup(\mathcal{F} \cap \mathcal{G})$. □

    (c) Find an example of families of sets $\mathcal{F}$ and $\mathcal{G}$ for which $\cup(\mathcal{F} \cap \mathcal{G}) \neq (\cup\mathcal{F}) \cap (\cup\mathcal{G})$.

PD18. Suppose $\mathcal{F}$ and $\mathcal{G}$ are families of sets. Prove that $(\cup\mathcal{F}) \cap (\cup\mathcal{G}) \subseteq \cup(\mathcal{F} \cap \mathcal{G})$ iff $\forall A \in \mathcal{F} \forall B \in \mathcal{G}(A \cap B \subseteq \cup(\mathcal{F} \cap \mathcal{G}))$.

PD19. Suppose $\mathcal{F}$ and $\mathcal{G}$ are families of sets. Prove that $\cup\mathcal{F}$ and $\cup\mathcal{G}$ are disjoint iff for all $A \in \mathcal{F}$ and $B \in \mathcal{G}$, A and B are disjoint.

PD20. Suppose $\mathcal{F}$ and $\mathcal{G}$ are families of sets.

    (a) Prove that $(\cup\mathcal{F}) \setminus (\cup\mathcal{G}) \subseteq \cup(\mathcal{F} \setminus \mathcal{G})$.

    (b) What's wrong with the following proof that $\cup(\mathcal{F} \setminus \mathcal{G}) \subseteq (\cup\mathcal{F}) \setminus (\cup\mathcal{G})$?

Proof. Suppose $x \in \cup(\mathcal{F} \setminus \mathcal{G})$. Then we can choose some $A \in \mathcal{F} \setminus \mathcal{G}$ such that $x \in A$. Since $A \in \mathcal{F} \setminus \mathcal{G}$, $A \in \mathcal{F}$ and $A \notin \mathcal{G}$. Since $x \in A$ and $A \in \mathcal{F}$, $x \in \cup\mathcal{F}$. Since $x \in A$ and $A \notin \mathcal{G}$, $x \notin \cup\mathcal{G}$. Therefore $x \in (\cup\mathcal{F}) \setminus (\cup\mathcal{G})$. □

(c) Prove that $\cup(\mathcal{F} \setminus \mathcal{G}) \subseteq (\cup\mathcal{F}) \setminus (\cup\mathcal{G})$ iff $\forall A \in (\mathcal{F} \setminus \mathcal{G})\forall B \in \mathcal{G}(A \cap B = \varnothing)$.

(d) Find an example of families of sets $\mathcal{F}$ and $\mathcal{G}$ for which $\cup(\mathcal{F} \setminus \mathcal{G}) \neq (\cup\mathcal{F}) \setminus (\cup\mathcal{G})$.

PD*21. Suppose $\mathcal{F}$ and $\mathcal{G}$ are families of sets. Prove that if $\cup\mathcal{F} \not\subseteq \cup\mathcal{G}$, then there is some $A \in \mathcal{F}$ such that for all $B \in \mathcal{G}$, $A \not\subseteq B$.

22. Suppose B is a set, $\{A_i \mid i \in I\}$ is an indexed family of sets, and $I \neq \varnothing$.

(a) What proof strategies are used in the following proof that $B \cap (\cup_{i\in I} A_i) = \cup_{i\in I}(B \cap A_i)$?

Proof. Let x be arbitrary. Suppose $x \in B \cap (\cup_{i\in I} A_i)$. Then $x \in B$ and $x \in \cup_{i\in I} A_i$, so we can choose some $i_0 \in I$ such that $x \in A_{i_0}$. Since $x \in B$ and $x \in A_{i_0}$, $x \in B \cap A_{i_0}$. Therefore $x \in \cup_{i\in I}(B \cap A_i)$.

Now suppose $x \in \cup_{i\in I}(B \cap A_i)$. Then we can choose some $i_0 \in I$ such that $x \in B \cap A_{i_0}$. Therefore $x \in B$ and $x \in A_{i_0}$. Since $x \in A_{i_0}$, $x \in \cup_{i\in I} A_i$. Since $x \in B$ and $x \in \cup_{i\in I} A_i$, $x \in B \cap (\cup_{i\in I} A_i)$.

Since x was arbitrary, we have shown that $\forall x[x \in B \cap (\cup_{i\in I} A_i) \leftrightarrow x \in \cup_{i\in I}(B \cap A_i)]$, so $B \cap (\cup_{i\in I} A_i) = \cup_{i\in I}(B \cap A_i)$. □

(b) Prove that $B \setminus (\cup_{i\in I} A_i) = \cap_{i\in I}(B \setminus A_i)$.

(c) Can you discover and prove a similar theorem about $B \setminus (\cap_{i\in I} A_i)$? (Hint: Try to guess the theorem, and then try to prove it. If you can't finish the proof, it might be because your guess was wrong. Change your guess and try again.)

*23. Suppose $\{A_i \mid i \in I\}$ and $\{B_i \mid i \in I\}$ are indexed families of sets and $I \neq \varnothing$.

(a) Prove that $\cup_{i\in I}(A_i \setminus B_i) \subseteq (\cup_{i\in I} A_i) \setminus (\cap_{i\in I} B_i)$.

(b) Find an example for which $\cup_{i\in I}(A_i \setminus B_i) \neq (\cup_{i\in I} A_i) \setminus (\cap_{i\in I} B_i)$.

24. Suppose $\{A_i \mid i \in I\}$ and $\{B_i \mid i \in I\}$ are indexed families of sets.

(a) Prove that $\cup_{i\in I}(A_i \cap B_i) \subseteq (\cup_{i\in I} A_i) \cap (\cup_{i\in I} B_i)$.

(b) Find an example for which $\cup_{i\in I}(A_i \cap B_i) \neq (\cup_{i\in I} A_i) \cap (\cup_{i\in I} B_i)$.

25. Prove that for all integers a and b there is an integer c such that $a \mid c$ and $b \mid c$.

26. (a) Prove that for every integer n, $15 \mid n$ iff $3 \mid n$ and $5 \mid n$.

(b) Prove that it is *not* true that for every integer n, $60 \mid n$ iff $6 \mid n$ and $10 \mid n$.

3.5. Proofs Involving Disjunctions

Suppose one of your givens in a proof has the form $P \vee Q$. This given tells you that either P or Q is true, but it doesn't tell you which. Thus, there are two possibilities that you must take into account. One way to do the proof would be to consider these two possibilities in turn. In other words, first assume that P is true and use this assumption to prove your goal. Then assume Q is true and give another proof that the goal is true. Although you don't know which of these assumptions is correct, the given $P \vee Q$ tells you that *one* of them must be correct. Whichever one it is, you have shown that it implies the goal. Thus, the goal must be true.

The two possibilities that are considered separately in this type of proof – the possibility that P is true and the possibility that Q is true – are called *cases*. The given $P \vee Q$ justifies the use of these two cases by guaranteeing that these cases cover all of the possibilities. Mathematicians say in this situation that the cases are *exhaustive*. Any proof can be broken into two or more cases at any time, as long as the cases are exhaustive.

To use a given of the form $P \vee Q$:

Break your proof into cases. For case 1, assume that P is true and use this assumption to prove the goal. For case 2, assume Q is true and give another proof of the goal.

Scratch work

Before using strategy:

Givens	*Goal*
$P \vee Q$	—
—	

After using strategy:

Case 1: Givens	*Goal*
P	—
—	
Case 2: Givens	*Goal*
Q	—
—	

Form of final proof:

Case 1. P is true.
[Proof of goal goes here.]

Case 2. Q is true.

[Proof of goal goes here.]

Since we know $P \lor Q$, these cases cover all the possibilities. Therefore the goal must be true.

Example 3.5.1. Suppose that A, B, and C are sets. Prove that if $A \subseteq C$ and $B \subseteq C$ then $A \cup B \subseteq C$.

Scratch work

We assume $A \subseteq C$ and $B \subseteq C$ and prove $A \cup B \subseteq C$. Writing out the goal using logical symbols gives us the following givens and goal:

Givens	*Goal*
$A \subseteq C$	$\forall x(x \in A \cup B \rightarrow x \in C)$
$B \subseteq C$	

To prove the goal we let x be arbitrary, assume $x \in A \cup B$, and try to prove $x \in C$. Thus, we now have a new given $x \in A \cup B$, which we write as $x \in A \lor x \in B$, and our goal is now $x \in C$.

Givens	*Goal*
$A \subseteq C$	$x \in C$
$B \subseteq C$	
$x \in A \lor x \in B$	

Because the goal cannot be analyzed any further at this point, we look more closely at the givens. The first given will be useful if we ever come across an object that is an element of A, since it would allow us to conclude immediately that this object must also be an element of C. Similarly, the second given will be useful if we come across an element of B. Keeping in mind that we should watch for any elements of A or B that might come up, we move on to the third given. Because this given has the form $P \lor Q$, we try proof by cases. For the first case we assume $x \in A$, and for the second we assume $x \in B$. In the first case we therefore have the following givens and goal:

Givens	*Goal*
$A \subseteq C$	$x \in C$
$B \subseteq C$	
$x \in A$	

We've already decided that if we ever come across an element of A, we can use the first given to conclude that it is also an element of C. Since we now have $x \in A$ as a given, we can conclude that $x \in C$, which is our goal. The reasoning for the second case is quite similar, using the second given instead of the first.

Solution

Theorem. *Suppose that A, B, and C are sets. If $A \subseteq C$ and $B \subseteq C$ then $A \cup B \subseteq C$.*

Proof. Suppose $A \subseteq C$ and $B \subseteq C$, and let x be an arbitrary element of $A \cup B$. Then either $x \in A$ or $x \in B$.

Case 1. $x \in A$. Then since $A \subseteq C$, $x \in C$.

Case 2. $x \in B$. Then since $B \subseteq C$, $x \in C$.

Since we know that either $x \in A$ or $x \in B$, these cases cover all the possibilities, so we can conclude that $x \in C$. Since x was an arbitrary element of $A \cup B$, this means that $A \cup B \subseteq C$. □

Note that the cases in this proof are not *exclusive*. In other words, it is possible for both $x \in A$ and $x \in B$ to be true, so some values of x might fall under both cases. There is nothing wrong with this. The cases in a proof by cases must cover all possibilities, but there is no harm in covering some possibilities more than once. In other words, the cases must be exhaustive, but they need not be exclusive.

Proof by cases is sometimes also helpful if you are proving a goal of the form $P \vee Q$. If you can prove P in some cases and Q in others, then as long as your cases are exhaustive you can conclude that $P \vee Q$ is true. This method is particularly useful if one of the givens also has the form of a disjunction, because then you can use the cases suggested by this given.

To prove a goal of the form $P \vee Q$:

Break your proof into cases. In each case, either prove P or prove Q.

Example 3.5.2. Suppose that A, B and C are sets. Prove that $A \setminus (B \setminus C) \subseteq (A \setminus B) \cup C$.

Scratch work

Because the goal is $\forall x(x \in A \setminus (B \setminus C) \rightarrow x \in (A \setminus B) \cup C)$, we let x be arbitrary, assume $x \in A \setminus (B \setminus C)$, and try to prove $x \in (A \setminus B) \cup C$. Writing these statements out in logical symbols gives us:

Givens	*Goal*
$x \in A \wedge \neg(x \in B \wedge x \notin C)$	$(x \in A \wedge x \notin B) \vee x \in C$

We split the given into two separate givens, $x \in A$ and $\neg(x \in B \wedge x \notin C)$, and since the second is a negated statement we use one of DeMorgan's laws to reexpress it as the positive statement $x \notin B \lor x \in C$.

Givens	*Goal*
$x \in A$	$(x \in A \land x \notin B) \lor x \in C$
$x \notin B \lor x \in C$	

Now the second given and the goal are both disjunctions, so we'll try considering the two cases $x \notin B$ and $x \in C$ suggested by the second given. According to our strategy for proving goals of the form $P \lor Q$, if in each case we can either prove $x \in A \land x \notin B$ or prove $x \in C$, then the proof will be complete. For the first case we assume $x \notin B$.

Givens	*Goal*
$x \in A$	$(x \in A \land x \notin B) \lor x \in C$
$x \notin B$	

In this case the goal is clearly true, because in fact we can conclude that $x \in A \land x \notin B$. For the second case we assume $x \in C$, and once again the goal is clearly true.

Solution

Theorem. *Suppose that A, B, and C are sets. Then $A \setminus (B \setminus C) \subseteq (A \setminus B) \cup C$.*

Proof. Suppose $x \in A \setminus (B \setminus C)$. Then $x \in A$ and $x \notin B \setminus C$. Since $x \notin B \setminus C$, it follows that either $x \notin B$ or $x \in C$. We will consider these cases separately.

Case 1. $x \notin B$. Then since $x \in A$, $x \in A \setminus B$, so $x \in (A \setminus B) \cup C$.

Case 2. $x \in C$. Then clearly $x \in (A \setminus B) \cup C$.

Since x was an arbitrary element of $A \setminus (B \setminus C)$, we can conclude that $A \setminus (B \setminus C) \subseteq (A \setminus B) \cup C$. □

Sometimes you may find it useful to break a proof into cases even if the cases are not suggested by a given of the form $P \lor Q$. Any proof can be broken into cases at any time, as long as the cases exhaust all of the possibilities.

Example 3.5.3. Prove that for every integer x, the remainder when x^2 is divided by 4 is either 0 or 1.

Scratch work

We start by letting x be an arbitrary integer and then try to prove that the remainder when x^2 is divided by 4 is either 0 or 1.

Givens	*Goal*
$x \in \mathbb{Z}$	$(x^2 \div 4 \text{ has remainder } 0) \lor (x^2 \div 4 \text{ has remainder } 1)$

Because the goal is a disjunction, breaking the proof into cases seems like a likely approach, but there is no given that suggests what cases to use. However, trying out a few values for x suggests the right cases:

x	x^2	quotient of $x^2 \div 4$	remainder of $x^2 \div 4$
1	1	0	1
2	4	1	0
3	9	2	1
4	16	4	0
5	25	6	1
6	36	9	0

It appears that the remainder is 0 when x is even and 1 when x is odd. These are the cases we will use. Thus, for case 1 we assume x is even and try to prove that the remainder is 0, and for case 2 we assume x is odd and prove that the remainder is 1. Because every integer is either even or odd, these cases are exhaustive.

Filling in the definition of *even*, here are our givens and goal for case 1:

Givens	*Goal*
$x \in \mathbb{Z}$	$x^2 \div 4$ has remainder 0
$\exists k \in \mathbb{Z}(x = 2k)$	

We immediately use the second given and let k stand for some particular integer for which $x = 2k$. Then $x^2 = (2k)^2 = 4k^2$, so clearly when we divide x^2 by 4 the quotient is k^2 and the remainder is 0.

Case 2 is quite similar:

Givens	*Goal*
$x \in \mathbb{Z}$	$x^2 \div 4$ has remainder 1
$\exists k \in \mathbb{Z}(x = 2k + 1)$	

Once again we use the second given immediately and let k stand for an integer for which $x = 2k + 1$. Then $x^2 = (2k + 1)^2 = 4k^2 + 4k + 1 = 4(k^2 + k) + 1$, so when x^2 is divided by 4 the quotient is $k^2 + k$ and the remainder is 1.

Solution

Theorem. *For every integer x, the remainder when x^2 is divided by 4 is either 0 or 1.*

Proof. Suppose x is an integer. We consider two cases.

Case 1. x is even. Then $x = 2k$ for some integer k, so $x^2 = 4k^2$. Clearly the remainder when x^2 is divided by 4 is 0.

Case 2. x is odd. Then $x = 2k + 1$ for some integer k, so $x^2 = 4k^2 + 4k + 1$. Clearly in this case the remainder when x^2 is divided by 4 is 1. □

Sometimes in a proof of a goal that has the form $P \lor Q$ it is hard to figure out how to break the proof into cases. Here's a way of doing it that is often helpful. Simply assume that P is true in case 1 and assume that it is false in case 2. Certainly P is either true or false, so these cases are exhaustive. In the first case you have assumed that P is true, so certainly the goal $P \lor Q$ is true. Thus, no further reasoning is needed in case 1. In the second case you have assumed that P is false, so the only way the goal $P \lor Q$ could be true is if Q is true. Thus, to complete this case you should try to prove Q.

To prove a goal of the form $P \lor Q$:
If P is true, then clearly the goal $P \lor Q$ is true, so you only need to worry about the case in which P is false. You can complete the proof in this case by proving that Q is true.

Scratch work

Before using strategy:

Givens	*Goal*
—	$P \lor Q$
—	

After using strategy:

Givens	*Goal*
—	Q
—	
$\neg P$	

Form of final proof:

If P is true, then of course $P \lor Q$ is true. Now suppose P is false.
[Proof of Q goes here.]
Thus, $P \lor Q$ is true.

Thus, this strategy for proving $P \lor Q$ suggests that you transform the problem by adding $\neg P$ as a new given and changing the goal to Q. It is interesting to note that this is exactly the same as the transformation you would use if you were proving the goal $\neg P \to Q$! This is not really surprising, because we already know that the statements $P \lor Q$ and $\neg P \to Q$ are equivalent. But we derived this equivalence before from the truth table for the conditional connective, and this truth table may have been hard to understand at first. Perhaps the reasoning we've given makes this equivalence, and therefore the truth table for the conditional connective, seem more natural.

Of course, the roles of P and Q could be reversed in using this strategy. Thus, you can also prove $P \vee Q$ by assuming that Q is false and proving P.

Example 3.5.4. Prove that for every real number x, if $x^2 \geq x$ then either $x \leq 0$ or $x \geq 1$.

Scratch work

Our goal is $\forall x(x^2 \geq x \rightarrow (x \leq 0 \vee x \geq 1))$, so to get started we let x be an arbitrary real number, assume $x^2 \geq x$, and set $x \leq 0 \vee x \geq 1$ as our goal:

Givens	*Goal*
$x^2 \geq x$	$x \leq 0 \vee x \geq 1$

According to our strategy, to prove this goal we can either assume $x > 0$ and prove $x \geq 1$ or assume $x < 1$ and prove $x \leq 0$. The assumption that x is positive seems more likely to be useful in reasoning about inequalities, so we take the first approach.

Givens	*Goal*
$x^2 \geq x$	$x \geq 1$
$x > 0$	

The proof is now easy. Since $x > 0$, we can divide the given inequality $x^2 \geq x$ by x to get the goal $x \geq 1$.

Solution

Theorem. *For every real number x, if $x^2 \geq x$ then either $x \leq 0$ or $x \geq 1$.*

Proof. Suppose $x^2 \geq x$. If $x \leq 0$, then of course $x \leq 0$ or $x \geq 1$. Now suppose $x > 0$. Then we can divide both sides of the inequality $x^2 \geq x$ by x to conclude that $x \geq 1$. Thus, either $x \leq 0$ or $x \geq 1$. □

The equivalence of $P \vee Q$ and $\neg P \rightarrow Q$ also suggests a rule of inference called *disjunctive syllogism* for using a given statement of the form $P \vee Q$:

To use a given of the form $P \vee Q$:

If you are also given $\neg P$, or you can prove that P is false, then you can use this given to conclude that Q is true. Similarly, if you are given $\neg Q$ or can prove that Q is false, then you can conclude that P is true.

In fact, this rule is the one we used in our first example of deductive reasoning in Chapter 1!

Once again, we end this section with a proof for you to read without the benefit of a preliminary scratch work analysis.

Theorem 3.5.5. *Suppose m and n are integers. If mn is even, then either m is even or n is even.*

Proof. Suppose mn is even. Then we can choose an integer k such that $mn = 2k$. If m is even then there is nothing more to prove, so suppose m is odd. Then $m = 2j + 1$ for some integer j. Substituting this into the equation $mn = 2k$, we get $(2j + 1)n = 2k$, so $2jn + n = 2k$, and therefore $n = 2k - 2jn = 2(k - jn)$. Since $k - jn$ is an integer, it follows that n is even. □

Commentary. The overall form of the proof is the following:

Suppose mn is even.
If m is even, then clearly either m is even or n is even. Now suppose m is not even. Then m is odd.
[Proof that n is even goes here.]
Therefore either m is even or n is even.
Therefore if mn is even then either m is even or n is even.

The assumptions that mn is even and m is odd lead, by existential instantiation, to the equations $mn = 2k$ and $m = 2j + 1$. Although the proof doesn't say so explicitly, you are expected to work out for yourself that in order to prove that n is even it suffices to find an integer c such that $n = 2c$. Straightforward algebra leads to the equation $n = 2(k - jn)$, so the choice $c = k - jn$ works.

Exercises

℞*1. Suppose A, B, and C are sets. Prove that $A \cap (B \cup C) \subseteq (A \cap B) \cup C$.

℞2. Suppose A, B, and C are sets. Prove that $(A \cup B) \setminus C \subseteq A \cup (B \setminus C)$.

℞3. Suppose A and B are sets. Prove that $A \setminus (A \setminus B) = A \cap B$.

℞*4. Suppose $A \cap C \subseteq B \cap C$ and $A \cup C \subseteq B \cup C$. Prove that $A \subseteq B$.

℞5. Recall from Section 1.4 that the symmetric difference of two sets A and B is the set $A \triangle B = (A \setminus B) \cup (B \setminus A) = (A \cup B) \setminus (A \cap B)$. Prove that if $A \triangle B \subseteq A$ then $B \subseteq A$.

℞6. Suppose A, B, and C are sets. Prove that $A \cup C \subseteq B \cup C$ iff $A \setminus C \subseteq B \setminus C$.

℞*7. Prove that for any sets A and B, $\mathscr{P}(A) \cup \mathscr{P}(B) \subseteq \mathscr{P}(A \cup B)$.

℔8. Prove that for any sets A and B, if $\mathscr{P}(A) \cup \mathscr{P}(B) = \mathscr{P}(A \cup B)$ then either $A \subseteq B$ or $B \subseteq A$.

9. Suppose x and y are real numbers and $x \neq 0$. Prove that $y + 1/x = 1 + y/x$ iff either $x = 1$ or $y = 1$.

10. Prove that for every real number x, if $|x - 3| > 3$ then $x^2 > 6x$. (Hint: According to the definition of $|x - 3|$, if $x - 3 \geq 0$ then $|x - 3| = x - 3$, and if $x - 3 < 0$ then $|x - 3| = 3 - x$. The easiest way to use this fact is to break your proof into cases. Assume that $x - 3 \geq 0$ in case 1, and $x - 3 < 0$ in case 2.)

*11. Prove that for every real number x, $|2x - 6| > x$ iff $|x - 4| > 2$. (Hint: Read the hint for exercise 10.)

12. (a) Prove that for all real numbers a and b, $|a| \leq b$ iff $-b \leq a \leq b$.
    (b) Prove that for any real number x, $-|x| \leq x \leq |x|$. (Hint: Use part (a).)
    (c) Prove that for all real numbers x and y, $|x + y| \leq |x| + |y|$. (This is called the *triangle inequality*. One way to prove this is to combine parts (a) and (b), but you can also do it by considering a number of cases.)

13. Prove that for every integer x, $x^2 + x$ is even.

14. Prove that for every integer x, the remainder when x^4 is divided by 8 is either 0 or 1.

*15. Suppose $\mathcal{F}$ and $\mathcal{G}$ are nonempty families of sets.
    ℔(a) Prove that $\cup(\mathcal{F} \cup \mathcal{G}) = (\cup\mathcal{F}) \cup (\cup\mathcal{G})$.
    (b) Can you discover and prove a similar theorem about $\cap(\mathcal{F} \cup \mathcal{G})$?

16. Suppose $\mathcal{F}$ is a nonempty family of sets and B is a set.
    ℔(a) Prove that $B \cup (\cup\mathcal{F}) = \cup(\mathcal{F} \cup \{B\})$.
    (b) Prove that $B \cup (\cap\mathcal{F}) = \cap_{A \in \mathcal{F}}(B \cup A)$.
    (c) Can you discover and prove a similar theorem about $B \cap (\cap\mathcal{F})$?

17. Suppose $\mathcal{F}$, $\mathcal{G}$, and $\mathcal{H}$ are nonempty families of sets and for every $A \in \mathcal{F}$ and every $B \in \mathcal{G}$, $A \cup B \in \mathcal{H}$. Prove that $\cap\mathcal{H} \subseteq (\cap\mathcal{F}) \cup (\cap\mathcal{G})$.

℔18. Suppose A and B are sets. Prove that $\forall x(x \in A \triangle B \leftrightarrow (x \in A \leftrightarrow x \notin B))$.

℔*19. Suppose A, B, and C are sets. Prove that $A \triangle B$ and C are disjoint iff $A \cap C = B \cap C$.

℔20. Suppose A, B, and C are sets. Prove that $A \triangle B \subseteq C$ iff $A \cup C = B \cup C$.

℔21. Suppose A, B, and C are sets. Prove that $C \subseteq A \triangle B$ iff $C \subseteq A \cup B$ and $A \cap B \cap C = \varnothing$.

℔*22. Suppose A, B, and C are sets.
    (a) Prove that $A \setminus C \subseteq (A \setminus B) \cup (B \setminus C)$.
    (b) Prove that $A \triangle C \subseteq (A \triangle B) \cup (B \triangle C)$.

℔*23. Suppose A, B, and C are sets.

(a) Prove that $(A \cup B) \triangle C \subseteq (A \triangle C) \cup (B \triangle C)$.

(b) Find an example of sets A, B, and C such that $(A \cup B) \triangle C \neq (A \triangle C) \cup (B \triangle C)$

℔24. Suppose A, B, and C are sets.

(a) Prove that $(A \triangle C) \cap (B \triangle C) \subseteq (A \cap B) \triangle C$.

(b) Is it always true that $(A \cap B) \triangle C \subseteq (A \triangle C) \cap (B \triangle C)$? Give either a proof or a counterexample.

℔25. Suppose A, B, and C are sets. Consider the sets $(A \setminus B) \triangle C$ and $(A \triangle C) \setminus (B \triangle C)$. Can you prove that either is a subset of the other? Justify your conclusions with either proofs or counterexamples.

*26. Consider the following putative theorem.

Theorem? *For every real number x, if $|x - 3| < 3$ then $0 < x < 6$.*

Is the following proof correct? If so, what proof strategies does it use? If not, can it be fixed? Is the theorem correct?

Proof. Let x be an arbitrary real number, and suppose $|x - 3| < 3$. We consider two cases:

Case 1. $x - 3 \geq 0$. Then $|x - 3| = x - 3$. Plugging this into the assumption that $|x - 3| < 3$, we get $x - 3 < 3$, so clearly $x < 6$.

Case 2. $x - 3 < 0$. Then $|x - 3| = 3 - x$, so the assumption $|x - 3| < 3$ means that $3 - x < 3$. Therefore $3 < 3 + x$, so $0 < x$.

Since we have proven both $0 < x$ and $x < 6$, we can conclude that $0 < x < 6$. □

27. Consider the following putative theorem.

Theorem? *For any sets A, B, and C, if $A \setminus B \subseteq C$ and $A \not\subseteq C$ then $A \cap B \neq \varnothing$.*

Is the following proof correct? If so, what proof strategies does it use? If not, can it be fixed? Is the theorem correct?

Proof. Since $A \not\subseteq C$, we can choose some x such that $x \in A$ and $x \notin C$. Since $x \notin C$ and $A \setminus B \subseteq C$, $x \notin A \setminus B$. Therefore either $x \notin A$ or $x \in B$. But we already know that $x \in A$, so it follows that $x \in B$. Since $x \in A$ and $x \in B$, $x \in A \cap B$. Therefore $A \cap B \neq \varnothing$. □

*28. Consider the following putative theorem.

Theorem? $\forall x \in \mathbb{R} \exists y \in \mathbb{R}(xy^2 \neq y - x)$.

Is the following proof correct? If so, what proof strategies does it use? If not, can it be fixed? Is the theorem correct?

Proof. Let x be an arbitrary real number.

Case 1. $x = 0$. Let $y = 1$. Then $xy^2 = 0$ and $y - x = 1 - 0 = 1$, so $xy^2 \neq y - x$.

Case 2. $x \neq 0$. Let $y = 0$. Then $xy^2 = 0$ and $y - x = -x \neq 0$, so $xy^2 \neq y - x$.

Since these cases are exhaustive, we have shown that $\exists y \in \mathbb{R}(xy^2 \neq y - x)$. Since x was arbitrary, this shows that $\forall x \in \mathbb{R} \exists y \in \mathbb{R}(xy^2 \neq y - x)$. □

29. Prove that if $\forall x P(x) \rightarrow \exists x Q(x)$ then $\exists x(P(x) \rightarrow Q(x))$. (Hint: Remember that $P \rightarrow Q$ is equivalent to $\neg P \vee Q$).

*30. Consider the following putative theorem.

Theorem? *Suppose A, B, and C are sets and $A \subseteq B \cup C$. Then either $A \subseteq B$ or $A \subseteq C$.*

Is the following proof correct? If so, what proof strategies does it use? If not, can it be fixed? Is the theorem correct?

Proof. Let x be an arbitrary element of A. Since $A \subseteq B \cup C$, it follows that either $x \in B$ or $x \in C$.

Case 1. $x \in B$. Since x was an arbitrary element of A, it follows that $\forall x \in A(x \in B)$, which means that $A \subseteq B$.

Case 2. $x \in C$. Similarly, since x was an arbitrary element of A, we can conclude that $A \subseteq C$.

Thus, either $A \subseteq B$ or $A \subseteq C$. □

31. Prove $\exists x(P(x) \rightarrow \forall y P(y))$.

3.6. Existence and Uniqueness Proofs

In this section we consider proofs in which the goal has the form $\exists! x P(x)$. As we saw in Section 2.2, this can be thought of as an abbreviation for the formula $\exists x(P(x) \wedge \neg\exists y(P(y) \wedge y \neq x))$. According to the proof strategies discussed in previous sections, we could therefore prove this goal by finding a particular value of x for which we could prove both $P(x)$ and $\neg\exists y(P(y) \wedge y \neq x)$. The last part of this proof would involve proving a negated statement, but we can reexpress it as an equivalent positive statement:

$\neg\exists y(P(y) \wedge y \neq x)$
is equivalent to $\forall y \neg(P(y) \wedge y \neq x)$ (quantifier negation law),
which is equivalent to $\forall y(\neg P(y) \vee y = x)$ (DeMorgan's law),
which is equivalent to $\forall y(P(y) \rightarrow y = x)$ (conditional law).

Thus, we see that $\exists!x\,P(x)$ could also be written as $\exists x(P(x) \wedge \forall y(P(y) \rightarrow y = x))$. In fact, as the next example shows, several other formulas are also equivalent to $\exists!x\,P(x)$, and they suggest other approaches to proving goals of this form.

Example 3.6.1. Prove that the following formulas are all equivalent:

1. $\exists x(P(x) \wedge \forall y(P(y) \rightarrow y = x))$.
2. $\exists x \forall y(P(y) \leftrightarrow y = x)$.
3. $\exists x\,P(x) \wedge \forall y \forall z((P(y) \wedge P(z)) \rightarrow y = z)$.

Scratch work

If we prove directly that each of these statements is equivalent to each of the others, then we will have three biconditionals to prove: statement 1 iff statement 2, statement 1 iff statement 3, and statement 2 iff statement 3. If we prove each biconditional by the methods of Section 3.4, then each will involve two conditional proofs, so we will need a total of six conditional proofs. Fortunately, there is an easier way. We will prove that statement 1 implies statement 2, statement 2 implies statement 3, and statement 3 implies statement 1 – just three conditionals. Although we will not give a separate proof that statement 2 implies statement 1, it will follow from the fact that statement 2 implies statement 3 and statement 3 implies statement 1. Similarly, the other two conditionals follow from the three we will prove. Mathematicians almost always use some such shortcut when proving that several statements are all equivalent. Because we'll be proving three conditional statements, our proof will have three parts, which we will label $1 \rightarrow 2$, $2 \rightarrow 3$, and $3 \rightarrow 1$. We'll need to work out our strategy for the three parts separately.

$1 \rightarrow 2$. We assume statement 1 and prove statement 2. Because statement 1 starts with an existential quantifier, we choose a name, say x_0, for some object for which both $P(x_0)$ and $\forall y(P(y) \rightarrow y = x_0)$ are true. Thus, we now have the following situation:

Givens	*Goal*
$P(x_0)$	$\exists x \forall y(P(y) \leftrightarrow y = x)$
$\forall y(P(y) \rightarrow y = x_0)$	

Our goal also starts with an existential quantifier, so to prove it we should try to find a value of x that makes the rest of the statement come out true. Of course, the obvious choice is $x = x_0$. Plugging in x_0 for x, we see that we must now prove $\forall y(P(y) \leftrightarrow y = x_0)$. We let y be arbitrary and prove both directions of the biconditional. The $\rightarrow$ direction is clear by the second given.

For the $\leftarrow$ direction, suppose $y = x_0$. We also have $P(x_0)$ as a given, and plugging in y for x_0 in this given we get $P(y)$.

$2 \rightarrow 3$. Statement 2 is an existential statement, so we let x_0 be some object such that $\forall y(P(y) \leftrightarrow y = x_0)$. The goal, statement 3, is a conjunction, so we treat it as two separate goals.

Givens	*Goals*
$\forall y(P(y) \leftrightarrow y = x_0)$	$\exists x P(x)$
	$\forall y \forall z((P(y) \land P(z)) \rightarrow y = z)$

To prove the first goal we must choose a value for x, and of course the obvious value is $x = x_0$ again. Thus, we must prove $P(x_0)$. The natural way to use our only given is to plug in something for y; and to prove the goal $P(x_0)$, the obvious thing to plug in is x_0. This gives us $P(x_0) \leftrightarrow x_0 = x_0$. Of course, $x_0 = x_0$ is true, so by the $\leftarrow$ direction of the biconditional, we get $P(x_0)$.

For the second goal, we let y and z be arbitrary, assume $P(y)$ and $P(z)$, and try to prove $y = z$.

Givens	*Goal*
$\forall y(P(y) \leftrightarrow y = x_0)$	$y = z$
$P(y)$	
$P(z)$	

Plugging in each of y and z in the first given we get $P(y) \leftrightarrow y = x_0$ and $P(z) \leftrightarrow z = x_0$. Since we've assumed $P(y)$ and $P(z)$, this time we use the $\rightarrow$ directions of these biconditionals to conclude that $y = x_0$ and $z = x_0$. Our goal $y = z$ clearly follows.

$3 \rightarrow 1$. Because statement 3 is a conjunction, we treat it as two separate givens. The first is an existential statement, so we let x_0 stand for some object such that $P(x_0)$ is true. To prove statement 1 we again let $x = x_0$, so we have this situation:

Givens	*Goal*
$P(x_0)$	$P(x_0) \land \forall y(P(y) \rightarrow y = x_0)$
$\forall y \forall z((P(y) \land P(z)) \rightarrow y = z)$	

We already know the first half of the goal, so we only need to prove the second. For this we let y be arbitrary, assume $P(y)$, and make $y = x_0$ our goal.

Givens	*Goal*
$P(x_0)$	$y = x_0$
$\forall y \forall z((P(y) \land P(z)) \rightarrow y = z)$	
$P(y)$	

But now we know both $P(y)$ and $P(x_0)$, so the goal $y = x_0$ follows from the second given.

Solution

Theorem. *The following are equivalent:*

1. $\exists x(P(x) \wedge \forall y(P(y) \rightarrow y = x))$.
2. $\exists x \forall y(P(y) \leftrightarrow y = x)$.
3. $\exists x P(x) \wedge \forall y \forall z((P(y) \wedge P(z)) \rightarrow y = z)$.

Proof. $1 \rightarrow 2$. By statement 1, we can let x_0 be some object such that $P(x_0)$ and $\forall y(P(y) \rightarrow y = x_0)$. To prove statement 2 we will show that $\forall y(P(y) \leftrightarrow y = x_0)$. We already know the $\rightarrow$ direction. For the $\leftarrow$ direction, suppose $y = x_0$. Then since we know $P(x_0)$, we can conclude $P(y)$.

$2 \rightarrow 3$. By statement 2, choose x_0 such that $\forall y(P(y) \leftrightarrow y = x_0)$. Then, in particular, $P(x_0) \leftrightarrow x_0 = x_0$, and since clearly $x_0 = x_0$, it follows that $P(x_0)$ is true. Thus, $\exists x P(x)$. To prove the second half of statement 3, let y and z be arbitrary and suppose $P(y)$ and $P(z)$. Then by our choice of x_0 (as something for which $\forall y(P(y) \leftrightarrow y = x_0)$ is true), it follows that $y = x_0$ and $z = x_0$, so $y = z$.

$3 \rightarrow 1$. By the first half of statement 3, let x_0 be some object such that $P(x_0)$. Statement 1 will follow if we can show that $\forall y(P(y) \rightarrow y = x_0)$, so suppose $P(y)$. Since we now have both $P(x_0)$ and $P(y)$, by the second half of statement 3 we can conclude that $y = x_0$, as required. □

Because all three of the statements in the theorem are equivalent to $\exists! x P(x)$, we can prove a goal of this form by proving any of the three statements in the theorem. Probably the most common technique for proving a goal of the form $\exists! x P(x)$ is to prove statement 3 of the theorem.

To prove a goal of the form $\exists! x P(x)$:

Prove $\exists x P(x)$ and $\forall y \forall z((P(y) \wedge P(z)) \rightarrow y = z)$. The first of these goals shows that there exists an x such that $P(x)$ is true, and the second shows that it is unique. The two parts of the proof are therefore sometimes labeled *existence* and *uniqueness*. Each part is proven using strategies discussed earlier.

Form of final proof:

Existence: [Proof of $\exists x P(x)$ goes here.]
Uniqueness: [Proof of $\forall y \forall z((P(y) \wedge P(z)) \rightarrow y = z)$ goes here.]

Example 3.6.2. Prove that there is a unique set A such that for every set B, $A \cup B = B$.

Scratch work

Our goal is $\exists! A P(A)$, where $P(A)$ is the statement $\forall B(A \cup B = B)$. According to our strategy, we can prove this by proving existence and uniqueness separately. For the existence half of the proof we must prove $\exists A P(A)$, so we try to find a value of A that makes $P(A)$ true. There is no formula for finding this set A, but if you think about what the statement $P(A)$ means, you should realize that the right choice is $A = \varnothing$. Plugging this value in for A, we see that to complete the existence half of the proof we must show that $\forall B(\varnothing \cup B = B)$. This is clearly true. (If you're not sure of this, work out the proof!)

For the uniqueness half of the proof we prove $\forall C \forall D((P(C) \wedge P(D)) \rightarrow C = D)$. To do this, we let C and D be arbitrary, assume $P(C)$ and $P(D)$, and prove $C = D$. Writing out what the statements $P(C)$ and $P(D)$ mean, we have the following givens and goal:

Givens	*Goal*
$\forall B(C \cup B = B)$	$C = D$
$\forall B(D \cup B = B)$	

To use the givens, we should try to find something to plug in for B in each of them. There is a clever choice that makes the rest of the proof easy: We plug in D for B in the first given, and C for B in the second. This gives us $C \cup D = D$ and $D \cup C = C$. But clearly $C \cup D = D \cup C$. (If you don't see why, prove it!) The goal $C = D$ follows immediately.

Solution

Theorem. *There is a unique set A such that for every set B, $A \cup B = B$.*

Proof. Existence: Clearly $\forall B(\varnothing \cup B = B)$, so $\varnothing$ has the required property.

Uniqueness: Suppose $\forall B(C \cup B = B)$ and $\forall B(D \cup B = B)$. Applying the first of these assumptions to D we see that $C \cup D = D$, and applying the second to C we get $D \cup C = C$. But clearly $C \cup D = D \cup C$, so $C = D$. □

Sometimes a statement of the form $\exists! x P(x)$ is proven by proving statement 1 from Example 3.6.1. This leads to the following proof strategy.

To prove a goal of the form $\exists! x P(x)$:

Prove $\exists x(P(x) \wedge \forall y(P(y) \rightarrow y = x))$, using strategies from previous sections.

Example 3.6.3. Prove that for every real number x, if $x \neq 2$ then there is a unique real number y such that $2y/(y + 1) = x$.

Scratch work

Our goal is $\forall x(x \neq 2 \rightarrow \exists! y(2y/(y+1) = x))$. We therefore let x be arbitrary, assume $x \neq 2$, and prove $\exists! y(2y/(y+1) = x)$. According to the preceding strategy, we can prove this goal by proving the equivalent statement $\exists y(2y/(y+1) = x \wedge \forall z(2z/(z+1) = x \rightarrow z = y))$. We start by trying to find a value of y that will make the equation $2y/(y+1) = x$ come out true. In other words, we solve this equation for y:

$$\frac{2y}{y+1} = x \Rightarrow 2y = x(y+1) \Rightarrow y(2-x) = x \Rightarrow y = \frac{x}{2-x}.$$

Note that we have $x \neq 2$ as a given, so the division by $2 - x$ in the last step makes sense. Of course, these steps will not appear in the proof. We simply let $y = x/(2-x)$ and try to prove both $2y/(y+1) = x$ and $\forall z(2z/(z+1) = x \rightarrow z = y)$.

Givens	*Goals*
$x \neq 2$	$\frac{2y}{y+1} = x$
$y = \frac{x}{2-x}$	$\forall z\left(\frac{2z}{z+1} = x \rightarrow z = y\right)$

The first goal is easy to verify by simply plugging in $x/(2-x)$ for y. For the second, we let z be arbitrary, assume $2z/(z+1) = x$, and prove $z = y$:

Givens	*Goal*
$x \neq 2$	$z = y$
$y = \frac{x}{2-x}$	
$\frac{2z}{z+1} = x$	

We can show that $z = y$ now by solving for z in the third given:

$$\frac{2z}{z+1} = x \Rightarrow 2z = x(z+1) \Rightarrow z(2-x) = x \Rightarrow z = \frac{x}{2-x} = y.$$

Note that the steps we used here are exactly the same as the steps we used earlier in solving for y. This is a common pattern in existence and uniqueness proofs. Although the scratch work for figuring out an existence proof should not appear in the proof, this scratch work, or reasoning similar to it, can sometimes be used to prove that the object shown to exist is unique.

Solution

Theorem. *For every real number x, if $x \neq 2$ then there is a unique real number y such that $2y/(y+1) = x$.*

Proof. Let x be an arbitrary real number, and suppose $x \neq 2$. Let $y = x/(2-x)$, which is defined since $x \neq 2$. Then

$$\frac{2y}{y+1} = \frac{\frac{2x}{2-x}}{\frac{x}{2-x}+1} = \frac{\frac{2x}{2-x}}{\frac{2}{2-x}} = \frac{2x}{2} = x.$$

To see that this solution is unique, suppose $2z/(z+1) = x$. Then $2z = x(z+1)$, so $z(2-x) = x$. Since $x \neq 2$ we can divide both sides by $2-x$ to get $z = x/(2-x) = y$. □

The theorem in Example 3.6.1 can also be used to formulate strategies for using givens of the form $\exists! x P(x)$. Once again, statement 3 of the theorem is the one used most often.

To use a given of the form $\exists! x P(x)$:

Treat this as two given statements, $\exists x P(x)$ and $\forall y \forall z((P(y) \wedge P(z)) \rightarrow y = z)$. To use the first statement you should probably choose a name, say x_0, to stand for some object such that $P(x_0)$ is true. The second tells you that if you ever come across two objects y and z such that $P(y)$ and $P(z)$ are both true, you can conclude that $y = z$.

Example 3.6.4. Suppose A, B, and C are sets, A and B are not disjoint, A and C are not disjoint, and A has exactly one element. Prove that B and C are not disjoint.

Scratch work

Givens	*Goal*
$A \cap B \neq \varnothing$	$B \cap C \neq \varnothing$
$A \cap C \neq \varnothing$	
$\exists! x(x \in A)$	

We treat the last given as two separate givens, as suggested by our strategy. Writing out the meanings of the other givens and the goal, we have the following situation:

Givens	*Goal*
$\exists x(x \in A \wedge x \in B)$	$\exists x(x \in B \wedge x \in C)$
$\exists x(x \in A \wedge x \in C)$	
$\exists x(x \in A)$	
$\forall y \forall z((y \in A \wedge z \in A) \rightarrow y = z)$	

To prove the goal, we must find something that is an element of both B and C. To do this, we turn to the givens. The first given tells us that we can choose a name, say b, for something such that $b \in A$ and $b \in B$. Similarly, by the second given we can let c be something such that $c \in A$ and $c \in C$. At this point the third given is redundant. We already know that there's something in A, because in fact we already know that $b \in A$ and $c \in A$. We may as well skip to the last given, which says that if we ever come across two objects that are elements of A, we can conclude that they are equal. But as we have just observed, we know that $b \in A$ and $c \in A$! We can therefore conclude that $b = c$. Since $b \in B$ and $b = c \in C$, we have found something that is an element of both B and C, as required to prove the goal.

Solution

Theorem. *Suppose A, B, and C are sets, A and B are not disjoint, A and C are not disjoint, and A has exactly one element. Then B and C are not disjoint.*

Proof. Since A and B are not disjoint, we can let b be something such that $b \in A$ and $b \in B$. Similarly, since A and C are not disjoint, there is some object c such that $c \in A$ and $c \in C$. Since A has only one element, we must have $b = c$. Thus $b = c \in B \cap C$ and therefore B and C are not disjoint. □

Exercises

*1. Prove that for every real number x there is a unique real number y such that $x^2y = x - y$.

2. Prove that there is a unique real number x such that for every real number y, $xy + x - 4 = 4y$.

3. Prove that for every real number x, if $x \neq 0$ and $x \neq 1$ then there is a unique real number y such that $y/x = y - x$.

*4. Prove that for every real number x, if $x \neq 0$ then there is a unique real number y such that for every real number z, $zy = z/x$.

5. Recall that if $\mathcal{F}$ is a family of sets, then $\cup\mathcal{F} = \{x \mid \exists A(A \in \mathcal{F} \wedge x \in A)\}$. Suppose we define a new set $\cup!\mathcal{F}$ by the formula $\cup!\mathcal{F} = \{x \mid \exists!A(A \in \mathcal{F} \wedge x \in A)\}$.
   (a) Prove that for any family of sets $\mathcal{F}$, $\cup!\mathcal{F} \subseteq \cup\mathcal{F}$.
   (b) A family of sets $\mathcal{F}$ is said to be *pairwise disjoint* if every pair of distinct elements of $\mathcal{F}$ are disjoint; that is, $\forall A \in \mathcal{F} \forall B \in \mathcal{F}(A \neq B \rightarrow A \cap B = \varnothing)$. Prove that for any family of sets $\mathcal{F}$, $\cup!\mathcal{F} = \cup\mathcal{F}$ iff $\mathcal{F}$ is pairwise disjoint.

P_D*6. Let U be any set.
   (a) Prove that there is a unique $A \in \mathscr{P}(U)$ such that for every $B \in \mathscr{P}(U)$, $A \cup B = B$.

(b) Prove that there is a unique $A \in \mathscr{P}(U)$ such that for every $B \in \mathscr{P}(U)$, $A \cup B = A$.

Pd7. Let U be any set.

(a) Prove that there is a unique $A \in \mathscr{P}(U)$ such that for every $B \in \mathscr{P}(U)$, $A \cap B = B$.

(b) Prove that there is a unique $A \in \mathscr{P}(U)$ such that for every $B \in \mathscr{P}(U)$, $A \cap B = A$.

Pd8. Let U be any set.

(a) Prove that for every $A \in \mathscr{P}(U)$ there is a unique $B \in \mathscr{P}(U)$ such that for every $C \in \mathscr{P}(U)$, $C \setminus A = C \cap B$.

(b) Prove that for every $A \in \mathscr{P}(U)$ there is a unique $B \in \mathscr{P}(U)$ such that for every $C \in \mathscr{P}(U)$, $C \cap A = C \setminus B$.

Pd9. Recall that you showed in exercise 12 of Section 1.4 that symmetric difference is associative; in other words, for all sets A, B, and C, $A \,\triangle\, (B \,\triangle\, C) = (A \,\triangle\, B) \,\triangle\, C$. You may also find it useful in this problem to note that symmetric difference is clearly commutative; in other words, for all sets A and B, $A \,\triangle\, B = B \,\triangle\, A$.

(a) Prove that there is a unique identity element for symmetric difference. In other words, there is a unique set X such that for every set A, $A \,\triangle\, X = A$.

(b) Prove that every set has a unique inverse for the operation of symmetric difference. In other words, for every set A there is a unique set B such that $A \,\triangle\, B = X$, where X is the identity element from part (a).

(c) Prove that for any sets A and B there is a unique set C such that $A \,\triangle\, C = B$.

(d) Prove that for every set A there is a unique set $B \subseteq A$ such that for every set $C \subseteq A$, $B \,\triangle\, C = A \setminus C$.

Pd10. Suppose A is a set, and for every family of sets $\mathcal{F}$, if $\cup\mathcal{F} = A$ then $A \in \mathcal{F}$. Prove that A has exactly one element. (Hint: For both the existence and uniqueness parts of the proof, try proof by contradiction.)

Pd*11. Suppose $\mathcal{F}$ is a family of sets that has the property that for every $\mathcal{G} \subseteq \mathcal{F}$, $\cup\mathcal{G} \in \mathcal{F}$. Prove that there is a unique set A such that $A \in \mathcal{F}$ and $\forall B \in \mathcal{F}(B \subseteq A)$.

12. (a) Suppose $P(x)$ is a statement with a free variable x. Find a formula, using the logical symbols we have studied, that means "there are exactly two values of x for which $P(x)$ is true."

(b) Based on your answer to part (a), design a proof strategy for proving a statement of the form "there are exactly two values of x for which $P(x)$ is true."

(c) Prove that there are exactly two solutions to the equation $x^3 = x^2$.

3.7. More Examples of Proofs

So far, most of our proofs have involved fairly straightforward applications of the proof techniques we've discussed. We end this chapter with a few examples of somewhat more difficult proofs. These proofs use the techniques of this chapter, but for various reasons they're a little harder than most of our earlier proofs. Some are simply longer, involving the application of more proof strategies. Some require clever choices of which strategies to use. In some cases it's clear what strategy to use, but some insight is required to see exactly how to use it. Our earlier examples, which were intended only to illustrate and clarify the proof techniques, may have made proof-writing seem mechanical and dull. We hope that by studying these more difficult examples you will begin to see that mathematical reasoning can also be surprising and beautiful.

Some proof techniques are particularly difficult to apply. For example, when you're proving a goal of the form $\exists x P(x)$, the obvious way to proceed is to try to find a value of x that makes the statement $P(x)$ true, but sometimes it will not be obvious how to find that value of x. Using a given of the form $\forall x P(x)$ is similar. You'll probably want to plug in a particular value for x, but to complete the proof you may have to make a clever choice of what to plug in. Proofs that must be broken down into cases are also sometimes difficult to figure out. It is sometimes hard to know when to use cases and what cases to use.

We begin by looking again at the proofs from the introduction. Some aspects of these proofs probably seemed somewhat mysterious when you read them in the introduction. See if they make more sense to you now that you have a better understanding of how proofs are constructed. We will present each proof exactly as it appeared in the introduction and then follow it with a commentary discussing the proof techniques used.

Theorem 3.7.1. *Suppose n is an integer larger than 1 and n is not prime. Then $2^n - 1$ is not prime.*

Proof. Since n is not prime, there are positive integers a and b such that $a < n$, $b < n$, and $n = ab$. Let $x = 2^b - 1$ and $y = 1 + 2^b + 2^{2b} + \cdots + 2^{(a-1)b}$. Then

$$\begin{aligned}
xy &= (2^b - 1) \cdot (1 + 2^b + 2^{2b} + \cdots + 2^{(a-1)b}) \\
&= 2^b \cdot (1 + 2^b + 2^{2b} + \cdots + 2^{(a-1)b}) - (1 + 2^b + 2^{2b} + \cdots + 2^{(a-1)b}) \\
&= (2^b + 2^{2b} + 2^{3b} + \cdots + 2^{ab}) - (1 + 2^b + 2^{2b} + \cdots + 2^{(a-1)b}) \\
&= 2^{ab} - 1 \\
&= 2^n - 1.
\end{aligned}$$

Since $b < n$, we can conclude that $x = 2^b - 1 < 2^n - 1$. Also, since $ab = n > a$, it follows that $b > 1$. Therefore, $x = 2^b - 1 > 2^1 - 1 = 1$, so $y < xy = 2^n - 1$. Thus, we have shown that $2^n - 1$ can be written as the product of two positive integers x and y, both of which are smaller than $2^n - 1$, so $2^n - 1$ is not prime. □

Commentary. We are given that n is not prime, and we must prove that $2^n - 1$ is not prime. Both of these are negative statements, but fortunately it is easy to reexpress them as positive statements. To say that an integer larger than 1 is not prime means that it can be written as a product of two smaller positive integers. Thus, the hypothesis that n is not prime means $\exists a \in \mathbb{Z}^+ \exists b \in \mathbb{Z}^+(ab = n \wedge a < n \wedge b < n)$, and what we must prove is that $2^n - 1$ is not prime, which means $\exists x \in \mathbb{Z}^+ \exists y \in \mathbb{Z}^+(xy = 2^n - 1 \wedge x < 2^n - 1 \wedge y < 2^n - 1)$. In the second sentence of the proof we apply existential instantiation to the hypothesis that n is not prime, and the rest of the proof is devoted to exhibiting numbers x and y with the properties required to prove that $2^n - 1$ is not prime.

As usual in proofs of existential statements, the proof doesn't explain how the values of x and y were chosen, it simply demonstrates that these values work. After the values of x and y have been given, the goal remaining to be proven is $xy = 2^n - 1 \wedge x < 2^n - 1 \wedge y < 2^n - 1$. Of course, this is treated as three separate goals, which are proven one at a time. The proofs of these three goals involve only elementary algebra.

One of the attractive features of this proof is the calculation used to show that $xy = 2^n - 1$. The formulas for x and y are somewhat complicated, and at first their product looks even more complicated. It is a pleasant surprise when most of the terms in this product cancel and, as if by magic, the answer $2^n - 1$ appears. Of course, we can see with hindsight that it was this calculation that motivated the choice of x and y. There is, however, one aspect of this calculation that may bother you. The use of "$\cdots$" in the formulas indicates that the proof depends on a pattern in the calculation that is not being spelled out. We'll give a more rigorous proof that $xy = 2^n - 1$ in Chapter 6, after we have introduced the method of proof by mathematical induction.

Theorem 3.7.2. *There are infinitely many prime numbers.*

Proof. Suppose there are only finitely many prime numbers. Let $p_1, p_2, \ldots, p_n$ be a list of all prime numbers. Let $m = p_1 p_2 \cdots p_n + 1$. Note that m is not divisible by p_1, since dividing m by p_1 gives a quotient of $p_2 p_3 \cdots p_n$ and a remainder of 1. Similarly, m is not divisible by any of $p_2, p_3 \ldots, p_n$.

We now use the fact that every integer larger than 1 is either prime or can be written as a product of primes. (We'll see a proof of this fact in Chapter 6.) Clearly m is larger than 1, so m is either prime or a product of primes. Suppose first that m is prime. Note that m is larger than all of the numbers in the list $p_1, p_2, \ldots, p_n$, so we've found a prime number not in this list. But this contradicts our assumption that this was a list of *all* prime numbers.

Now suppose m is a product of primes. Let q be one of the primes in this product. Then m is divisible by q. But we've already seen that m is not divisible by any of the numbers in the list $p_1, p_2, \ldots, p_n$, so once again we have a contradiction with the assumption that this list included all prime numbers.

Since the assumption that there are finitely many prime numbers has led to a contradiction, there must be infinitely many prime numbers. □

Commentary. Because *infinite* means *not finite*, the statement of the theorem might be considered to be a negative statement. It is therefore not surprising that the proof proceeds by contradiction. The assumption that there are finitely many primes means that there exists a natural number n such that there are n primes, and the statement that there are n primes means that there is a list of distinct numbers $p_1, p_2, \ldots, p_n$ such that every number in the list is prime, and there are no primes that are not in the list. Thus, the second sentence of the proof applies existential instantiation to introduce the numbers n and $p_1, p_2, \ldots, p_n$ into the proof. At this point in the proof we have the following situation:

Givens	*Goal*
$p_1, p_2, \ldots, p_n$ are all prime	Contradiction
$\neg\exists q(q \text{ is prime} \wedge q \notin \{p_1, p_2 \ldots, p_n\})$	

The second given could be reexpressed as a positive statement, but since we are doing a proof by contradiction, another reasonable approach would be to try to reach a contradiction by proving that $\exists q(q \text{ is prime} \wedge q \notin \{p_1, p_2, \ldots, p_n\})$. This is the strategy used in the proof. Thus, the goal for the rest of the proof is to show that there is a prime number not in the list $p_1, p_2, \ldots, p_n$ – an "unlisted prime."

Because our goal is now an existential statement, it is not surprising that the next step in the proof is to introduce the new number m, without any explanation of how m was chosen. What *is* surprising is that m may or may not be the unlisted prime we are looking for. The problem is that m might not be prime. All we can be sure of is that m is either prime or a product of primes. Because this statement is a disjunction, it suggests proof by cases, and this is the method used in the rest of the proof. Although the cases are not explicitly labeled as cases in the proof, it is important to realize that the rest of the proof has the form of a proof by cases. In case 1 we assume that m is prime, and in case 2 we assume that it is a product of primes. In both cases we are able to produce an unlisted prime as required to complete the proof.

Theorem 3.7.3. *For every positive integer n, there is a sequence of n consecutive positive integers containing no primes.*

Proof. Suppose n is a positive integer. Let $x = (n+1)! + 2$. We will show that none of the numbers $x, x+1, x+2, \cdots, x+(n-1)$ is prime. Since this is a sequence of n consecutive positive integers, this will prove the theorem.

To see that x is not prime, note that

$$\begin{aligned} x &= 1 \cdot 2 \cdot 3 \cdot 4 \cdots (n+1) + 2 \\ &= 2 \cdot (1 \cdot 3 \cdot 4 \cdots (n+1) + 1). \end{aligned}$$

Thus, x can be written as a product of two smaller positive integers, so x is not prime.

Similarly, we have

$$\begin{aligned} x+1 &= 1 \cdot 2 \cdot 3 \cdot 4 \cdots (n+1) + 3 \\ &= 3 \cdot (1 \cdot 2 \cdot 4 \cdots (n+1) + 1), \end{aligned}$$

so $x+1$ is also not prime. In general, consider any number $x+i$, where $0 \leq i \leq n-1$. Then we have

$$\begin{aligned} x+i &= 1 \cdot 2 \cdot 3 \cdot 4 \cdots (n+1) + (i+2) \\ &= (i+2) \cdot (1 \cdot 2 \cdot 3 \cdots (i+1) \cdot (i+3) \cdots (n+1) + 1), \end{aligned}$$

so $x+i$ is not prime. □

Commentary. A sequence of n consecutive positive integers is a sequence of the form $x, x+1, x+2, \ldots, x+(n-1)$, where x is a positive integer. Thus, the logical form of the statement to be proven is $\forall n > 0 \exists x > 0 \forall i (0 \leq i \leq n-1 \rightarrow x+i$ is not prime), where all variables range over the integers. The overall plan of the proof is exactly what one would expect for a proof of a statement of this form: We let $n > 0$ be arbitrary, specify a value for x, let i be arbitrary, and then assume that $0 \leq i \leq n-1$ and prove that $x+i$ is not prime. As in the proof of Theorem 3.7.1, to prove that $x+i$ is not prime we show how to write it as a product of two smaller integers.

Before the demonstration that $x+i$ is not prime, where i is an arbitrary integer between 0 and $n-1$, the proof includes verifications that x and $x+1$

are not prime. These are completely unnecessary and are only included to make the proof easier to read.

For readers who are familiar with the definition of limits from calculus, we give one more example, showing how proofs involving limits can be worked out using the techniques in this chapter. Readers who are not familiar with this definition should skip this example.

Example 3.7.4. Show that $\lim_{x \to 3} \frac{2x^2 - 5x - 3}{x - 3} = 7$.

Scratch work

According to the definition of limits, our goal means that for every positive number ε there is a positive number δ such that if x is any number such that $0 < |x - 3| < \delta$, then $\left|\frac{2x^2-5x-3}{x-3} - 7\right| < \varepsilon$. Translating this into logical symbols, we have

$$\forall \varepsilon > 0 \exists \delta > 0 \forall x \left(0 < |x - 3| < \delta \rightarrow \left|\frac{2x^2 - 5x - 3}{x - 3} - 7\right| < \varepsilon\right).$$

We therefore start by letting ε be an arbitrary positive number and then try to find a positive number δ for which we can prove

$$\forall x \left(0 < |x - 3| < \delta \rightarrow \left|\frac{2x^2 - 5x - 3}{x - 3} - 7\right| < \varepsilon\right).$$

The scratch work involved in finding δ will not appear in the proof, of course. In the final proof we'll just write "Let δ = (some positive number)" and then proceed to prove

$$\forall x \left(0 < |x - 3| < \delta \rightarrow \left|\frac{2x^2 - 5x - 3}{x - 3} - 7\right| < \varepsilon\right).$$

Before working out the value of δ, let's figure out what the rest of the proof will look like. Based on the form of the goal at this point, we should proceed by letting x be arbitrary, assuming $0 < |x - 3| < \delta$, and then proving $\left|\frac{2x^2-5x-3}{x-3} - 7\right| < \varepsilon$. Thus, the entire proof will have the following form:

Let ε be an arbitrary positive number.
  Let δ = (some positive number).
    Let x be arbitrary.
      Suppose $0 < |x - 3| < \delta$.
        [Proof of $\left|\frac{2x^2-5x-3}{x-3} - 7\right| < \varepsilon$ goes here.]
      Therefore $0 < |x - 3| < \delta \rightarrow \left|\frac{2x^2-5x-3}{x-3} - 7\right| < \varepsilon$.

Since x was arbitrary, we can conclude that $\forall x(0 < |x-3| < \delta \rightarrow \left|\frac{2x^2-5x-3}{x-3} - 7\right| < \varepsilon)$.

Therefore $\exists\delta > 0 \forall x\left(0 < |x-3| < \delta \rightarrow \left|\frac{2x^2-5x-3}{x-3} - 7\right| < \varepsilon\right)$.

Since ε was arbitrary, it follows that $\forall\varepsilon > 0 \exists\delta > 0 \forall x\left(0 < |x-3| < \delta \rightarrow \left|\frac{2x^2-5x-3}{x-3} - 7\right| < \varepsilon\right)$.

Two steps remain to be worked out. We must decide what value to assign to δ, and we must fill in the proof of $\left|\frac{2x^2-5x-3}{x-3} - 7\right| < \varepsilon$. We'll work on the second of these steps first, and in the course of working out this step it will become clear what value we should use for δ. The givens and goal for this second step are as follows:

Givens	*Goal*
$\varepsilon > 0$	$\left\|\frac{2x^2-5x-3}{x-3} - 7\right\| < \varepsilon$
$\delta =$ (some positive number)	
$0 < \|x-3\| < \delta$	

First of all, note that we have $0 < |x-3|$ as a given, so $x \neq 3$ and therefore the fraction $\frac{2x^2-5x-3}{x-3}$ is defined. Factoring the numerator, we find that

$$\begin{aligned}\left|\frac{2x^2-5x-3}{x-3} - 7\right| &= \left|\frac{(2x+1)(x-3)}{x-3} - 7\right| \\ &= |2x+1-7| = |2x-6| = 2|x-3|.\end{aligned}$$

Now we also have as a given that $|x-3| < \delta$, so $2|x-3| < 2\delta$. Combining this with the previous equation, we get $\left|\frac{2x^2-5x-3}{x-3} - 7\right| < 2\delta$, and our goal is $\left|\frac{2x^2-5x-3}{x-3} - 7\right| < \varepsilon$. Thus, if we chose δ so that $2\delta = \varepsilon$, we'd be done. In other words, we should let $\delta = \varepsilon/2$. Note that since $\varepsilon > 0$, this is a positive number, as required.

Solution

Theorem. $\lim_{x \to 3} \frac{2x^2-5x-3}{x-3} = 7$.

Proof. Suppose $\varepsilon > 0$. Let $\delta = \varepsilon/2$, which is also clearly positive. Let x be an arbitrary real number, and suppose that $0 < |x-3| < \delta$. Then

$$\begin{aligned}\left|\frac{2x^2-5x-3}{x-3} - 7\right| &= \left|\frac{(2x+1)(x-3)}{x-3} - 7\right| \\ &= |2x+1-7| = |2x-6| \\ &= 2|x-3| < 2\delta = 2\left(\frac{\varepsilon}{2}\right) = \varepsilon.\end{aligned}$$

□

Exercises

PD*1. Suppose $\mathcal{F}$ is a family of sets. Prove that there is a unique set A that has the following two properties:
(a) $\mathcal{F} \subseteq \mathscr{P}(A)$.
(b) $\forall B(\mathcal{F} \subseteq \mathscr{P}(B) \rightarrow A \subseteq B)$.
(Hint: First try an example. Let $\mathcal{F} = \{\{1, 2, 3\}, \{2, 3, 4\}, \{3, 4, 5\}\}$. Can you find the set A that has properties (a) and (b)?)

PD2. Suppose A and B are sets. What can you prove about $\mathscr{P}(A \setminus B) \setminus (\mathscr{P}(A) \setminus \mathscr{P}(B))$? (No, it's not equal to $\varnothing$. Try some examples and see what you get.)

PD3. Suppose that A, B, and C are sets. Prove that the following statements are equivalent:
(a) $(A \mathbin{\triangle} C) \cap (B \mathbin{\triangle} C) = \varnothing$.
(b) $A \cap B \subseteq C \subseteq A \cup B$.
(c) $A \mathbin{\triangle} C \subseteq A \mathbin{\triangle} B$.

*4. Suppose $\{A_i \mid i \in I\}$ is a family of sets. Prove that if $\mathscr{P}(\bigcup_{i \in I} A_i) \subseteq \bigcup_{i \in I} \mathscr{P}(A_i)$, then there is some $i \in I$ such that $\forall j \in I(A_j \subseteq A_i)$.

5. Suppose $\mathcal{F}$ is a nonempty family of sets. Let $I = \bigcup\mathcal{F}$ and $J = \bigcap\mathcal{F}$. Suppose also that $J \neq \varnothing$, and notice that it follows that for every $X \in \mathcal{F}$, $X \neq \varnothing$, and also that $I \neq \varnothing$. Finally, suppose that $\{A_i \mid i \in I\}$ is an indexed family of sets.
(a) Prove that $\bigcup_{i \in I} A_i = \bigcup_{X \in \mathcal{F}}(\bigcup_{i \in X} A_i)$.
(b) Prove that $\bigcap_{i \in I} A_i = \bigcap_{X \in \mathcal{F}}(\bigcap_{i \in X} A_i)$.
(c) Prove that $\bigcup_{i \in J} A_i \subseteq \bigcap_{X \in \mathcal{F}}(\bigcup_{i \in X} A_i)$. Is it always true that $\bigcup_{i \in J} A_i = \bigcap_{X \in \mathcal{F}}(\bigcup_{i \in X} A_i)$? Give either a proof or a counterexample to justify your answer.
(d) Discover and prove a theorem relating $\bigcap_{i \in J} A_i$ and $\bigcup_{X \in \mathcal{F}}(\bigcap_{i \in X} A_i)$.

6. Prove that $\lim\limits_{x \to 2} \frac{3x^2 - 12}{x - 2} = 12$.

*7. Prove that if $\lim\limits_{x \to c} f(x) = L$ and $L > 0$, then there is some number $\delta > 0$ such that for all x, if $0 < |x - c| < \delta$ then $f(x) > 0$.

8. Prove that if $\lim\limits_{x \to c} f(x) = L$ then $\lim\limits_{x \to c} 7f(x) = 7L$.

*9. Consider the following putative theorem.

Theorem. *There are irrational numbers a and b such that a^b is rational.*

Is the following proof correct? If so, what proof strategies does it use? If not, can it be fixed? Is the theorem correct? (Note: The proof uses the fact that $\sqrt{2}$ is irrational, which we'll prove in Chapter 6.)

Proof. Either $\sqrt{2}^{\sqrt{2}}$ is rational or it's irrational.

Case 1. $\sqrt{2}^{\sqrt{2}}$ is rational. Let $a = b = \sqrt{2}$. Then a and b are irrational, and $a^b = \sqrt{2}^{\sqrt{2}}$, which we are assuming in this case is rational.

Case 2. $\sqrt{2}^{\sqrt{2}}$ is irrational. Let $a = \sqrt{2}^{\sqrt{2}}$ and $b = \sqrt{2}$. Then a is irrational by assumption, and we know that b is also irrational. Also, $a^b = \left(\sqrt{2}^{\sqrt{2}}\right)^{\sqrt{2}} = \sqrt{2}^{(\sqrt{2}\cdot\sqrt{2})} = (\sqrt{2})^2 = 2$, which is rational. $\square$

4

Relations

4.1. Ordered Pairs and Cartesian Products

In Chapter 1 we discussed truth sets for statements containing a single free variable. In this chapter we extend this idea to include statements with more than one free variable.

For example, suppose $P(x, y)$ is a statement with two free variables x and y. We can't speak of this statement as being true or false until we have specified two values – one for x and one for y. Thus, if we want the truth set to identify which assignments of values to free variables make the statement come out true, then the truth set will have to contain not individual values, but pairs of values. We will specify a pair of values by writing the two values in parentheses separated by a comma. For example, let $D(x, y)$ mean "x divides y." Then $D(6, 18)$ is true, since $6 \mid 18$, so the pair of values $(6, 18)$ is an assignment of values to the variables x and y that makes the statement $D(x, y)$ come out true. Note that 18 does not divide 6, so the pair of values $(18, 6)$ makes the statement $D(x, y)$ false. We must therefore distinguish between the pairs $(18, 6)$ and $(6, 18)$. Because the order of the values in the pair makes a difference, we will refer to a pair (a, b) as an *ordered pair*, with *first coordinate a* and *second coordinate b*.

You have probably seen ordered pairs before when studying points in the xy plane. The use of x and y coordinates to identify points in the plane works by assigning to each point in the plane an ordered pair, whose coordinates are the x and y coordinates of the point. The pairs must be ordered because, for example, the points $(2, 5)$ and $(5, 2)$ are different points in the plane. In this case the coordinates of the ordered pairs are real numbers, but ordered pairs can have anything at all as their coordinates. For example, suppose we let $C(x, y)$ stand for the statement "x has y children." In this statement the variable x ranges over the set of all people, and y ranges over the set of all natural numbers. Thus, the only ordered pairs it makes sense to consider when discussing assignments of values to the variables x and y in this statement are pairs in which the first coordinate is a person and the second is a natural number. For example, the assignment (Prince Charles, 2) makes the statement $C(x, y)$ come out true, because Prince Charles does have two children, whereas the assignment (Johnny Carson, 37) makes the statement false. Note that the assignment (2, Prince Charles) makes no sense, because it would lead to the nonsensical statement "2 has Prince Charles children."

In general, if $P(x, y)$ is a statement in which x ranges over some set A and y ranges over a set B, then the only assignments of values to x and y that will make sense in $P(x, y)$ will be ordered pairs in which the first coordinate is an element of A and the second comes from B. We therefore make the following definition:

Definition 4.1.1. Suppose A and B are sets. Then the *Cartesian product* of A and B, denoted $A \times B$, is the set of all ordered pairs in which the first coordinate is an element of A and the second is an element of B. In other words,

$$A \times B = \{(a, b) \mid a \in A \text{ and } b \in B\}.$$

Example 4.1.2.

1. If $A = \{\text{red, green}\}$ and $B = \{2, 3, 5\}$ then $A \times B = \{(\text{red}, 2), (\text{red}, 3), (\text{red}, 5), (\text{green}, 2), (\text{green}, 3), (\text{green}, 5)\}$.
2. If $P =$ the set of all people then $P \times \mathbb{N} = \{(p, n) \mid p$ is a person and n is a natural number$\} = \{$(Prince Charles, 0), (Prince Charles, 1), (Prince Charles, 2), ..., (Johnny Carson, 0), (Johnny Carson, 1), ...$\}$. These are the ordered pairs that make sense as assignments of values to the free variables x and y in the statement $C(x, y)$.
3. $\mathbb{R} \times \mathbb{R} = \{(x, y) \mid x$ and y are real numbers$\}$. These are the coordinates of all the points in the plane. For obvious reasons, this set is sometimes written $\mathbb{R}^2$.

The introduction of a new mathematical concept gives us an opportunity to practice our proof-writing techniques by proving some basic properties of the new concept. Here's a theorem giving some basic properties of Cartesian products.

Theorem 4.1.3. *Suppose A, B, C, and D are sets.*

1. $A \times (B \cap C) = (A \times B) \cap (A \times C)$.
2. $A \times (B \cup C) = (A \times B) \cup (A \times C)$.
3. $(A \times B) \cap (C \times D) = (A \cap C) \times (B \cap D)$.
4. $(A \times B) \cup (C \times D) \subseteq (A \cup C) \times (B \cup D)$.
5. $A \times \emptyset = \emptyset \times A = \emptyset$.

Proof of 1. Let p be an arbitrary element of $A \times (B \cap C)$. Then by the definition of Cartesian product, p must be an ordered pair whose first coordinate is an element of A and second coordinate is an element of $B \cap C$. In other words, $p = (x, y)$ for some $x \in A$ and $y \in B \cap C$. Since $y \in B \cap C$, $y \in B$ and $y \in C$. Since $x \in A$ and $y \in B$, $p = (x, y) \in A \times B$, and similarly $p \in A \times C$. Thus, $p \in (A \times B) \cap (A \times C)$. Since p was an arbitrary element of $A \times (B \cap C)$, it follows that $A \times (B \cap C) \subseteq (A \times B) \cap (A \times C)$.

Now let p be an arbitrary element of $(A \times B) \cap (A \times C)$. Then $p \in A \times B$, so $p = (x, y)$ for some $x \in A$ and $y \in B$. Also, $(x, y) = p \in A \times C$, so $y \in C$. Since $y \in B$ and $y \in C$, $y \in B \cap C$. Thus, $p = (x, y) \in A \times (B \cap C)$. Since p was an arbitrary element of $(A \times B) \cap (A \times C)$ we can conclude that $(A \times B) \cap (A \times C) \subseteq A \times (B \cap C)$, so $A \times (B \cap C) = (A \times B) \cap (A \times C)$. □

Commentary. Before continuing with the proofs of the other parts, we give a brief commentary on the proof just given. Statement 1 is an equation between two sets, so as we saw in Example 3.4.4, there are two natural approaches we could take to prove it. We could prove $\forall p[p \in A \times (B \cap C) \leftrightarrow p \in (A \times B) \cap (A \times C)]$ or we could prove both $A \times (B \cap C) \subseteq (A \times B) \cap (A \times C)$ and $(A \times B) \cap (A \times C) \subseteq A \times (B \cap C)$. In this proof, we have taken the second approach. The first paragraph gives the proof that $A \times (B \cap C) \subseteq (A \times B) \cap (A \times C)$ and the second gives the proof that $(A \times B) \cap (A \times C) \subseteq A \times (B \cap C)$.

In the first of these proofs we take the usual approach of letting p be an arbitrary element of $A \times (B \cap C)$ and then proving $p \in (A \times B) \cap (A \times C)$. Because $p \in A \times (B \cap C)$ means $\exists x \exists y(x \in A \wedge y \in B \cap C \wedge p = (x, y))$, we immediately introduce the variables x and y by existential instantiation. The rest of the proof involves simply working out the definitions of the set theory operations involved. The proof of the opposite inclusion in the second paragraph is similar.

Note that in both parts of this proof we introduced an arbitrary object p that turned out to be an ordered pair, and we were therefore able to say that $p = (x, y)$ for some objects x and y. In most proofs involving Cartesian products mathematicians suppress this step. If it is clear from the beginning that an object will turn out to be an ordered pair, it is usually just called (x, y) from the outset. We will follow this practice in our proofs.

We leave the proofs of statements 2 and 3 as exercises (see exercise 5).

Proof of 4. Let (x, y) be an arbitrary element of $(A \times B) \cup (C \times D)$. Then either $(x, y) \in A \times B$ or $(x, y) \in C \times D$.

Case 1. $(x, y) \in A \times B$. Then $x \in A$ and $y \in B$, so clearly $x \in A \cup C$ and $y \in B \cup D$. Therefore $(x, y) \in (A \cup C) \times (B \cup D)$.

Case 2. $(x, y) \in C \times D$. A similar argument shows that $(x, y) \in (A \cup C) \times (B \cup D)$.

Since (x, y) was an arbitrary element of $(A \times B) \cup (C \times D)$, it follows that $(A \times B) \cup (C \times D) \subseteq (A \cup C) \times (B \cup D)$. □

Proof of 5. Suppose $A \times \varnothing \neq \varnothing$. Then $A \times \varnothing$ has at least one element, and by the definition of Cartesian product this element must be an ordered pair (x, y) for some $x \in A$ and $y \in \varnothing$. But this is impossible, because $\varnothing$ has no elements. Thus, $A \times \varnothing = \varnothing$. The proof that $\varnothing \times A = \varnothing$ is similar. □

Commentary. Statement 4 says that one set is a subset of another, and the proof follows the usual pattern for statements of this form: We start with an arbitrary element of the first set and then prove that it's an element of the second. It is clear that the arbitrary element of the first set must be an ordered pair, so we have written it as an ordered pair from the beginning.

Thus, for the rest of the proof we have $(x, y) \in (A \times B) \cup (C \times D)$ as a given, and the goal is to prove that $(x, y) \in (A \cup C) \times (B \cup D)$. The given means $(x, y) \in A \times B \vee (x, y) \in C \times D$, so proof by cases is an appropriate strategy. In each case it is easy to prove the goal.

Statement 5 means $A \times \varnothing = \varnothing \wedge \varnothing \times A = \varnothing$, so we treat this as two goals and prove $A \times \varnothing = \varnothing$ and $\varnothing \times A = \varnothing$ separately. To say that a set equals the empty set is actually a negative statement, although it may not look like it on the surface, because it means that the set does *not* have any elements. Thus, it is not surprising that the proof that $A \times \varnothing = \varnothing$ proceeds by contradiction. The assumption that $A \times \varnothing \neq \varnothing$ means $\exists p(p \in A \times \varnothing)$, so our next step is to introduce a name for an element of $A \times \varnothing$. Once again, it is clear that the new object being introduced in the proof is an ordered pair, so we have written it as an ordered pair (x, y) from the beginning. Writing out the meaning of $(x, y) \in A \times \varnothing$ leads immediately to a contradiction.

The proof that $\varnothing \times A = \varnothing$ is similar, but simply *saying* this doesn't *prove* it. Thus, the claim in the proof that this part of the proof is similar is really an indication that the second half of the proof is being left as an exercise. You should work through the details of this proof in your head (or if necessary write them out on paper) to make sure that a proof similar to the proof in the first half will really work.

Because the order of the coordinates in an ordered pair matters, $A \times B$ and $B \times A$ mean different things. Does it ever happen that $A \times B = B \times A$? Well, one way this could happen is if $A = B$. Clearly if $A = B$ then $A \times B = A \times A = B \times A$. Are there any other possibilities?

Here's an incorrect proof that $A \times B = B \times A$ only if $A = B$: The first coordinates of the ordered pairs in $A \times B$ come from A, and the first coordinates of the ordered pairs in $B \times A$ come from B. But if $A \times B = B \times A$, then the first coordinates in these two sets must be the same, so $A = B$.

This is a good example of why it's important to stick to the rules of proof-writing we've studied rather than allowing yourself to be convinced by any reasoning that looks plausible. The informal reasoning in the preceding paragraph is incorrect, and we can find the error by trying to reformulate this reasoning as a formal proof. Suppose $A \times B = B \times A$. To prove that $A = B$ we could let x be arbitrary and then try to prove $x \in A \rightarrow x \in B$ and $x \in B \rightarrow x \in A$. For the first of these we assume $x \in A$ and try to prove $x \in B$. Now the incorrect proof suggests that we should try to show that x is the first coordinate of some ordered pair in $A \times B$ and then use the fact that $A \times B = B \times A$. We could do this by trying to find some object $y \in B$ and then forming the ordered pair (x, y). Then we would have $(x, y) \in A \times B$ and $A \times B = B \times A$, and it would follow that $(x, y) \in B \times A$ and therefore $x \in B$. But how can we find an object $y \in B$? We don't have any given information about B, other than the fact that $A \times B = B \times A$. In fact, *B could be the empty set!* This is the flaw in the proof. If $B = \emptyset$, then it will be impossible to choose $y \in B$, and the proof will fall apart. For similar reasons, the other half of the proof won't work if $A = \emptyset$.

Not only have we found the flaw in the proof, but we can now figure out what to do about it. We must take into account the possibility that A or B might be the empty set.

Theorem 4.1.4. *Suppose A and B are sets. Then $A \times B = B \times A$ iff either $A = \emptyset$, $B = \emptyset$, or $A = B$.*

Proof. ($\rightarrow$) Suppose $A \times B = B \times A$. If either $A = \emptyset$ or $B = \emptyset$, then there is nothing more to prove, so suppose $A \neq \emptyset$ and $B \neq \emptyset$. We will show that $A = B$. Let x be arbitrary, and suppose $x \in A$. Since $B \neq \emptyset$ we can choose some $y \in B$. Then $(x, y) \in A \times B = B \times A$, so $x \in B$.

Now suppose $x \in B$. Since $A \neq \emptyset$ we can choose some $z \in A$. Therefore $(x, z) \in B \times A = A \times B$, so $x \in A$. Thus $A = B$, as required.

($\leftarrow$) Suppose either $A = \emptyset$, $B = \emptyset$, or $A = B$.

Case 1. $A = \emptyset$. Then $A \times B = \emptyset \times B = \emptyset = B \times \emptyset = B \times A$.

Case 2. $B = \emptyset$. Similar to case 1.

Case 3. $A = B$. Then $A \times B = A \times A = B \times A$. □

Commentary. Of course, the statement to be proven is an iff statement, so we prove both directions separately. For the $\rightarrow$ direction, our goal is $A = \emptyset \vee B = \emptyset \vee A = B$, which could be written as $(A = \emptyset \vee B = \emptyset) \vee A = B$, so by one of our strategies for disjunctions from Chapter 3 we can assume $\neg(A = \varnothing \vee B = \varnothing)$ and prove $A = B$. Note that by one of DeMorgan's laws, $\neg(A = \varnothing \vee B = \varnothing)$ is equivalent to $A \neq \varnothing \wedge B \neq \varnothing$, so we treat this as two assumptions, $A \neq \varnothing$ and $B \neq \varnothing$. Of course we could also have proceeded differently, for example by assuming $A \neq B$ and $B \neq \varnothing$ and then proving $A = \varnothing$. But recall from the commentary on part 5 of Theorem 4.1.3 that $A = \varnothing$ and $B = \varnothing$ are actually negative statements, so because it is generally better to work with positive than negative statements, we're better off negating both of them to get the assumptions $A \neq \varnothing$ and $B \neq \varnothing$ and then proving the positive statement $A = B$. The assumptions $A \neq \varnothing$ and $B \neq \varnothing$ are existential statements, so they are used in the proof to justify the introduction of y and z. The proof that $A = B$ proceeds in the obvious way, by introducing an arbitrary object x and then proving $x \in A \leftrightarrow x \in B$.

For the $\leftarrow$ direction of the proof, we have $A = \varnothing \vee B = \varnothing \vee A = B$ as a given, so it is natural to use proof by cases. In each case, the goal is easy to prove.

This theorem is a better illustration of how mathematics is really done than most of the examples we've seen so far. Usually when you're trying to find the answer to a mathematical question you won't know in advance what the answer is going to be. You might be able to take a guess at the answer and you might have an idea for how the proof might go, but your guess might be wrong and your idea for the proof might be flawed. It is only by turning your idea into a formal proof, according to the rules in Chapter 3, that you can be sure your answer is right. Often in the course of trying to construct a formal proof you will discover a flaw in your reasoning, as we did earlier, and you may have to revise your ideas to overcome the flaw. The final theorem and proof are often the result of repeated mistakes and corrections. Of course, when mathematicians write up their theorems and proofs, they follow our rule that proofs are for justifying theorems, not for explaining thought processes, and so they don't describe all the mistakes they made. But just because mathematicians don't explain their mistakes in their proofs, you shouldn't be fooled into thinking they don't make any!

Now that we know how to use ordered pairs and Cartesian products to talk about assigning values to free variables, we're ready to define truth sets for statements containing two free variables.

Definition 4.1.5. Suppose $P(x, y)$ is a statement with two free variables in which x ranges over a set A and y ranges over another set B. Then $A \times B$ is the set of all assignments to x and y that make sense in the statement $P(x, y)$. The *truth set* of $P(x, y)$ is the subset of $A \times B$ consisting of those assignments that make the statement come out true. In other words,

$$\text{truth set of } P(x, y) = \{(a, b) \in A \times B \mid P(a, b)\}.$$

Example 4.1.6. What are the truth sets of the following statements?

1. "x has y children," where x ranges over the set P of all people and y ranges over $\mathbb{N}$.
2. "x is located in y," where x ranges over the set C of all cities and y ranges over the set N of all countries.
3. "$y = 2x - 3$," where x and y range over $\mathbb{R}$.

Solutions

1. $\{(p, n) \in P \times \mathbb{N} \mid \text{the person } p \text{ has } n \text{ children}\} = \{(\text{Prince Charles}, 2), \ldots\}$.
2. $\{(c, n) \in C \times N \mid \text{the city } c \text{ is located in the country } n\} = \{(\text{New York, United States}), (\text{Tokyo, Japan}), (\text{Paris, France}), \ldots\}$.
3. $\{(x, y) \in \mathbb{R} \times \mathbb{R} \mid y = 2x - 3\} = \{(0, -3), (1, -1), (2, 1), \ldots\}$. You are probably already familiar with the fact that the ordered pairs in this set are the coordinates of points in the plane that lie along a certain straight line, called the *graph* of the equation $y = 2x - 3$. Thus, you can think of the graph of the equation as a picture of its truth set!

Many of the facts about truth sets for statements with one free variable that we discussed in Chapter 1 carry over to truth sets for statements with two free variables. For example, suppose T is the truth set of a statement $P(x, y)$, where x ranges over some set A and y ranges over B. Then for any $a \in A$ and $b \in B$ the statement $(a, b) \in T$ means the same thing as $P(a, b)$. Also, if $P(x, y)$ is true for every $x \in A$ and $y \in B$, then $T = A \times B$, and if $P(x, y)$ is false for every $x \in A$ and $y \in B$, then $T = \varnothing$. If S is the truth set of another statement $Q(x, y)$, then the truth set of the statement $P(x, y) \wedge Q(x, y)$ is $T \cap S$, and the truth set of $P(x, y) \vee Q(x, y)$ is $T \cup S$.

Although we'll be concentrating on ordered pairs for the rest of this chapter, it is possible to work with ordered triples, ordered quadruples, and so on. These might be used to talk about truth sets for statements containing three or more free variables. For example, let $L(x, y, z)$ be the statement "x has lived in y for z years," where x ranges over the set P of all people, y ranges over the set C of all cities, and z ranges over $\mathbb{N}$. Then the assignments of values to the free variables that make sense in this statement would be ordered triples (p, c, n), where p is a person, c is a city, and n is a natural number. The set of all such ordered triples would be written $P \times C \times \mathbb{N}$, and the truth set of the statement

$L(x, y, z)$ would be the set $\{(p, c, n) \in P \times C \times \mathbb{N} \mid$ the person p has lived in the city c for n years$\}$.

Exercises

*1. What are the truth sets of the following statements? List a few elements of each truth set.
   (a) "x is a parent of y," where x and y both range over the set P of all people.
   (b) "There is someone who lives in x and attends y," where x ranges over the set C of all cities and y ranges over the set U of all universities.

2. What are the truth sets of the following statements? List a few elements of each truth set.
   (a) "x lives in y," where x ranges over the set P of all people and y ranges over the set C of all cities.
   (b) "The population of x is y," where x ranges over the set C of all cities and y ranges over $\mathbb{N}$.

3. The truth sets of the following statements are subsets of $\mathbb{R}^2$. List a few elements of each truth set. Draw a picture showing all the points in the plane whose coordinates are in the truth set.
   (a) $y = x^2 - x - 2$.
   (b) $y < x$.
   (c) Either $y = x^2 - x - 2$ or $y = 3x - 2$.
   (d) $y < x$, and either $y = x^2 - x - 2$ or $y = 3x - 2$.

*4. Let $A = \{1, 2, 3\}$, $B = \{1, 4\}$, $C = \{3, 4\}$, and $D = \{5\}$. Compute all the sets mentioned in Theorem 4.1.3 and verify that all parts of the theorem are true.

5. Prove parts 2 and 3 of Theorem 4.1.3.

*6. What's wrong with the following proof that for any sets A, B, C, and D, $(A \cup C) \times (B \cup D) \subseteq (A \times B) \cup (C \times D)$? (Note that this is the reverse of the inclusion in part 4 of Theorem 4.1.3.)

   Proof. Suppose $(x, y) \in (A \cup C) \times (B \cup D)$. Then $x \in A \cup C$ and $y \in B \cup D$, so either $x \in A$ or $x \in C$, and either $y \in B$ or $y \in D$. We consider these cases separately.

   Case 1. $x \in A$ and $y \in B$. Then $(x, y) \in A \times B$.

   Case 2. $x \in C$ and $y \in D$. Then $(x, y) \in C \times D$.

   Thus, either $(x, y) \in A \times B$ or $(x, y) \in C \times D$, so $(x, y) \in (A \times B) \cup (C \times D)$. □

7. If A has m elements and B has n elements, how many elements does $A \times B$ have?

℔*8. Is it true that for any sets A, B, and C, $A \times (B \setminus C) = (A \times B) \setminus (A \times C)$? Give either a proof or a counterexample to justify your answer.

℔9. Prove that for any sets A, B, C, and D, $(A \times B) \setminus (C \times D) = [A \times (B \setminus D)] \cup [(A \setminus C) \times B]$.

℔10. Prove that for any sets A, B, C, and D, if $A \times B$ and $C \times D$ are disjoint, then either A and C are disjoint or B and D are disjoint.

11. Suppose $\{A_i \mid i \in I\}$ and $\{B_i \mid i \in I\}$ are indexed families of sets.
    (a) Prove that $\cup_{i \in I}(A_i \times B_i) \subseteq (\cup_{i \in I} A_i) \times (\cup_{i \in I} B_i)$.
    (b) For each $(i, j) \in I \times I$ let $C_{(i,j)} = A_i \times B_j$, and let $P = I \times I$. Prove that $\cup_{p \in P} C_p = (\cup_{i \in I} A_i) \times (\cup_{i \in I} B_i)$.

*12. This problem was suggested by Prof. Alan Taylor of Union College. Consider the following putative theorem.

Theorem? *For any sets A, B, C, and D, if $A \times B \subseteq C \times D$ then $A \subseteq C$ and $B \subseteq D$.*

Is the following proof correct? If so, what proof strategies does it use? If not, can it be fixed? Is the theorem correct?

Proof. Suppose $A \times B \subseteq C \times D$. Let a be an arbitrary element of A and let b be an arbitrary element of B. Then $(a, b) \in A \times B$, so since $A \times B \subseteq C \times D$, $(a, b) \in C \times D$. Therefore $a \in C$ and $b \in D$. Since a and b were arbitrary elements of A and B, respectively, this shows that $A \subseteq C$ and $B \subseteq D$. □

4.2. Relations

Suppose $P(x, y)$ is a statement with two free variables x and y. Often such a statement can be thought of as expressing a *relationship* between x and y. The truth set of the statement $P(x, y)$ is a set of ordered pairs that records when this relationship holds. In fact, it is often useful to think of any set of ordered pairs in this way, as a record of when some relationship holds. This is the motivation behind the following definition.

Definition 4.2.1. Suppose A and B are sets. Then a set $R \subseteq A \times B$ is called a *relation from A to B*.

If x ranges over A and y ranges over B, then clearly the truth set of any statement $P(x, y)$ will be a relation from A to B. However, note that

Definition 4.2.1 does not require that a set of ordered pairs be defined as the truth set of some statement for the set to be a relation. Although thinking about truth sets was the motivation for this definition, the definition says nothing explicitly about truth sets. According to the definition, *any* subset of $A \times B$ is to be called a relation from A to B.

Example 4.2.2. Here are some examples of relations from one set to another.

1. Let $A = \{1, 2, 3\}$, $B = \{3, 4, 5\}$, and $R = \{(1, 3), (1, 5), (3, 3)\}$. Then $R \subseteq A \times B$, so R is a relation from A to B.
2. Let $G = \{(x, y) \in \mathbb{R} \times \mathbb{R} \mid x > y\}$. Then G is a relation from $\mathbb{R}$ to $\mathbb{R}$.
3. Let $A = \{1, 2\}$ and $B = \mathscr{P}(A) = \{\varnothing, \{1\}, \{2\}, \{1, 2\}\}$. Let $E = \{(x, y) \in A \times B \mid x \in y\}$. Then E is a relation from A to B. In this case, $E = \{(1, \{1\}), (1, \{1, 2\}), (2, \{2\}), (2, \{1, 2\})\}$.

   For the next three examples, let S be the set of all students at your school, R the set of all dorm rooms, P the set of all professors, and C the set of all courses.
4. Let $L = \{(s, r) \in S \times R \mid \text{the student } s \text{ lives in the dorm room } r\}$. Then L is a relation from S to R.
5. Let $E = \{(s, c) \in S \times C \mid \text{the student } s \text{ is enrolled in the course } c\}$. Then E is a relation from S to C.
6. Let $T = \{(c, p) \in C \times P \mid \text{the course } c \text{ is taught by the professor } p\}$. Then T is a relation from C to P.

So far we have concentrated mostly on developing your proof-writing skills. Another important skill in mathematics is the ability to understand and apply new definitions. Here are the definitions for several new concepts involving relations. We'll soon give examples illustrating these concepts, but first see if you can understand the concepts based on their definitions.

Definition 4.2.3. Suppose R is a relation from A to B. Then the *domain* of R is the set

$$\text{Dom}(R) = \{a \in A \mid \exists b \in B((a, b) \in R)\}.$$

The *range* of R is the set

$$\text{Ran}(R) = \{b \in B \mid \exists a \in A((a, b) \in R)\}.$$

The *inverse* of R is the relation R^{-1} from B to A defined as follows:

$$R^{-1} = \{(b, a) \in B \times A \mid (a, b) \in R\}.$$

Finally, suppose R is a relation from A to B and S is a relation from B to C. Then the *composition* of S and R is the relation $S \circ R$ from A to C defined as follows:

$$S \circ R = \{(a, c) \in A \times C \mid \exists b \in B((a, b) \in R \text{ and } (b, c) \in S)\}.$$

Notice that we have assumed that the second coordinates of pairs in R and the first coordinates of pairs in S both come from the same set, B. If these sets were not the same, the composition $S \circ R$ would be undefined.

According to Definition 4.2.3, the domain of a relation from A to B is the set containing all the first coordinates of ordered pairs in the relation. This will in general be a subset of A, but it need not be all of A. For example, consider the relation L from part 4 of Example 4.2.2, which pairs up students with the dorm rooms in which they live. The domain of L would contain all students who appear as the first coordinate in some ordered pair in L – in other words, all students who live in some dorm room – but would not contain, for example, students who live in apartments off campus. Working it out more carefully from the definition as stated, we have

$$\begin{aligned}
\text{Dom}(L) &= \{s \in S \mid \exists r \in R((s, r) \in L)\} \\
&= \{s \in S \mid \exists r \in R \text{ (the student } s \text{ lives in the dorm room } r)\} \\
&= \{s \in S \mid \text{ the student } s \text{ lives in some dorm room}\}.
\end{aligned}$$

Similarly, the range of a relation is the set containing all the second coordinates of its ordered pairs. For example, the range of the relation L would be the set of all dorm rooms in which some student lives. Any dorm rooms that are unoccupied would not be in the range of L.

The inverse of a relation contains exactly the same ordered pairs as the original relation, but with the order of the coordinates of each pair reversed. Thus, in the case of the relation L, if Joe Smith lives in room 213 Davis Hall, then (Joe Smith, 213 Davis Hall) $\in L$ and (213 Davis Hall, Joe Smith) $\in L^{-1}$. In general, for any student s and dorm room r, we would have $(r, s) \in L^{-1}$ iff $(s, r) \in L$. For another example, consider the relation G from part 2 of Example 4.2.2. It contains all ordered pairs of real numbers (x, y) in which x is greater than y. We might call it the "greater-than" relation. Its inverse is

$$\begin{aligned}
G^{-1} &= \{(x, y) \in \mathbb{R} \times \mathbb{R} \mid (y, x) \in G\} \\
&= \{(x, y) \in \mathbb{R} \times \mathbb{R} \mid y > x\} \\
&= \{(x, y) \in \mathbb{R} \times \mathbb{R} \mid x < y\}.
\end{aligned}$$

In other words, the inverse of the greater-than relation is the less-than relation!

The most difficult concept introduced in Definition 4.2.3 is the concept of the composition of two relations. For an example of this concept, consider the relations E and T from parts 5 and 6 of Example 4.2.2. Recall that E is a relation from the set S of all students to the set C of all courses, and T is a relation from C to the set P of all professors. According to Definition 4.2.3, the composition $T \circ E$ will be the relation from S to P defined as follows:

$$\begin{aligned} T \circ E &= \{(s, p) \in S \times P \mid \exists c \in C((s, c) \in E \text{ and } (c, p) \in T)\} \\ &= \{(s, p) \in S \times P \mid \exists c \in C(\text{the student } s \text{ is enrolled in the course } c \\ &\qquad \text{and the course } c \text{ is taught by the professor } p)\} \\ &= \{(s, p) \in S \times P \mid \text{ the student } s \text{ is enrolled in some course} \\ &\qquad \text{taught by the professor } p\}. \end{aligned}$$

Thus, if Joe Smith is enrolled in Biology 12 and Biology 12 is taught by Professor Evans, then (Joe Smith, Biology 12) $\in E$ and (Biology 12, Professor Evans) $\in T$, and therefore (Joe Smith, Professor Evans) $\in T \circ E$. In general, if s is some particular student and p is a particular professor, then $(s, p) \in T \circ E$ iff there is some course c such that $(s, c) \in E$ and $(c, p) \in T$. This notation may seem backward at first. If $(s, c) \in E$ and $(c, p) \in T$, then you might be tempted to write $(s, p) \in E \circ T$, but according to our definition, the proper notation is $(s, p) \in T \circ E$. In fact, $E \circ T$ is undefined, because the second coordinates of ordered pairs in T and the first coordinates of pairs in E do not come from the same set. The reason we've chosen to write compositions of relations in this way will become clear in Chapter 5. For the moment, you'll just have to be careful about this notational detail when working with compositions of relations.

Example 4.2.4. Let S, R, C, and P be the sets of students, dorm rooms, courses, and professors at your school, as before, and let L, E, and T be the relations defined in parts 4–6 of Example 4.2.2. Describe the following relations.

1. E^{-1}.
2. $E \circ L^{-1}$.
3. $E^{-1} \circ E$.
4. $E \circ E^{-1}$.
5. $T \circ (E \circ L^{-1})$.
6. $(T \circ E) \circ L^{-1}$.

Solutions

1. $E^{-1} = \{(c, s) \in C \times S \mid (s, c) \in E\} = \{(c, s) \in C \times S \mid$ the student s is enrolled in the course $c\}$. For example, if Joe Smith is enrolled in Biology 12, then (Joe Smith, Biology 12) $\in E$ and (Biology 12, Joe Smith) $\in E^{-1}$.
2. Because L^{-1} is a relation from R to S and E is a relation from S to C, $E \circ L^{-1}$ will be the relation from R to C defined as follows.

$$\begin{aligned} E \circ L^{-1} &= \{(r, c) \in R \times C \mid \exists s \in S((r, s) \in L^{-1} \text{ and } (s, c) \in E)\} \\ &= \{(r, c) \in R \times C \mid \exists s \in S((s, r) \in L \text{ and } (s, c) \in E)\} \\ &= \{(r, c) \in R \times C \mid \exists s \in S(\text{the student } s \text{ lives in the dorm} \\ &\qquad \text{room } r \text{ and is enrolled in the course } c)\} \\ &= \{(r, c) \in R \times C \mid \text{ some student who lives in the room } r \\ &\qquad \text{is enrolled in the course } c\}. \end{aligned}$$

   Returning to our favorite student Joe Smith, who is enrolled in Biology 12 and lives in room 213 Davis Hall, we have (213 Davis Hall, Joe Smith) $\in L^{-1}$ and (Joe Smith, Biology 12) $\in E$, and therefore (213 Davis Hall, Biology 12) $\in E \circ L^{-1}$.
3. Because E is a relation from S to C and E^{-1} is a relation from C to S, $E^{-1} \circ E$ is the relation from S to S defined as follows.

$$\begin{aligned} E^{-1} \circ E &= \{(s, t) \in S \times S \mid \exists c \in C((s, c) \in E \text{ and } (c, t) \in E^{-1})\} \\ &= \{(s, t) \in S \times S \mid \exists c \in C(\text{the student } s \text{ is enrolled in the} \\ &\qquad \text{course } c, \text{ and so is the student } t)\} \\ &= \{(s, t) \in S \times S \mid \text{ there is some course that the students } s \\ &\qquad \text{and } t \text{ are both enrolled in}\}. \end{aligned}$$

   (Note that an arbitrary element of $S \times S$ is written (s, t), not (s, s), because we don't want to assume that the two coordinates are equal.)
4. This is not the same as the last example! Because E^{-1} is a relation from C to S and E is a relation from S to C, $E \circ E^{-1}$ is a relation from C to C. It is defined as follows.

$$\begin{aligned} E \circ E^{-1} &= \{(c, d) \in C \times C \mid \exists s \in S((c, s) \in E^{-1} \text{ and } (s, d) \in E)\} \\ &= \{(c, d) \in C \times C \mid \exists s \in S(\text{the student } s \text{ is enrolled in the} \\ &\qquad \text{course } c, \text{ and he is also enrolled in the course } d)\} \\ &= \{(c, d) \in C \times C \mid \text{ there is some student who is enrolled in} \\ &\qquad \text{both of the courses } c \text{ and } d\}. \end{aligned}$$

5. We saw in part 2 that $E \circ L^{-1}$ is a relation from R to C, and T is a relation from C to P, so $T \circ (E \circ L^{-1})$ is the relation from R to P defined as follows.

$$
\begin{aligned}
T \circ (E \circ L^{-1}) &= \{(r, p) \in R \times P \mid \exists c \in C((r, c) \in E \circ L^{-1} \text{ and } \\
&\qquad (c, p) \in T)\} \\
&= \{(r, p) \in R \times P \mid \exists c \in C \text{ (some student who lives} \\
&\qquad \text{in the room } r \text{ is enrolled in the course } c, \text{ and } c \\
&\qquad \text{is taught by the professor } p)\} \\
&= \{(r, p) \in R \times P \mid \text{ some student who lives in} \\
&\qquad \text{the room } r \text{ is enrolled in some course taught by} \\
&\qquad \text{the professor } p\}.
\end{aligned}
$$

6.
$$
\begin{aligned}
(T \circ E) \circ L^{-1} &= \{(r, p) \in R \times P \mid \exists s \in S((r, s) \in L^{-1} \text{ and } \\
&\qquad (s, p) \in T \circ E)\} \\
&= \{(r, p) \in R \times P \mid \exists s \in S(\text{the student } s \text{ lives in the} \\
&\qquad \text{room } r, \text{ and is enrolled in some course taught} \\
&\qquad \text{by the professor } p)\} \\
&= \{(r, p) \in R \times P \mid \text{ some student who lives in} \\
&\qquad \text{the room } r \text{ is enrolled in some course taught by} \\
&\qquad \text{the professor } p\}.
\end{aligned}
$$

Notice that our answers for parts 3 and 4 of Example 4.2.4 were different. so composition of relations is not commutative. However, our answers for parts 5 and 6 turned out to be the same. Is this a coincidence, or is it true in general that composition of relations is associative? Often, looking at examples of a new concept will suggest general rules that might apply to it. Although one counterexample is enough to show that a rule is incorrect, we should never accept a rule as correct without a proof. The next theorem summarizes some of the basic properties of the new concepts we have introduced.

Theorem 4.2.5. *Suppose R is a relation from A to B, S is a relation from B to C, and T is a relation from C to D. Then:*

1. $(R^{-1})^{-1} = R$.
2. $\text{Dom}(R^{-1}) = \text{Ran}(R)$.
3. $\text{Ran}(R^{-1}) = \text{Dom}(R)$.
4. $T \circ (S \circ R) = (T \circ S) \circ R$.
5. $(S \circ R)^{-1} = R^{-1} \circ S^{-1}$.

Proof. We will prove 1, 2, and half of 4, and leave the rest as exercises. (See exercise 6.)

1. First of all, note that R^{-1} is a relation from B to A, so $(R^{-1})^{-1}$ is a relation from A to B, just like R. To see that $(R^{-1})^{-1} = R$, let (a, b) be an arbitrary ordered pair in $A \times B$. Then

$$(a, b) \in (R^{-1})^{-1} \text{ iff } (b, a) \in R^{-1} \text{ iff } (a, b) \in R.$$

2. First note that $\text{Dom}(R^{-1})$ and $\text{Ran}(R)$ are both subsets of B. Now let b be an arbitrary element of B. Then

$$\begin{aligned} b \in \text{Dom}(R^{-1}) & \text{ iff } \exists a \in A((b, a) \in R^{-1}) \\ & \text{ iff } \exists a \in A((a, b) \in R) \text{ iff } b \in \text{Ran}(R). \end{aligned}$$

4. Clearly $T \circ (S \circ R)$ and $(T \circ S) \circ R$ are both relations from A to D. Let (a, d) be an arbitrary element of $A \times D$.

   First, suppose $(a, d) \in T \circ (S \circ R)$. By the definition of composition, this means that we can choose some $c \in C$ such that $(a, c) \in S \circ R$ and $(c, d) \in T$. Since $(a, c) \in S \circ R$, we can again use the definition of composition and choose some $b \in B$ such that $(a, b) \in R$ and $(b, c) \in S$. Now since $(b, c) \in S$ and $(c, d) \in T$, we can conclude that $(b, d) \in T \circ S$. Similarly, since $(a, b) \in R$ and $(b, d) \in T \circ S$, it follows that $(a, d) \in (T \circ S) \circ R$.

   Now suppose $(a, d) \in (T \circ S) \circ R$. A similar argument, which is left to the reader, shows that $(a, d) \in T \circ (S \circ R)$. Thus, $T \circ (S \circ R) = (T \circ S) \circ R$. □

Commentary. Statement 1 means $\forall p(p \in (R^{-1})^{-1} \leftrightarrow p \in R)$, so the proof should proceed by introducing an arbitrary object p and then proving $p \in (R^{-1})^{-1} \leftrightarrow p \in R$. But because R and $(R^{-1})^{-1}$ are both relations from A to B, we could think of the universe over which p ranges as being $A \times B$, so p must be an ordered pair. Thus, in the preceding proof we've written it as an ordered pair (a, b) from the start. The proof of the biconditional statement $(a, b) \in (R^{-1})^{-1} \leftrightarrow (a, b) \in R$ uses the method, introduced in Example 3.4.4, of stringing together a sequence of equivalences.

The proofs of statements 2 and 4 are similar, except that the biconditional proof for statement 4 cannot easily be done by stringing together equivalences, so we prove the two directions separately. Only one direction was proven. The key to this proof is to recognize that the given $(a, d) \in T \circ (S \circ R)$ is an existential statement, since it means $\exists c \in C((a, c) \in S \circ R \text{ and } (c, d) \in T)$, so we should introduce a new variable c into the proof to stand for some element of C such that $(a, c) \in S \circ R$ and $(c, d) \in T$. Similarly, $(a, c) \in S \circ R$ is an existential statement, so it suggests introducing the variable b. Once these new variables have been introduced, it is easy to prove the goal $(a, d) \in (T \circ S) \circ R$.

Statement 5 of Theorem 4.2.5 perhaps deserves some comment. First of all, notice that the right-hand side of the equation is $R^{-1} \circ S^{-1}$, *not* $S^{-1} \circ R^{-1}$; the order of the relations has been reversed. You are asked to prove statement 5 in exercise 6, but it might be worthwhile to try an example first. We've already seen that, for the relations E and T from parts 5 and 6 of Example 4.2.2,

$$T \circ E = \{(s, p) \in S \times P \mid \text{the student } s \text{ is enrolled in some course taught by the professor } p\}.$$

It follows that

$$(T \circ E)^{-1} = \{(p, s) \in P \times S \mid \text{the student } s \text{ is enrolled in some course taught by the professor } p\}.$$

To compute $E^{-1} \circ T^{-1}$, first note that T^{-1} is a relation from P to C and E^{-1} is a relation from C to S, so $E^{-1} \circ T^{-1}$ is a relation from P to S. Now, applying the definition of composition, we get

$$\begin{aligned}
E^{-1} \circ T^{-1} &= \{(p, s) \in P \times S \mid \exists c \in C((p, c) \in T^{-1} \text{ and } (c, s) \in E^{-1})\} \\
&= \{(p, s) \in P \times S \mid \exists c \in C((c, p) \in T \text{ and } (s, c) \in E)\} \\
&= \{(p, s) \in P \times S \mid \exists c \in C(\text{the course } c \text{ is taught by the professor } p \text{ and the student } s \text{ is enrolled in the course } c)\} \\
&= \{(p, s) \in P \times S \mid \text{the student } s \text{ is enrolled in some course taught by the professor } p\}.
\end{aligned}$$

Thus, $(T \circ E)^{-1} = E^{-1} \circ T^{-1}$.

Exercises

*1. Find the domains and ranges of the following relations.
   (a) $\{(p, q) \in P \times P \mid \text{the person } p \text{ is a parent of the person } q\}$, where P is the set of all living people.
   (b) $\{(x, y) \in \mathbb{R}^2 \mid y > x^2\}$.
2. Find the domains and ranges of the following relations.
   (a) $\{(p, q) \in P \times P \mid \text{the person } p \text{ is a brother of the person } q\}$, where P is the set of all living people.
   (b) $\{(x, y) \in \mathbb{R}^2 \mid y^2 = 1 - 2/(x^2 + 1)\}$.

3. Let L and E be the relations defined in parts 4 and 5 of Example 4.2.2. Describe the following relations:
   (a) $L^{-1} \circ L$.
   (b) $E \circ (L^{-1} \circ L)$.

*4. Suppose that $A = \{1, 2, 3\}$, $B = \{4, 5, 6\}$, $R = \{(1, 4), (1, 5), (2, 5), (3, 6)\}$, and $S = \{(4, 5), (4, 6), (5, 4), (6, 6)\}$. Note that R is a relation from A to B and S is a relation from B to B. Find the following relations:
   (a) $S \circ R$.
   (b) $S \circ S^{-1}$.

5. Suppose that $A = \{1, 2, 3\}$, $B = \{4, 5\}$, $C = \{6, 7, 8\}$, $R = \{(1, 7), (3, 6), (3, 7)\}$, and $S = \{(4, 7), (4, 8), (5, 6)\}$. Note that R is a relation from A to C and S is a relation from B to C. Find the following relations:
   (a) $S^{-1} \circ R$.
   (b) $R^{-1} \circ S$.

6. (a) Prove part 3 of Theorem 4.2.5 by imitating the proof of part 2 in the text.
   (b) Give an alternate proof of part 3 of Theorem 4.2.5 by showing that it follows from parts 1 and 2.
   (c) Complete the proof of part 4 of Theorem 4.2.5.
   (d) Prove part 5 of Theorem 4.2.5.

*7. Let $E = \{(p, q) \in P \times P \mid$ the person p is an enemy of the person $q\}$, and $F = \{(p, q) \in P \times P \mid$ the person p is a friend of the person $q\}$, where P is the set of all people. What does the saying "an enemy of one's enemy is one's friend" mean about the relations E and F?

8. Suppose R is a relation from A to B and S is a relation from B to C.
   (a) Prove that $\text{Dom}(S \circ R) \subseteq \text{Dom}(R)$.
   (b) Prove that if $\text{Ran}(R) \subseteq \text{Dom}(S)$ then $\text{Dom}(S \circ R) = \text{Dom}(R)$.
   (c) Formulate and prove similar theorems about $\text{Ran}(S \circ R)$.

9. Suppose R and S are relations from A to B. Must the following statements be true? Justify your answers with proofs or counterexamples.
   (a) $R \subseteq \text{Dom}(R) \times \text{Ran}(R)$.
   (b) If $R \subseteq S$ then $R^{-1} \subseteq S^{-1}$.
   (c) $(R \cup S)^{-1} = R^{-1} \cup S^{-1}$.

*10. Suppose R is a relation from A to B and S is a relation from B to C. Prove that $S \circ R = \varnothing$ iff Ran (R) and Dom (S) are disjoint.

$^{P}_{D}$11. Suppose R is a relation from A to B and S and T are relations from B to C.
   (a) Prove that $(S \circ R) \setminus (T \circ R) \subseteq (S \setminus T) \circ R$.

(b) What's wrong with the following proof that $(S \setminus T) \circ R \subseteq (S \circ R) \setminus (T \circ R)$?

Proof. Suppose $(a, c) \in (S \setminus T) \circ R$. Then we can choose some $b \in B$ such that $(a, b) \in R$ and $(b, c) \in S \setminus T$, so $(b, c) \in S$ and $(b, c) \notin T$. Since $(a, b) \in R$ and $(b, c) \in S$, $(a, c) \in S \circ R$. Similarly, since $(a, b) \in R$ and $(b, c) \notin T$, $(a, c) \notin T \circ R$. Therefore $(a, c) \in (S \circ R) \setminus (T \circ R)$. Since (a, c) was arbitrary, this shows that $(S \setminus T) \circ R \subseteq (S \circ R) \setminus (T \circ R)$. □

(c) Must it be true that $(S \setminus T) \circ R \subseteq (S \circ R) \setminus (T \circ R)$? Justify your answer with either a proof or a counterexample.

12. Suppose R is a relation from A to B and S and T are relations from B to C. Must the following statements be true? Justify your answers with proofs or counterexamples.
(a) If $S \subseteq T$ then $S \circ R \subseteq T \circ R$.
(b) $(S \cap T) \circ R \subseteq (S \circ R) \cap (T \circ R)$.
(c) $(S \cap T) \circ R = (S \circ R) \cap (T \circ R)$.
(d) $(S \cup T) \circ R = (S \circ R) \cup (T \circ R)$.

4.3. More About Relations

Although we have defined relations to be sets of ordered pairs, it is sometimes useful to be able to think about them in other ways. Often even a small change in notation can help us see things differently. One alternative notation that mathematicians sometimes use with relations is motivated by the fact that in mathematics we often express a relationship between two objects x and y by putting some symbol between them. For example, the notations $x = y$, $x < y$, $x \in y$, and $x \subseteq y$ express four important mathematical relationships between x and y. Imitating these notations, if R is a relation from A to B, $x \in A$, and $y \in B$, mathematicians sometimes write xRy to mean $(x, y) \in R$.

For example, if L is the relation defined in part 4 of Example 4.2.2, then for any student s and dorm room r, sLr means $(s, r) \in L$, or in other words, the student s lives in the dorm room r. Similarly, if E and T are the relations defined in parts 5 and 6 of Example 4.2.2, then sEc means that the student s is enrolled in the course c, and cTp means that the course c is taught by the professor p. The definition of composition of relations could have been stated by saying that if R is a relation from A to B and S is a relation from B to C, then $S \circ R = \{(a, c) \in A \times C \mid \exists b \in B(aRb \text{ and } bSc)\}$.

Another way to think about relations is to draw pictures of them. Figure 1 shows a picture of the relation $R = \{(1,3), (1,5), (3,3)\}$ from part 1 of Example 4.2.2. Recall that this was a relation from the set $A = \{1, 2, 3\}$ to the set $B = \{3, 4, 5\}$. In the figure, each of these sets is represented by an oval, with the elements of the set represented by dots inside the oval. Each ordered pair $(a, b) \in R$ is represented by an arrow from the dot representing a to the dot representing b. For example, there is an arrow from the dot inside A labeled 1 to the dot inside B labeled 5 because the ordered pair $(1, 5)$ is an element of R.

In general, any relation R from a set A to a set B can be represented by such a picture. The dots representing the elements of A and B in such a picture are called *vertices*, and the arrows representing the ordered pairs in R are called *edges*. It doesn't matter exactly how the vertices representing elements of A and B are arranged on the page; what's important is that the edges correspond precisely to the ordered pairs in R. Drawing these pictures may help you understand the concepts discussed in the last section. For example, you should be able to convince yourself that you could find the domain of R by locating those vertices in A that have edges pointing away from them. Similarly, the range of R would consist of those elements of B whose vertices have edges pointing toward them. For the relation R shown in Figure 1, we have $\text{Dom}(R) = \{1, 3\}$ and $\text{Ran}(R) = \{3, 5\}$. A picture of R^{-1} would look just like a picture of R but with the directions of all the arrows reversed.

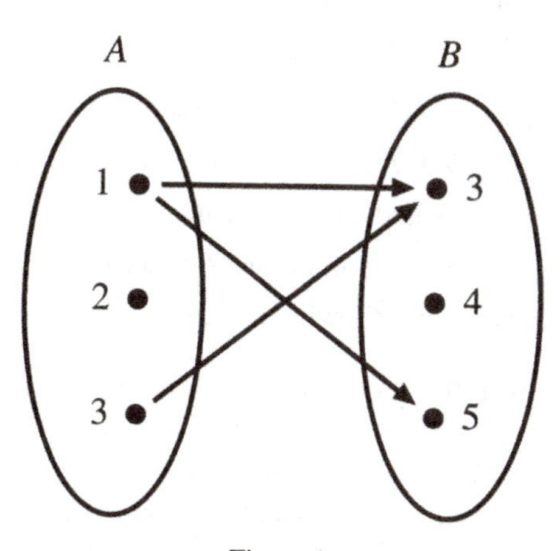

Figure 1

Pictures illustrating the composition of two relations are a little harder to understand. For example, consider again the relations E and T from parts 5 and 6 of Example 4.2.2. Figure 2 shows what part of both relations might look like. (The complete picture might be quite large if there are many students, courses, and professors at your school.) We can see in this picture that, for example, Joe Smith is taking Biology 12 and Math 21, that Biology 12 is taught by Prof. Evans, and that Math 21 is taught by Prof. Andrews. Thus, applying the definition of composition, we can see that the pairs (Joe Smith, Prof. Evans) and (Joe Smith, Prof. Andrews) are both elements of the relation $T \circ E$.

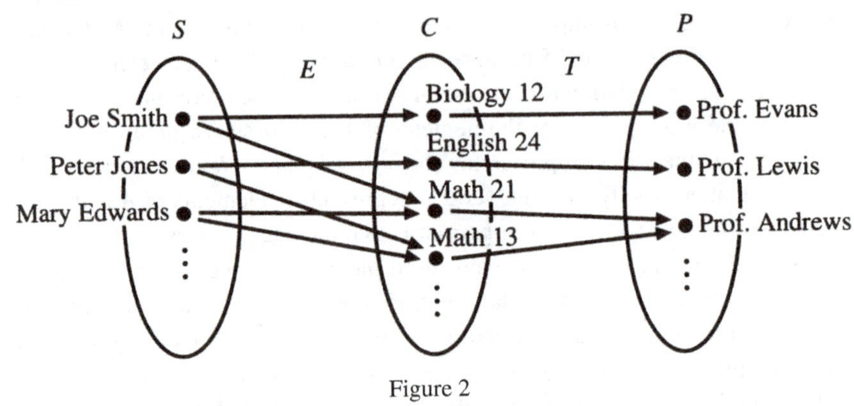

Figure 2

To see more clearly how the composition $T \circ E$ is represented in this picture, first note that for any student s, course c, and professor p, there is an arrow from s to c iff sEc, and there is an arrow from c to p iff cTp. Thus, according to the definition of composition,

$$\begin{aligned} T \circ E &= \{(s, p) \in S \times P \mid \exists c \in C(sEc \text{ and } cTp)\} \\ &= \{(s, p) \in S \times P \mid \exists c \in C(\text{in Figure 2, there is an arrow} \\ &\qquad \text{from } s \text{ to } c \text{ and an arrow from } c \text{ to } p)\} \\ &= \{(s, p) \in S \times P \mid \text{ in Figure 2, you can get from } s \text{ to } p \text{ in} \\ &\qquad \text{two steps by following the arrows}\}. \end{aligned}$$

For example, starting at the vertex labeled Mary Edwards, we can get to Prof. Andrews in two steps (going by way of either Math 21 or Math 13), so we can conclude that (Mary Edwards, Prof. Andrews) $\in T \circ E$.

In some situations we draw pictures of relations in a slightly different way. For example, if A is a set and $R \subseteq A \times A$, then according to Definition 4.2.1, R would be called a relation from A to A. Such a relation is also sometimes called a *relation on A* (or a *binary relation on A*). Relations of this type come up often in mathematics; in fact, we have already seen a few of them. For example, we described the relation G in part 2 of Example 4.2.2 as a relation from $\mathbb{R}$ to $\mathbb{R}$, but in our new terminology we could call it a relation (or a binary relation) on

$\mathbb{R}$. The relation $E^{-1} \circ E$ from Example 4.2.4 was a relation on the set S, and $E \circ E^{-1}$ was a relation on C.

Example 4.3.1. Here are some more examples of relations on sets.

1. Let $A = \{1, 2\}$ and $B = \mathscr{P}(A) = \{\varnothing, \{1\}, \{2\}, \{1, 2\}\}$ as in part 3 of Example 4.2.2. Let $S = \{(x, y) \in B \times B \mid x \subseteq y\} = \{(\varnothing, \varnothing), (\varnothing, \{1\}), (\varnothing, \{2\}), (\varnothing, \{1, 2\}), (\{1\}, \{1\}), (\{1\}, \{1, 2\}), (\{2\}, \{2\}), (\{2\}, \{1, 2\}), (\{1, 2\}, \{1, 2\})\}$. Then S is a relation on B.
2. Suppose A is a set. Let $i_A = \{(x, y) \in A \times A \mid x = y\}$. Then i_A is a relation on A. (It is sometimes called the *identity relation* on A.) For example, if $A = \{1, 2, 3\}$, then $i_A = \{(1, 1), (2, 2), (3, 3)\}$. Note that i_A could also be defined by writing $i_A = \{(x, x) \mid x \in A\}$.
3. For each positive real number r, let $D_r = \{(x, y) \in \mathbb{R} \times \mathbb{R} \mid x$ and y differ by less than r, or in other words $|x - y| < r\}$. Then D_r is a relation on $\mathbb{R}$.

Suppose R is a relation on a set A. If we used the method described earlier to draw a picture of R, then we would have to draw two copies of the set A and then draw edges from one copy of A to the other to represent the ordered pairs in R. An easier way to draw the picture would be to draw just one copy of A and then connect the vertices representing the elements of A with edges to represent the ordered pairs in R. For example, Figure 3 shows a picture of the relation S from part 1 of Example 4.3.1. Pictures like the one in Figure 3 are called *directed graphs*.

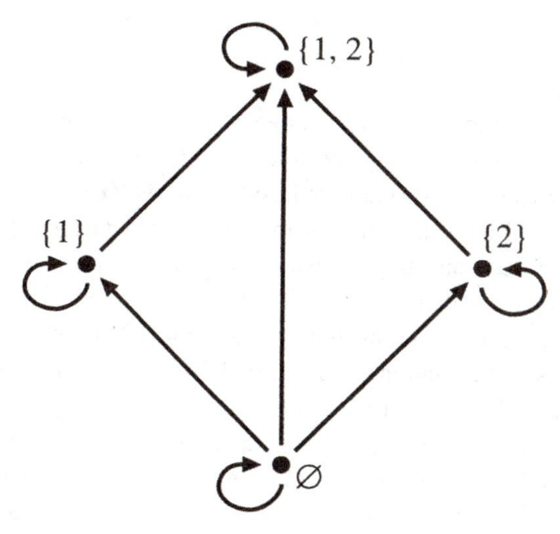

Figure 3

Note that in this directed graph there is an edge from $\varnothing$ to itself, because $(\varnothing, \varnothing) \in S$. Edges such as this one that go from a vertex to itself are called *loops*. In fact, in Figure 3 there is a loop at every vertex, because S has the property that $\forall x \in B((x, x) \in S)$. We describe this situation by saying that S is *reflexive*.

Definition 4.3.2. Suppose R is a relation on A.

1. R is said to be *reflexive on* A (or just *reflexive*, if A is clear from context) if $\forall x \in A(xRx)$, or in other words $\forall x \in A((x, x) \in R)$.
2. R is *symmetric* if $\forall x \in A \forall y \in A(xRy \rightarrow yRx)$.
3. R is *transitive* if $\forall x \in A \forall y \in A \forall z \in A((xRy \wedge yRz) \rightarrow xRz)$.

As we saw in Example 4.3.1, if R is reflexive on A, then the directed graph representing R will have loops at all vertices. If R is symmetric, then whenever there is an edge from x to y, there will also be an edge from y to x. If x and y are distinct, it follows that there will be two edges connecting x and y, one pointing in each direction. Thus, if R is symmetric, then all edges except loops will come in such pairs. If R is transitive, then whenever there is an edge from x to y and y to z, there is also an edge from x to z.

Example 4.3.3. Is the relation G from part 2 of Example 4.2.2 reflexive? Is it symmetric? Transitive? Are the relations in Example 4.3.1 reflexive, symmetric, or transitive?

Solution

Recall that the relation G from Example 4.2.2 is a relation on $\mathbb{R}$ and that for any real numbers x and y, xGy means $x > y$. Thus, to say that G is reflexive would mean that $\forall x \in \mathbb{R}(xGx)$, or in other words $\forall x \in \mathbb{R}(x > x)$, and this is clearly false. To say that G is symmetric would mean that $\forall x \in \mathbb{R} \forall y \in \mathbb{R}(x > y \rightarrow y > x)$, and this is also clearly false. Finally, to say that G is transitive would mean that $\forall x \in \mathbb{R} \forall y \in \mathbb{R} \forall z \in \mathbb{R}((x > y \wedge y > z) \rightarrow x > z)$, and this is true. Thus, G is transitive, but not reflexive or symmetric.

The analysis of the relations in Example 4.3.1 is similar. For the relation S in part 1 we use the fact that for any x and y in B, xSy means $x \subseteq y$. As we have already observed, S is reflexive, since $\forall x \in B(x \subseteq x)$, but it is not true that $\forall x \in B \forall y \in B(x \subseteq y \rightarrow y \subseteq x)$. For example, $\{1\} \subseteq \{1, 2\}$, but $\{1, 2\} \not\subseteq \{1\}$. You can see this in Figure 3 by noting that there is an edge from $\{1\}$ to $\{1, 2\}$ but not from $\{1, 2\}$ to $\{1\}$. Thus, S is not symmetric. S is transitive, because the statement $\forall x \in B \forall y \in B \forall z \in B((x \subseteq y \wedge y \subseteq z) \rightarrow x \subseteq z)$ is true.

For any set A the identity relation i_A will be reflexive, symmetric, and transitive, because the statements $\forall x \in A(x = x)$, $\forall x \in A \forall y \in A(x = y \rightarrow y = x)$, and $\forall x \in A \forall y \in A \forall z \in A((x = y \wedge y = z) \rightarrow x = z)$ are all clearly true. Finally, suppose r is a positive real number and consider the relation D_r. For any real number x, $|x - x| = 0 < r$, so $(x, x) \in D_r$. Thus, D_r is reflexive. Also, for any real numbers x and y, $|x - y| = |y - x|$, so if $|x - y| < r$ then $|y - x| < r$. Therefore, if $(x, y) \in D_r$ then $(y, x) \in D_r$, so D_r is symmetric. But D_r is not transitive. To see why, let x be any real number. Let $y = x + 2r/3$ and $z = y + 2r/3 = x + 4r/3$. Then $|x - y| = 2r/3 < r$ and $|y - z| = 2r/3 < r$, but $|x - z| = 4r/3 > r$. Thus, $(x, y) \in D_r$ and $(y, z) \in D_r$, but $(x, z) \notin D_r$.

Perhaps you've already guessed that the properties of relations defined in Definition 4.3.2 are related to the operations defined in Definition 4.2.3. To say that a relation R is symmetric involves reversing the roles of two variables in a way that may remind you of the definition of R^{-1}. The definition of transitivity of a relation involves stringing together two ordered pairs, just as the definition of composition of relations does. The following theorem spells these connections out more carefully.

Theorem 4.3.4. *Suppose R is a relation on a set A.*

1. *R is reflexive iff $i_A \subseteq R$, where as before i_A is the identity relation on A.*
2. *R is symmetric iff $R = R^{-1}$.*
3. *R is transitive iff $R \circ R \subseteq R$.*

Proof. We will prove 2 and leave the proofs of 1 and 3 as exercises (see exercises 7 and 8).

2. ($\rightarrow$) Suppose R is symmetric. Let (x, y) be an arbitrary element of R. Then xRy, so since R is symmetric, yRx. Thus, $(y, x) \in R$, so by the definition of R^{-1}, $(x, y) \in R^{-1}$. Since (x, y) was arbitrary, it follows that $R \subseteq R^{-1}$.

Now suppose $(x, y) \in R^{-1}$. Then $(y, x) \in R$, so since R is symmetric, $(x, y) \in R$. Thus, $R^{-1} \subseteq R$, so $R = R^{-1}$.

($\leftarrow$) Suppose $R = R^{-1}$, and let x and y be arbitrary elements of A. Suppose xRy. Then $(x, y) \in R$, so since $R = R^{-1}$, $(x, y) \in R^{-1}$. By the definition of R^{-1} this means $(y, x) \in R$, so yRx. Thus, $\forall x \in A \forall y \in A(xRy \rightarrow yRx)$, so R is symmetric. □

Commentary. This proof is fairly straightforward. The statement to be proven is an iff statement, so we prove both directions separately. In the $\rightarrow$ half we must prove that $R = R^{-1}$, and this is done by proving both $R \subseteq R^{-1}$ and $R^{-1} \subseteq R$. Each of these goals is proven by taking an arbitrary element of the first set and showing that it is in the second set. In the $\leftarrow$ half we must prove that R is symmetric, which means $\forall x \in A \forall y \in A(xRy \to yRx)$. We use the obvious strategy of letting x and y be arbitrary elements of A, assuming xRy, and proving yRx.

Exercises

*1. Let $L = \{a, b, c, d, e\}$ and $W = \{\text{bad, bed, cab}\}$. Let $R = \{(l, w) \in L \times W \mid \text{the letter } l \text{ occurs in the word } w\}$. Draw a diagram (like the one in Figure 1) of R.

2. Let $A = \{\text{cat, dog, bird, rat}\}$, and let $R = \{(x, y) \in A \times A \mid \text{there is at least one letter that occurs in both of the words } x \text{ and } y\}$. Draw a directed graph (like the one in Figure 3) for the relation R. Is R reflexive? symmetric? transitive?

*3. Let $A = \{1, 2, 3, 4\}$. Draw a directed graph for the identity relation on A, i_A.

4. List the ordered pairs in the relations represented by the following directed graphs. Determine whether each relation is reflexive, symmetric, or transitive.

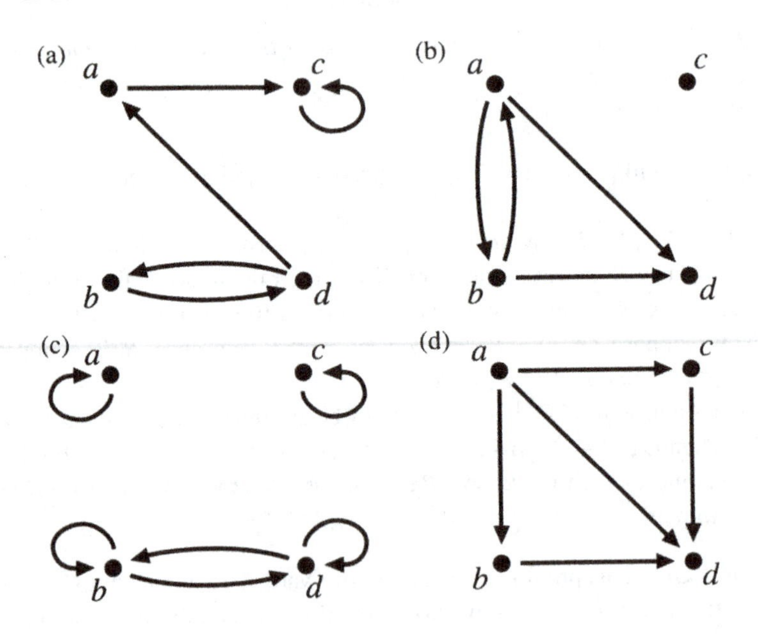

*5. The following diagram shows two relations R and S. Find $S \circ R$.

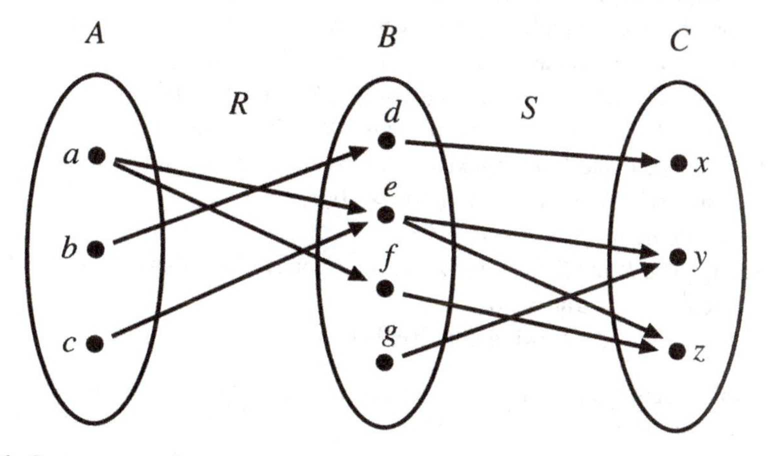

6. Suppose r and s are two positive real numbers. Let D_r and D_s be defined as in part 3 of Example 4.3.1. What is $D_r \circ D_s$? Justify your answer with a proof. (Hint: In your proof, you may find it helpful to use the triangle inequality; see exercise 12(c) of Section 3.5.)

*7. Prove part 1 of Theorem 4.3.4.

8. Prove part 3 of Theorem 4.3.4.

9. Suppose A and B are two sets.
   (a) Show that for every relation R from A to B, $R \circ i_A = R$.
   (b) Show that for every relation R from A to B, $i_B \circ R = R$.

*10. Suppose S is a relation on A. Let $D = \text{Dom}(S)$ and $R = \text{Ran}(S)$. Prove that $i_D \subseteq S^{-1} \circ S$ and $i_R \subseteq S \circ S^{-1}$.

11. Suppose R is a relation on A. Prove that if R is reflexive then $R \subseteq R \circ R$.

12. Suppose R is a relation on A.
   (a) Prove that if R is reflexive, then so is R^{-1}.
   (b) Prove that if R is symmetric, then so is R^{-1}.
   (c) Prove that if R is transitive, then so is R^{-1}.

*13. Suppose R_1 and R_2 are relations on A. For each part, give either a proof or a counterexample to justify your answer.
   (a) If R_1 and R_2 are reflexive, must $R_1 \cup R_2$ be reflexive?
   (b) If R_1 and R_2 are symmetric, must $R_1 \cup R_2$ be symmetric?
   (c) If R_1 and R_2 are transitive, must $R_1 \cup R_2$ be transitive?

14. Suppose R_1 and R_2 are relations on A. For each part, give either a proof or a counterexample to justify your answer.
   (a) If R_1 and R_2 are reflexive, must $R_1 \cap R_2$ be reflexive?
   (b) If R_1 and R_2 are symmetric, must $R_1 \cap R_2$ be symmetric?
   (c) If R_1 and R_2 are transitive, must $R_1 \cap R_2$ be transitive?

15. Suppose R_1 and R_2 are relations on A. For each part, give either a proof or a counterexample to justify your answer.
    (a) If R_1 and R_2 are reflexive, must $R_1 \setminus R_2$ be reflexive?
    (b) If R_1 and R_2 are symmetric, must $R_1 \setminus R_2$ be symmetric?
    (c) If R_1 and R_2 are transitive, must $R_1 \setminus R_2$ be transitive?
16. Suppose R and S are reflexive relations on A. Prove that $R \circ S$ is reflexive.
*17. Suppose R and S are symmetric relations on A. Prove that $R \circ S$ is symmetric iff $R \circ S = S \circ R$.
18. Suppose R and S are transitive relations on A. Prove that if $S \circ R \subseteq R \circ S$ then $R \circ S$ is transitive.
19. Consider the following putative theorem.

    Theorem? *Suppose R is a relation on A, and define a relation S on $\mathscr{P}(A)$ as follows:*

    $$S = \{(X, Y) \in \mathscr{P}(A) \times \mathscr{P}(A) \mid \exists x \in X \exists y \in Y(xRy)\}.$$

    If R is transitive, then so is S.

    (a) What's wrong with the following proof of the theorem?

    Proof. Suppose R is transitive. Suppose $(X, Y) \in S$ and $(Y, Z) \in S$. Then by the definition of S, xRy and yRz, where $x \in X$, $y \in Y$, and $z \in Z$. Since xRy, yRz, and R is transitive, xRz. But then since $x \in X$ and $z \in Z$, it follows from the definition of S that $(X, Z) \in S$. Thus, S is transitive. □

    (b) Is the theorem correct? Justify your answer with either a proof or a counterexample.
*20. Suppose R is a relation on A. Let $B = \{X \in \mathscr{P}(A) \mid X \neq \varnothing\}$, and define a relation S on B as follows:

    $$S = \{(X, Y) \in B \times B \mid \forall x \in X \forall y \in Y(xRy)\}.$$

    Prove that if R is transitive, then so is S. Why did the empty set have to be excluded from the set B to make this proof work?
21. Suppose R is a relation on A, and define a relation S on $\mathscr{P}(A)$ as follows:

    $$S = \{(X, Y) \in \mathscr{P}(A) \times \mathscr{P}(A) \mid \forall x \in X \exists y \in Y(xRy)\}.$$

    For each part, give either a proof or a counterexample to justify your answer.
    (a) If R is reflexive, must S be reflexive?
    (b) If R is symmetric, must S be symmetric?
    (c) if R is transitive, must S be transitive?

22. Consider the following putative theorem:

    Theorem? *Suppose R is a relation on A. If R is symmetric and transitive, then R is reflexive.*

    Is the following proof correct? If so, what proof strategies does it use? If not, can it be fixed? Is the theorem correct?

    Proof. Let x be an arbitrary element of A. Let y be any element of A such that xRy. Since R is symmetric, it follows that yRx. But then by transitivity, since xRy and yRx we can conclude that xRx. Since x was arbitrary, we have shown that $\forall x \in A(xRx)$, so R is reflexive. □

*23. This problem was suggested by Prof. William Zwicker of Union College. Suppose A is a set, and $\mathcal{F} \subseteq \mathscr{P}(A)$. Let $R = \{(a, b) \in A \times A \mid$ for every $X \subseteq A \setminus \{a, b\}$, if $X \cup \{a\} \in \mathcal{F}$ then $X \cup \{b\} \in \mathcal{F}\}$. Show that R is transitive.

4.4. Ordering Relations

Consider the relation $L = \{(x, y) \in \mathbb{R} \times \mathbb{R} \mid x \leq y\}$. You should be able to check for yourself that it is reflexive and transitive, but not symmetric. It fails to be symmetric in a rather extreme way because there are many pairs (x, y) such that xLy is true but yLx is false. In fact, the only way xLy and yLx can both be true is if $x \leq y$ and $y \leq x$, and thus $x = y$. We therefore say that L is *antisymmetric*. Here is the general definition.

Definition 4.4.1. Suppose R is a relation on a set A. Then R is said to be *antisymmetric* if $\forall x \in A \forall y \in A((xRy \wedge yRx) \rightarrow x = y)$.

We have already seen a relation with many of the same properties as L. Look again at the relation S defined in part 1 of Example 4.3.1. Recall that in that example we let $A = \{1, 2\}$, $B = \mathscr{P}(A)$, and $S = \{(x, y) \in B \times B \mid x \subseteq y\}$. Thus, if x and y are elements of B, then xSy means $x \subseteq y$. We checked in the last section that S is reflexive and transitive, but not symmetric. In fact, S is also antisymmetric, because for any sets x and y, if $x \subseteq y$ and $y \subseteq x$ then $x = y$. You may find it useful to look back at Figure 3 in the last section, which shows the directed graph representing S.

Intuitively, L and S are both relations that have something to do with comparing the sizes of two objects. Each of the statements $x \leq y$ and $x \subseteq y$ can be thought of as saying that, in some sense, y is "at least as large as" x. You might say that each of these statements specifies what *order* x and y come in. This motivates the following definition.

Definition 4.4.2. Suppose R is a relation on a set A. Then R is called a *partial order on A* (or just a *partial order* if A is clear from context) if it is reflexive, transitive, and antisymmetric. It is called a *total order on A* (or just a *total order*) if it is a partial order, and in addition it has the following property:

$$\forall x \in A \forall y \in A(xRy \lor yRx).$$

The relations L and S just considered are both partial orders. S is not a total order, because it is not true that $\forall x \in B \forall y \in B(x \subseteq y \lor y \subseteq x)$. For example, if we let $x = \{1\}$ and $y = \{2\}$, then $x \not\subseteq y$ and $y \not\subseteq x$. Thus, although we can think of the relation S as indicating a sense in which one element of B might be at least as large as another, it does not give us a way of comparing *every* pair of elements of B. For some pairs, such as $\{1\}$ and $\{2\}$, S doesn't pick out either one as being at least as large as the other. This is the sense in which the ordering is *partial*. On the other hand, L is a total order, because if x and y are any two real numbers, then either $x \leq y$ or $y \leq x$. Thus, L does give us a way of comparing *any* two real numbers.

Example 4.4.3. Which of the following relations are partial orders? Which are total orders?

1. Let A be any set, and let $B = \mathscr{P}(A)$ and $S = \{(x, y) \in B \times B \mid x \subseteq y\}$.
2. Let $A = \{1, 2\}$ and $B = \mathscr{P}(A)$ as before. Let $R = \{(x, y) \in B \times B \mid y$ has at least as many elements as $x\} = \{(\varnothing, \varnothing), (\varnothing, \{1\}), (\varnothing, \{2\}), (\varnothing, \{1, 2\}), (\{1\}, \{1\}), (\{1\}, \{2\}), (\{1\}, \{1, 2\}), (\{2\}, \{1\}), (\{2\}, \{2\}), (\{2\}, \{1, 2\}), (\{1, 2\}, \{1, 2\})\}$.
3. $D = \{(x, y) \in \mathbb{Z}^+ \times \mathbb{Z}^+ \mid x \text{ divides } y\}$.
4. $G = \{(x, y) \in \mathbb{R} \times \mathbb{R} \mid x \geq y\}$.

Solutions

1. This is just a generalization of one of the examples discussed earlier, and it is easy to check that it is a partial order. As long as A has at least two elements, it will not be a total order. To see why, just note that if a and b are distinct elements of A, then $\{a\}$ and $\{b\}$ are elements of B for which $\{a\} \not\subseteq \{b\}$ and $\{b\} \not\subseteq \{a\}$.
2. Note that $(\{1\}, \{2\}) \in R$ and $(\{2\}, \{1\}) \in R$, but of course $\{1\} \neq \{2\}$. Thus, R is not antisymmetric, so it is not a partial order. Although R was defined by

picking out pairs (x, y) in which y is, in a certain sense, at least as large as x, it does not satisfy the definition of partial order. This example shows that our description of partial orders as relations that indicate a sense in which one object is at least as large as another should not be taken too seriously. This was the *motivation* for the definition of partial order, but it is not the definition itself.

3. Clearly every positive integer is divisible by itself, so D is reflexive. Also, as we showed in Theorem 3.3.6, if $x \mid y$ and $y \mid z$ then $x \mid z$. Thus, if $(x, y) \in D$ and $(y, z) \in D$ then $(x, z) \in D$, so D is transitive. Finally, suppose $(x, y) \in D$ and $(y, x) \in D$. Then $x \mid y$ and $y \mid x$, and because x and y are positive it follows that $x \leq y$ and $y \leq x$, so $x = y$. Thus, D is antisymmetric, so it is a partial order. It is easy to find examples illustrating that D is not a total order. For example, $(3, 5) \notin D$ and $(5, 3) \notin D$.

   Perhaps you were surprised to discover that D is a partial order. It doesn't seem to involve comparing the sizes of things, like the other partial orders we've seen. But we have shown that it does share with these other relations the important properties of reflexivity, transitivity, and antisymmetry. In fact, this is one of the reasons for formulating definitions such as Definition 4.4.2. They help us to see similarities between things that, on the surface, might not seem similar at all.
4. You should be able to check for yourself that G is a total order. Notice that in this case it seems more reasonable to think of xGy as meaning that y is as least as *small* as x rather than at least as *large*. The definition of partial order, though motivated by thinking about orderings that go in one direction, actually applies to orderings in either direction. In fact, this example might lead you to conjecture that if R is a partial order on A, then so is R^{-1}. You are asked to prove this conjecture in exercise 13.

Here's another example of a partial order. Let A be the set of all words in English, and let $R = \{(x, y) \in A \times A \mid$ all the letters in the word x appear, consecutively and in the right order, in the word $y\}$. For example, (can, cannot), (tar, start), and (ball, ball) are all elements of R, but (can, anchor) and (can, carnival) are not. You should be able to check that R is reflexive, transitive, and antisymmetric, so R is a partial order. Now consider the set $B = \{$me, men, tame, mental$\} \subseteq$ A. Clearly many ordered pairs of words in B are in the relation R, but note in particular that the ordered pairs (me, me), (me, men), (me, tame), and (me, mental) are all in R. If we think of xRy as meaning that y is in some sense at least as large as x, then we could say that the word *me* is the *smallest* element of B, in the sense that it is smaller than everything else in the set.

Not every set of words will have an element that is smallest in this sense. For example, consider the set $C = \{\text{a, me, men, tame, mental}\} \subseteq \text{A}$. Each of the words *men*, *tame*, and *mental* is larger than at least one other word in the set, but neither *a* nor *me* is larger than anything else in the set. We'll call *a* and *me* *minimal* elements of C. But note that neither *a* nor *me* is the smallest element of C in the sense described in the last paragraph, because neither is smaller than the other. The set C has two minimal elements but no smallest element.

These examples might raise a number of questions in your mind about smallest and minimal elements. The set C has two minimal elements, but B has only one smallest element. Can a set ever have more than one smallest element? Until we have settled this question, we should only talk about an object being *a* smallest element of a set, rather than *the* smallest element. If a set has only one minimal element, must it be a smallest element? Can a set have a smallest element and a minimal element that are different? Would the answers to these questions be different if we restricted our attention to *total* orders rather than all partial orders? Before we try to answer any of these questions, we should state the definitions of the terms *smallest* and *minimal* more carefully.

Definition 4.4.4. Suppose R is a partial order on a set A, $B \subseteq A$, and $b \in B$. Then b is called an R-*smallest* element of B (or just a *smallest* element if R is clear from the context) if $\forall x \in B(bRx)$. It is called an R-*minimal* element (or just a *minimal* element) if $\neg\exists x \in B(xRb \land x \neq b)$.

Example 4.4.5.

1. Let $L = \{(x, y) \in \mathbb{R} \times \mathbb{R} \mid x \leq y\}$, as before. Let $B = \{x \in \mathbb{R} \mid x \geq 7\}$. Does B have any L-smallest or L-minimal elements? What about the set $C = \{x \in \mathbb{R} \mid x > 7\}$?
2. Let D be the divisibility relation defined in part 3 of Example 4.4.3. Let $B = \{3, 4, 5, 6, 7, 8, 9\}$. Does B have any D-smallest or D-minimal elements?
3. Let $S = \{(X, Y) \in \mathscr{P}(\mathbb{N}) \times \mathscr{P}(\mathbb{N}) \mid X \subseteq Y\}$, which is a partial order on the set $\mathscr{P}(\mathbb{N})$. Let $\mathcal{F} = \{X \in \mathscr{P}(\mathbb{N}) \mid 2 \in X \text{ and } 3 \in X\}$. Note that the elements of F are not natural numbers, but *sets* of natural numbers. For example, $\{1, 2, 3\}$ and $\{n \in \mathbb{N} \mid n \text{ is prime}\}$ are both elements of $\mathcal{F}$. Does $\mathcal{F}$ have any S-smallest or S-mimimal elements? What about the set $\mathcal{G} = \{X \in \mathscr{P}(\mathbb{N}) \mid \text{either } 2 \in X \text{ or } 3 \in X\}$?

Solutions

1. Clearly $7 \leq x$ for every $x \in B$, so $\forall x \in B(7Lx)$ and therefore 7 is a smallest element of B. It is also a minimal element, since nothing in B is smaller

than 7, so $\neg\exists x \in B(x \leq 7 \wedge x \neq 7)$. There are no other smallest or minimal elements. Note that 7 is *not* a smallest or minimal element of C, since $7 \notin C$. According to Definition 4.4.4, a smallest or minimal element of a set must actually be an element of the set. In fact, C has no smallest or minimal elements.

2. First of all, note that 6 and 9 are not minimal because both are divisible by 3, and 8 is not minimal because it is divisible by 4. All the other elements of B are minimal elements, but none is a smallest element.
3. The set $\{2, 3\}$ is a smallest element of $\mathcal{F}$, since 2 and 3 are elements of every set in $\mathcal{F}$, and therefore $\forall X \in \mathcal{F}(\{2, 3\} \subseteq X)$. It is also a minimal element, since no other element of $\mathcal{F}$ is a subset of it, and there are no other smallest or minimal elements. The get $\mathcal{G}$ has two minimal elements, $\{2\}$ and $\{3\}$. Every other set in $\mathcal{G}$ must contain one of these two as a subset, so no other set can be minimal. Neither set is smallest, since neither is a subset of the other.

We are now ready to answer some of the questions we raised before Definition 4.4.4.

Theorem 4.4.6. *Suppose R is a partial order on a set A, and $B \subseteq A$.*

1. *If B has a smallest element, then this smallest element is unique. Thus, we can speak of* the *smallest element of B rather than* a *smallest element.*
2. *Suppose b is the smallest element of B. Then b is also a minimal element of B, and it is the only minimal element.*
3. *If R is a total order and b is a minimal element of B, then b is the smallest element of B.*

Scratch work

These proofs are somewhat harder than earlier ones in this chapter, so we do some scratch work before the proofs.

1. Of course, we start by assuming that B has a smallest element, and because this is an existential statement, we immediately introduce a name, say b, for a smallest element of B. We must prove that b is the only smallest element. As we saw in Section 3.6, this can be written $\forall c(c$ is a smallest element of $B \to b = c)$, so our next step should be to let c be arbitrary, assume it is also a smallest element, and prove $b = c$.

At this point, we don't know much about b and c. We know they're both elements of B, but we don't even know what kinds of objects are in B – whether they're numbers, or sets, or some other type of object – so this doesn't help us much in deciding how to prove that $b = c$. The only other fact we know about b and c is that they are both smallest elements of B, which means $\forall x \in B(bRx)$ and $\forall x \in B(cRx)$. The most promising way to use these statements is to plug something in for x in each statement. What we plug in should be an element of B, and we only know of two elements of B at this point, b and c. Plugging in both of them in both statements, we get bRb, bRc, cRb, and cRc. Of course, we already knew bRb and cRc, since R is reflexive. But when you see that bRc and cRb, you should think of antisymmetry. Since R is a partial order, it is antisymmetric, so from bRc and cRb it follows that $b = c$.

2. Our first goal is to prove that b is a minimal element of B, which means $\neg\exists x \in B(xRb \wedge x \neq b)$. Because this is a negative statement, it might help to reexpress it as an equivalent positive statement:

$$\begin{aligned}\neg\exists x \in B(xRb \wedge x \neq b) &\text{ iff } \forall x \in B\neg(xRb \wedge x \neq b)\\ &\text{ iff } \forall x \in B(\neg xRb \vee x = b)\\ &\text{ iff } \forall x \in B(xRb \rightarrow x = b).\end{aligned}$$

Thus, to prove that b is minimal we could let x be an arbitrary element of B, assume that xRb, and prove $x = b$.

Once again, it's a good idea to take stock of what we know at this point about b and x. We know xRb, and we know that b is the smallest element of B, which means $\forall x \in B(bRx)$. If we apply this last fact to our arbitrary x, then as in part 1 we can use antisymmetry to complete the proof.

We still must prove that b is the only minimal element, and as in part 1 this means $\forall c(c$ is a minimal element of $B \rightarrow b = c)$. So we let c be arbitrary and assume that c is a minimal element of B, and we must prove that $b = c$. The assumption that c is a minimal element of B means that $c \in B$ and $\neg\exists x \in B(xRc \wedge x \neq c)$, but as before, we can reexpress this last statement in the equivalent positive form $\forall x \in B(xRc \rightarrow x = c)$. To use this statement we should plug in something for x, and because our goal is to show that $b = c$, plugging in b for x seems like a good idea. This gives us $bRc \rightarrow b = c$, so if only we could show bRc, we could complete the proof by using modus ponens to conclude that $b = c$. But we know b is the smallest element of B, so of course bRc is true.

3. Of course, we start by assuming that R is a total order and b is a minimal element of B. We must prove that b is the smallest element of B, which means $\forall x \in B(bRx)$, so we let x be an arbitrary element of B and try to prove bRx.

We know from examples we've looked at that minimal elements in *partial* orders are not always smallest elements, so the assumption that R is a *total* order must be crucial. The assumption that R is total means $\forall x \in A\forall y \in A(xRy \vee$

yRx), so to use it we should plug in something for x and y. The only likely candidates for what to plug in are b and our arbitrary object x, and plugging these in we get $xRb \lor bRx$. Our goal is bRx, so this certainly looks like progress. If only we could rule out the possibility that xRb, we'd be done. So let's see if we can prove $\neg xRb$.

Because this is a negative statement, we try proof by contradiction. Suppose xRb. What given statement can we contradict? The only given we haven't used yet is the fact that b is minimal, and since this is a negative statement, it is the natural place to look for a contradiction. To contradict the fact that b is minimal, we should try to show that $\exists x \in B(xRb \land x \neq b)$. But we've already assumed xRb, so if we could show $x \neq b$ we'd be done.

You should try proving $x \neq b$ at this point. You won't get anywhere. The fact is, we started out by letting x be an arbitrary element of B, and this means that it could be any element of B, including b. We then assumed that xRb, but since R is reflexive, this still doesn't rule out the possibility that $x = b$. There really isn't any hope of proving $x \neq b$. We seem to be stuck.

Let's review our overall plan for the proof. We needed to show $\forall x \in B(bRx)$, so we let x be an arbitrary element of B, and we're trying to show bRx. We've now run into problems because of the possibility that $x = b$. But if our ultimate goal is to prove bRx, then the possibility that $x = b$ really isn't a problem after all. Since R is reflexive, if $x = b$ then of course bRx will be true!

Now, how should we structure the final write-up of the proof? It appears that our reasoning to establish bRx will have to be different depending on whether or not $x = b$. This suggests *proof by cases*. In case 1 we assume that $x = b$, and use the fact that R is reflexive to complete the proof. In case 2 we assume that $x \neq b$, and then we can use our original line of attack, starting with the fact that R is total.

Proof.

1. Suppose b is a smallest element of B, and suppose c is also a smallest element of B. Since b is a smallest element, $\forall x \in B(bRx)$, so in particular bRc. Similarly, since c is a smallest element, cRb. But now since R is a partial order, it must be antisymmetric, so from bRc and cRb we can conclude $b = c$.
2. Let x be an arbitrary element of B and suppose that xRb. Since b is the smallest element of B, we must have bRx, and now by antisymmetry it follows that $x = b$. Thus, b must be a minimal element.

   To see that it is the only one, suppose c is also a minimal element. Since b is the smallest element of B, bRc. But then since c is minimal, we must have $b = c$. Thus b is the only minimal element of B.

3. Suppose R is a total order and b is a minimal element of B. Let x be an arbitrary element of B. If $x = b$, then since R is reflexive, bRx. Now suppose $x \neq b$. Since R is a total order, we know that either xRb or bRx. But xRb can't be true, since by combining xRb with our assumption that $x \neq b$ we could conclude that b is not minimal, thereby contradicting our assumption that it is minimal. Thus, bRx must be true. Since x was arbitrary, we can conclude that $\forall x \in B(bRx)$, so b is the smallest element of B. □

When comparing subsets of some set A, mathematicians often use the partial order $S = \{(X, Y) \in \mathscr{P}(A) \times \mathscr{P}(A) \mid X \subseteq Y\}$, although this is not always made explicit. Recall that if $\mathcal{F} \subseteq \mathscr{P}(A)$ and $X \in \mathcal{F}$, then according to Definition 4.4.4, X is the S-smallest element of $\mathcal{F}$ iff $\forall Y \in \mathcal{F}(X \subseteq Y)$. In other words, to say that an element of $\mathcal{F}$ is the smallest element means that it is a subset of every element of $\mathcal{F}$. Similarly, mathematicians sometimes talk of a set being the smallest one with a certain property. Generally this means that the set has the property in question, and furthermore it is a subset of every set that has the property. For example, we might describe our conclusion in part 3 of Example 4.4.5 by saying that $\{2, 3\}$ is the smallest set $X \subseteq \mathbb{N}$ with the property that $2 \in X$ and $3 \in X$. We will see more examples of this idea in the next section and in later chapters.

Example 4.4.7.

1. Find the smallest set of real numbers X such that $5 \in X$ and for all real numbers x and y, if $x \in X$ and $x < y$ then $y \in X$.
2. Find the smallest set of real numbers X such that $X \neq \varnothing$ and for all real numbers x and y, if $x \in X$ and $x < y$ then $y \in X$.

Solutions

1. Another way to phrase the question would be to say that we are looking for the smallest element of the family of sets $\mathcal{F} = \{X \subseteq \mathbb{R} \mid 5 \in X$ and $\forall x \forall y((x \in X \wedge x < y) \rightarrow y \in X)\}$, where it is understood that *smallest* means smallest with respect to the subset partial order. Now for any set $X \in \mathcal{F}$ we know that $5 \in X$, and we know that $\forall x \forall y((x \in X \wedge x < y) \rightarrow y \in X)$. In particular, since $5 \in X$ we can say that $\forall y(5 < y \rightarrow y \in X)$. Thus, if we let $A = \{y \in \mathbb{R} \mid 5 \leq y\}$, then we can conclude that $\forall X \in \mathcal{F}(A \subseteq X)$. But it is easy to see that $A \in \mathcal{F}$, so A is the smallest element of $\mathcal{F}$.
2. We must find the smallest element of the family of sets $\mathcal{F} = \{X \subseteq \mathbb{R} \mid X \neq \varnothing$ and $\forall x \forall y((x \in X \wedge x < y) \rightarrow y \in X)\}$. The set $A = \{y \in \mathbb{R} \mid 5 \leq y\}$ from part 1 is an element of $\mathcal{F}$, but it is not the smallest element, or even a

minimal element, because the set $A' = \{y \in \mathbb{R} \mid 6 \leq y\}$ is smaller – in other words, $A' \subseteq A$ and $A' \neq A$. But A' is also not the smallest element, since $A'' = \{y \in \mathbb{R} \mid 7 \leq y\}$ is still smaller. In fact, this family has no smallest, or even minimal, element. You're asked to verify this in exercise 12. This example shows that we must be careful when talking about the smallest set with some property. There may be no such smallest set!

You have probably already guessed how to define maximal and largest elements in partially ordered sets. Suppose R is a partial order on A, $B \subseteq A$, and $b \in B$. We say that b is the *largest* element of B if $\forall x \in B(xRb)$, and it is a *maximal* element of B if $\neg\exists x \in B(bRx \wedge b \neq x)$. Of course, these definitions are quite similar to the ones in Definition 4.4.4. You are asked in exercise 14 to work out some of the connections among these ideas. Another useful related idea is the concept of an upper or lower bound for a set.

Definition 4.4.8. Suppose R is a partial order on A, $B \subseteq A$, and $a \in A$. Then a is called a *lower bound* for B if $\forall x \in B(aRx)$. Similarly, it is an *upper bound* for B if $\forall x \in B(xRa)$.

Note that a lower bound for B need not be an element of B. This is the only difference between lower bounds and smallest elements. A smallest element of B is just a lower bound that is also an element of B. For example, in part 1 of Example 4.4.5, we concluded that 7 was not a smallest element of the set $C = \{x \in \mathbb{R} \mid x > 7\}$ because $7 \notin C$. But 7 *is* a lower bound for C. In fact, so is every real number smaller than 7, but not any number larger than 7. Thus, the set of all lower bounds of C is the set $\{x \in \mathbb{R} \mid x \leq 7\}$, and 7 is its largest element. We say that 7 is the *greatest lower bound* of the set C.

Definition 4.4.9. Suppose R is a partial order on A and $B \subseteq A$. Let U be the set of all upper bounds for B, and let L be the set of all lower bounds. If U has a smallest element, then this smallest element is called the *least upper bound* of B. If L has a largest element, then this largest element is called the *greatest lower bound* of B. The phrases *least upper bound* and *greatest lower bound* are sometimes abbreviated *l.u.b.* and *g.l.b.*

Example 4.4.10.

1. Let $L = \{(x, y) \in \mathbb{R} \times \mathbb{R} \mid x \leq y\}$, a total order on $\mathbb{R}$. Let $B = \{1/n \mid n \in \mathbb{Z}^+\} = \{1, 1/2, 1/3, 1/4, 1/5, \ldots\} \subseteq \mathbb{R}$. Does B have any upper or lower bounds? Does it have a least upper bound or greatest lower bound?

2. Let A be the set of all English words, and let R be the partial order on A described after Example 4.4.3. Let $B = \{\text{house, boat}\}$. Does B have any upper or lower bounds? Does it have a least upper bound or a greatest lower bound?

Solutions

1. Clearly the largest element of B is 1. It is also an upper bound for B, as is any number larger than 1. By definition, an upper bound for B must be at least as large as every element of B, so in particular it must be at least as large as 1. Thus, no number smaller than 1 is an upper bound for B, so the set of upper bounds for B is $\{x \in \mathbb{R} \mid x \geq 1\}$. Clearly the smallest element of this set is 1, so 1 is the l.u.b. of B.

   Clearly 0 is a lower bound for B, as is any negative number. On the other hand, suppose a is a positive number. Then for a large enough integer n we will have $1/n < a$. (You should convince yourself that any integer n larger than $1/a$ would do.) Thus, it is not the case that $\forall x \in B(a \leq x)$, and therefore a is not a lower bound for B. So the set of all lower bounds for B is $\{x \in \mathbb{R} \mid x \leq 0\}$, and the g.l.b. of B is 0.
2. Clearly *houseboat* and *boathouse* are upper bounds for B. In fact, no shorter word could be an upper bound, so they are both minimal elements of the set of all upper bounds. According to part 2 of Theorem 4.4.6, a set that has more than one minimal element can have no smallest element, so the set of all upper bounds for B does not have a smallest element, and therefore B doesn't have a l.u.b.

   The only letter that the words *house* and *boat* have in common is *o*, which is not a word of English. Thus, B has no lower bounds.

Notice that in part 1 of Example 4.4.10, the largest element of B also turned out to be its least upper bound. You might wonder whether largest elements are always least upper bounds and whether smallest elements are always greatest lower bounds. You are asked to prove that they are in exercise 20. Another interesting fact about this example is that, although B did not have a smallest element, it did have a greatest lower bound. This was not a coincidence. It is an important fact about the real numbers that *every* nonempty set of real numbers that has a lower bound has a greatest lower bound and, similarly, every nonempty set of real numbers that has an upper bound has a least upper bound. The proof of this fact is beyond the scope of this book, but it is important to realize that it is a special fact about the real numbers; it does not apply to all partial orders or even to all total orders. For example, the set B in the second part of Example 4.4.10 had upper bounds but no least upper bound.

We end this section by looking once again at how these new concepts apply to the subset partial order on $\mathscr{P}(A)$, for any set A. It turns out that in this partial order, least upper bounds and greatest lower bounds are our old friends unions and intersections.

Theorem 4.4.11. *Suppose A is a set, $\mathcal{F} \subseteq \mathscr{P}(A)$, and $\mathcal{F} \neq \varnothing$. Then the least upper bound of $\mathcal{F}$ (in the subset partial order) is $\cup\mathcal{F}$ and the greatest lower bound of $\mathcal{F}$ is $\cap\mathcal{F}$.*

Proof. See exercise 23. □

Exercises

*1. In each case, say whether or not R is a partial order on A. If so, is it a total order?
   (a) $A = \{a, b, c\}$, $R = \{(a, a), (b, a), (b, b), (b, c), (c, c)\}$.
   (b) $A = \mathbb{R}$, $R = \{(x, y) \in \mathbb{R} \times \mathbb{R} \mid |x| \leq |y|\}$.
   (c) $A = \mathbb{R}$, $R = \{(x, y) \in \mathbb{R} \times \mathbb{R} \mid |x| < |y| \text{ or } x = y\}$.
2. In each case, say whether or not R is a partial order on A. If so, is it a total order?
   (a) $A =$ the set of all words of English, $R = \{(x, y) \in A \times A \mid$ the word y occurs at least as late in alphabetical order as the word $x\}$.
   (b) $A =$ the set of all words of English, $R = \{(x, y) \in A \times A \mid$ the first letter of the word y occurs at least as late in the alphabet as the first letter of the word $x\}$.
   (c) $A =$ the set of all countries in the world, $R = \{(x, y) \in A \times A \mid$ the population of the country y is at least as large as the population of the country $x\}$.
3. In each case find all minimal and maximal elements of B. Also find, if they exist, the largest and smallest elements of B, and the least upper bound and greatest lower bound of B.
   (a) $R =$ the relation shown in the following directed graph, $B = \{2, 3, 4\}$.

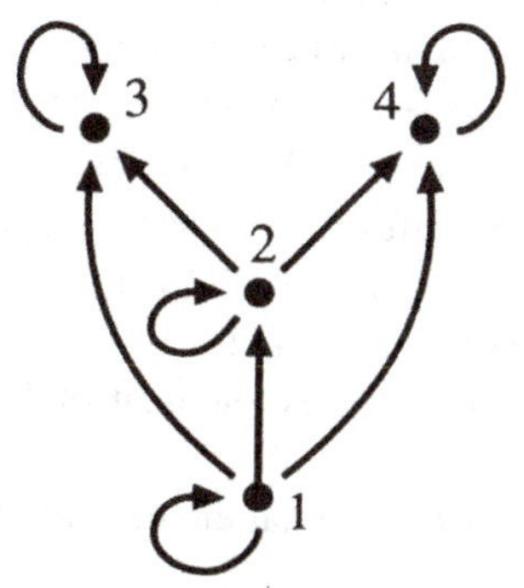

(b) $R = \{(x, y) \in \mathbb{R} \times \mathbb{R} \mid x \leq y\}$, $B = \{x \in \mathbb{R} \mid 1 \leq x < 2\}$.
(c) $R = \{(x, y) \in \mathscr{P}(\mathbb{N}) \times \mathscr{P}(\mathbb{N}) \mid x \subseteq y\}$, $B = \{x \in \mathscr{P}(\mathbb{N}) \mid x$ has at most 5 elements$\}$.

*4. Suppose R is a relation on A. You might think that R could not be both antisymmetric and symmetric, but this isn't true. Prove that R is both antisymmetric and symmetric iff $R \subseteq i_A$.

5. Suppose R is a partial order on A and $B \subseteq A$. Prove that $R \cap (B \times B)$ is a partial order on B.

6. Suppose R_1 and R_2 are partial orders on A. For each part, give either a proof or a counterexample to justify your answer.
(a) Must $R_1 \cap R_2$ be a partial order on A?
(b) Must $R_1 \cup R_2$ be a partial order on A?

7. Suppose R_1 is a partial order on A_1, R_2 is a partial order on A_2, and $A_1 \cap A_2 = \varnothing$.
(a) Prove that $R_1 \cup R_2$ is a partial order on $A_1 \cup A_2$.
(b) Prove that $R_1 \cup R_2 \cup (A_1 \times A_2)$ is a partial order on $A_1 \cup A_2$.
(c) Suppose that R_1 and R_2 are total orders. Are the partial orders in parts (a) and (b) also total orders?

*8. Suppose R is a partial order on A and S is a partial order on B. Define a relation T on $A \times B$ as follows: $T = \{((a, b), (a', b')) \in (A \times B) \times (A \times B) \mid aRa'$ and $bSb'\}$. Show that T is a partial order on $A \times B$. If both R and S are total orders, will T also be a total order?

9. Suppose R is a partial order on A and S is a partial order on B. Define a relation L on $A \times B$ as follows: $L = \{((a, b), (a', b')) \in (A \times B) \times (A \times B) \mid aRa'$, and if $a = a'$ then $bSb'\}$. Show that L is a partial order on $A \times B$. If both R and S are total orders, will L also be a total order?

10. Suppose R is a partial order on A. For each $x \in A$, let $P_x = \{a \in A \mid aRx\}$. Prove that $\forall x \in A \forall y \in A(xRy \leftrightarrow P_x \subseteq P_y)$.

*11. Let D be the divisibility relation defined in part 3 of Example 4.4.3. Let $B = \{x \in \mathbb{Z} \mid x > 1\}$. Does B have any minimal elements? If so, what are they? Does B have a smallest element? If so, what is it?

12. Show that, as was stated in part 2 of Example 4.4.7, $\{X \subseteq \mathbb{R} \mid X \neq \varnothing$ and $\forall x \forall y((x \in X \land x < y) \to y \in X)\}$ has no minimal element.

13. Suppose R is a partial order on A. Prove that R^{-1} is also a partial order on A. If R is a total order, will R^{-1} also be a total order?

*14. Suppose R is a partial order on A, $B \subseteq A$, and $b \in B$. Exercise 13 shows that R^{-1} is also a partial order on A.
(a) Prove that b is the R-largest element of B iff it is the R^{-1}-smallest element of B.
(b) Prove that b is an R-maximal element of B iff it is an R^{-1}-minimal element of B.

15. Suppose R_1 and R_2 are partial orders on A, $R_1 \subseteq R_2$, $B \subseteq A$, and $b \in B$.
    (a) Prove that if b is the R_1-smallest element of B, then it is also the R_2-smallest element of B.
    (b) Prove that if b is an R_2-minimal element of B, then it is also an R_1-minimal element of B.
16. Suppose R is a partial order on A, $B \subseteq A$, and $b \in B$. Prove that if b is the largest element of B, then b is also a maximal element of B, and it's the only maximal element.
*17. If a subset of a partially ordered set has exactly one minimal element, must that element be a smallest element? Give either a proof or a counter example to justify your answer.
18. Suppose R is a partial order on A, $B_1 \subseteq A$, $B_2 \subseteq A$, $\forall x \in B_1 \exists y \in B_2(xRy)$, and $\forall x \in B_2 \exists y \in B_1(xRy)$.
    (a) Prove that for all $x \in A$, x is an upper bound of B_1 iff x is an upper bound of B_2.
    (b) Prove that if B_1 and B_2 are disjoint then neither of them has a maximal element.
19. Consider the following putative theorem.

    Theorem? *Suppose R is a total order on A and $B \subseteq A$. Then every element of B is either the smallest element of B or the largest element of B.*

    (a) What's wrong with the following proof of the theorem?

    Proof. Suppose $b \in B$. Let x be an arbitrary element of B. Since R is a total order, either bRx or xRb.

    Case 1. bRx. Since x was arbitrary, we can conclude that $\forall x \in B(bRx)$, so b is the smallest element of R.

    Case 2. xRb. Since x was arbitrary, we can conclude that $\forall x \in B(xRb)$, so b is the largest element of R.

    Thus, b is either the smallest element of B or the largest element of B. Since b was arbitrary, every element of B is either its smallest element or its largest element. □

    (b) Is the theorem correct? Justify your answer with either a proof or a counterexample.

20. Suppose R is a partial order on A, $B \subseteq A$, and $b \in B$.
    (a) Prove that if b is the smallest element of B, then it is also the greatest lower bound of B.
    (b) Prove that if b is the largest element of B, then it is also the least upper bound of B.
*21. Suppose R is a partial order on A and $B \subseteq A$. Let U be the set of all upper bounds for B.

(a) Prove that U is *closed upward*; that is, prove that if $x \in U$ and xRy, then $y \in U$.
(b) Prove that every element of B is a lower bound for U.
(c) Prove that if x is the greatest lower bound of U, then x is the least upper bound of B.

22. Suppose that R is a partial order on A, $B_1 \subseteq A$, $B_2 \subseteq A$, x_1 is the least upper bound of B_1, and x_2 is the least upper bound of B_2. Prove that if $B_1 \subseteq B_2$ then $x_1 R x_2$.
23. Prove Theorem 4.4.11.

4.5. Closures

According to the definition we gave in the last section, the relation $L = \{(x, y) \in \mathbb{R} \times \mathbb{R} \mid x \leq y\}$ is a total order on $\mathbb{R}$, but the relation $M = \{(x, y) \in \mathbb{R} \times \mathbb{R} \mid x < y\}$ is not because it is not reflexive. Of course, these relations are closely related. It's clear that $M \subseteq L$, and the only ordered pairs in L that are not in M are pairs of the form (x, x), for some $x \in \mathbb{R}$. Note that all of these ordered pairs must be in any reflexive relation on $\mathbb{R}$. Thus, you could think of L as being formed by starting with M and then adding just those ordered pairs that must be added to create a reflexive relation. It follows that L is the smallest relation on $\mathbb{R}$ that is reflexive and contains M as a subset. We are using the word *smallest* here in exactly the way we defined it in the last section. If we let $\mathcal{F} = \{T \subseteq \mathbb{R} \times \mathbb{R} \mid M \subseteq T \text{ and } T \text{ is reflexive}\}$, then L is the smallest element of $\mathcal{F}$, where as usual it is understood that we mean smallest in the sense of the subset partial order. In other words, L is an element of $\mathcal{F}$, and it's a subset of every element of $\mathcal{F}$. We will say that L is the *reflexive closure* of M.

Definition 4.5.1. Suppose R is a relation on a set A. Then the *reflexive closure* of R is the smallest set $S \subseteq A \times A$ such that $R \subseteq S$ and S is reflexive, if there is such a smallest set. In other words, a relation $S \subseteq A \times A$ is the reflexive closure of R if it has the following three properties:

1. $R \subseteq S$.
2. S is reflexive.
3. For every relation $T \subseteq A \times A$, if $R \subseteq T$ and T is reflexive, then $S \subseteq T$.

According to Theorem 4.3.6, if a set has a smallest element, then it can have only one smallest element. Thus, if a relation R has a reflexive closure, then this reflexive closure must be unique, so it makes sense to call it *the* reflexive closure of R rather than *a* reflexive closure. However, as we saw in

Example 4.4.7, some families of sets don't have smallest elements, so it might not be clear at first whether every relation has a reflexive closure. In fact, every relation does have a reflexive closure.

Theorem 4.5.2. *Suppose R is a relation on A. Then R has a reflexive closure.*

Proof. Let $S = R \cup i_A$, where as usual i_A is the identity relation on A. We will show that S is the reflexive closure of A. Thus, we must show that S has the three properties listed in Definition 4.5.1. The first property is obviously true, since clearly $R \subseteq R \cup i_A = S$. For the second and third, we use Theorem 4.3.4. Clearly $i_A \subseteq R \cup i_A = S$, so by part 1 of Theorem 4.3.4, S is reflexive. Finally, to prove the third property, suppose T is a relation on A, $R \subseteq T$, and T is reflexive. Then by Theorem 4.3.4, since T is reflexive, $i_A \subseteq T$. Combining this with the fact that $R \subseteq T$, we can conclude that $S = R \cup i_A \subseteq T$, as required. □

Commentary. Our goal is the existential statement $\exists S$(S is the reflexive closure of R), so we start by specifying a value for S. Our earlier discussion suggests that to get S we should start with R and then add just those ordered pairs that must be added to create a reflexive relation. The ordered pairs that must be added are the elements of i_A that are not already in R, so we let $S = R \cup i_A$. Our goal now is to prove that S is the reflexive closure of R, and by definition this means that we must prove statements 1–3 of Definition 4.5.1. We prove these one at a time. Statements 1 and 2 are easy, and the logical form of statement 3 suggests the strategy of letting T be an arbitrary relation on A, assuming $R \subseteq T$ and T is reflexive, and then proving $S \subseteq T$.

For another example of a reflexive closure, let A be any set and consider the relation $P = \{(x, y) \in \mathscr{P}(A) \times \mathscr{P}(A) \mid x \subseteq y \text{ and } x \neq y\}$. Thus, if x and y are any two subsets of A, then xPy means that $x \subseteq y$ and $x \neq y$. If xPy, then we will say that x is a *proper* subset of y, which is written $x \subset y$. The reflexive closure of P would be the relation

$$\begin{aligned} P \cup i_{\mathscr{P}(A)} &= \{(x, y) \in \mathscr{P}(A) \times \mathscr{P}(A) \mid (x, y) \in P \text{ or } (x, y) \in i_{\mathscr{P}(A)}\} \\ &= \{(x, y) \in \mathscr{P}(A) \times \mathscr{P}(A) \mid x \subset y \text{ or } x = y\} \\ &= \{(x, y) \in \mathscr{P}(A) \times \mathscr{P}(A) \mid x \subseteq y\}. \end{aligned}$$

Thus, the reflexive closure of the proper subset relation is the subset relation.

The relations M and P in these examples are similar to partial orders except that they are not reflexive. Rather than expressing a sense in which one object can be "at least as large as" another, these relations seem to represent a sense in which one object can be "strictly larger" than another. They are therefore sometimes called *strict* partial orders.

Definition 4.5.3. Suppose R is a relation on A. Then R is said to be *irreflexive* if $\forall x \in A((x, x) \notin R)$. R is called a *strict partial order* if it is irreflexive and transitive. It is called a *strict total order* if it is a strict partial order, and in addition it satisfies the following requirement, called *trichotomy*:

$$\forall x \in A \forall y \in A(xRy \vee yRx \vee x = y).$$

Note that the terminology here is slightly misleading. A strict partial order isn't a special kind of partial order. It's not a partial order at all, since it's not reflexive! You may be surprised that we did not include antisymmetry in the definition of strict partial order, since it was part of the definition of partial order, but it turns out that antisymmetry is implied by the definition. For more on this, see exercise 3.

You should be able to check for yourself that P is a strict partial order and M is a strict total order. Perhaps you've already guessed from these examples that the reflexive closure of a strict partial order is always a partial order, and the reflexive closure of a strict total order is always a total order. You are asked to prove this in exercise 4.

Reflexivity is not the only property for which we can define a closure. Exactly the same idea could be applied to symmetry and transitivity.

Definition 4.5.4. Suppose R is a relation on A. The *symmetric closure* of R is the smallest set $S \subseteq A \times A$ such that $R \subseteq S$ and S is symmetric, if there is such a smallest set. In other words, a relation $S \subseteq A \times A$ is the symmetric closure of R if it has the following properties:

1. $R \subseteq S$.
2. S is symmetric.
3. For every relation $T \subseteq A \times A$, if $R \subseteq T$ and T is symmetric, then $S \subseteq T$.

The *transitive closure* of R is the smallest set $S \subseteq A \times A$ such that $R \subseteq S$ and S is transitive, if there is such a smallest set. In other words, a relation $S \subseteq A \times A$ is the transitive closure of R if it has the following properties:

1. $R \subseteq S$.
2. S is transitive.
3. For every relation $T \subseteq A \times A$, if $R \subseteq T$ and T is transitive, then $S \subseteq T$.

As we saw in the case of reflexive closure, it is not immediately obvious that these closures will always exist. It turns out that they do, although this will require proof. But before proving these theorems, let's look at a few examples of symmetric and transitive closures.

Let P be the set of all people, and let $H = \{(x, y) \in P \times P \mid x \text{ hates } y\}$. Then H may fail to be symmetric, because if x hates y, it doesn't necessarily follow that y hates x. To find the symmetric closure of H we would have to find the smallest relation on P that is symmetric and contains every ordered pair in H. We could do this by starting with H and then adding only those ordered pairs that *must* be added to make the relation symmetric. Now clearly if we want to create a symmetric relation, then we will have to add the ordered pair (x, y) whenever $(y, x) \in H$. In other words, if y hates x, then we'll have to include the pair (x, y) in the relation we are constructing. Adding these ordered pairs to H, we can see that any relation on P that is symmetric and contains H must contain all the ordered pairs in the set $S = H \cup \{(x, y) \in P \times P \mid y \text{ hates } x\} = \{(x, y) \in P \times P \mid \text{ either } x \text{ hates } y \text{ or } y \text{ hates } x\}$. Now it is not hard to check that S is a symmetric relation, so it must be the symmetric closure of H. If you were planning the guest list for a party, you might want to know about the relation S. If you are inviting some person x and you know that $x S y$, you probably shouldn't invite y!

For an example of a transitive closure, let C be the set of all cities in the world and let $B = \{(x, y) \in C \times C \mid \text{ there is a nonstop bus from } x \text{ to } y\}$. Now if $(a, b) \in B$ and $(b, c) \in B$, it does not necessarily follow that $(a, c) \in B$, since there might be nonstop buses from a to b and b to c, but no nonstop bus from a to c. Thus, if we want to add new ordered pairs to B in order to construct a transitive relation, we must add the pair (a, c). But notice that once we have added (a, c), we may be forced to add even more ordered pairs if we want to end up with a transitive relation. For example, if there is a nonstop bus from c to some other city d, then we will have to add (a, d). We were forced to add (a, d) because (a, b), (b, c), and (c, d) were all elements of B. In other words, you could go by bus from a to b, from b to c, and from c to d. In fact, it should be clear now that for any two cities x and y, if there is a way to get from x to y by bus, changing buses any number of times at other cities, then we will eventually be forced by the transitivity requirement to add the pair (x, y). Thus, any transitive relation on C that contains all the ordered pairs in B must contain the relation $T = \{(x, y) \in C \times C \mid \text{ it is possible to get from } x \text{ to } y \text{ by bus (possibly changing buses several times at other cities)}\}$. But if you can get from x to y by bus and you can get from y to z by bus, then by combining the two bus trips you can get from x to z by bus. Thus, T is transitive, so it is the transitive closure of B.

It might be helpful in thinking about this last example to draw the directed graph of the relation B. This is illustrated in Figure 1 for a small set of cities. A bus company might draw such a diagram to represent all its bus routes. Now we can describe the transitive closure T as consisting of all ordered pairs (x, y) such that you can get from x to y in Figure 1 by following the arrows. For example, the pair (Dallas, New York) would be in T, because you can get from Dallas to New York by changing buses in Washington. Note that although it might be convenient to position the dots in this diagram as they would appear on a map, the precise positions of the dots have nothing to do with the relations B and T. To read from the diagram which ordered pairs are in B or T, we only need to know which dots are connected by arrows and which aren't, not precisely where the dots are located.

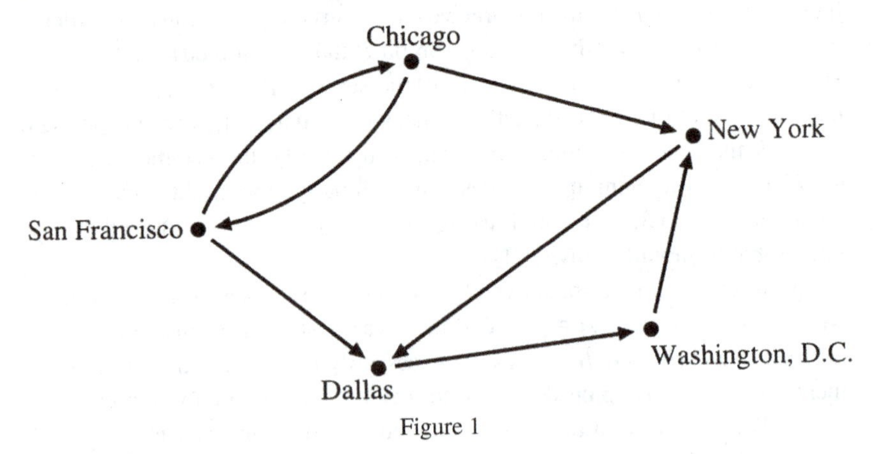

Figure 1

Directed graphs can often be helpful in thinking about symmetric and transitive closures of relations. Let's work out one more example. Suppose $A = \{1, 2, 3, 4\}$ and $R = \{(1, 2), (1, 3), (2, 1), (2, 2), (3, 4)\}$. The directed graph representing R is shown in Figure 2.

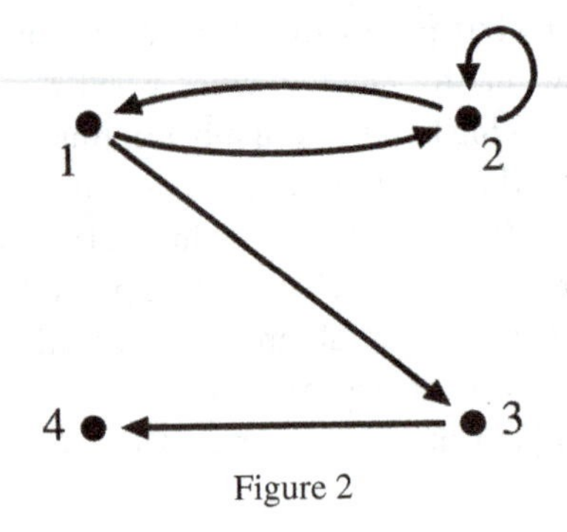

Figure 2

We can find the symmetric and transitive closures of R by imitating the reasoning we used for the relations H and B. To find the symmetric closure we start with R, and for each ordered pair (x, y) in R we add the ordered pair (y, x). This gives us the relation $S = \{(1, 2), (1, 3), (2, 1), (2, 2), (3, 1), (3, 4), (4, 3)\}$, whose directed graph is shown in Figure 3. Note that the only difference between this graph and the one in Figure 2 is that single arrows connecting distinct vertices have been changed to pairs of arrows pointing in opposite directions. You should be able to check that S is symmetric, and since the only ordered pairs we added to R are those we were *forced* to add by the definition of symmetry, S must be the symmetric closure of R.

To find the transitive closure of R, we let $T = \{(x, y) \in A \times A \mid$ you can get from vertex x to vertex y in Figure 2 by following the arrows$\}$ $= \{(1, 1), (1, 2), (1, 3), (1, 4), (2, 1), (2, 2), (2, 3), (2, 4), (3, 4)\}$. The graph of T is also shown in Figure 3. Note in particular that $(1, 1) \in T$, because you can get from vertex 1 to itself by following the arrows in Figure 2, going first from vertex 1 to vertex 2 and then back to vertex 1. Rephrasing this in terms of the definition of transitivity, since $(1, 2) \in R$ and $(2, 1) \in R$, we must add the ordered pair $(1, 1)$ to R if we want to create a transitive relation. Once again, you should be able to verify that T is the transitive closure of R by checking that T is transitive and that the only ordered pairs in T that are not in R are those we were *forced* to add by the definition of transitivity.

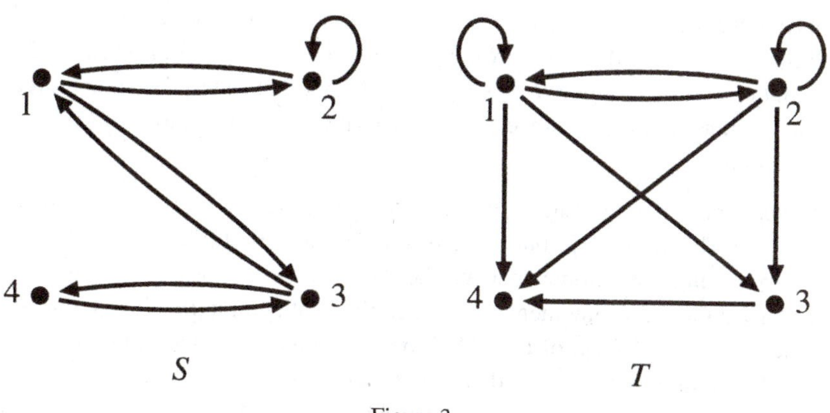

Figure 3

Let's return now to the problem of proving that every relation has a symmetric closure and a transitive closure. Both of these proofs will involve proving that something with a certain property exists, and as we saw in Chapter 3, proofs of this kind are sometimes difficult. The most straightforward way to proceed is to try to find something that has the required property, but sometimes this object is difficult to find. For the proofs we are concerned with we will need to be able to find, for any given relation R, relations that satisfy the definitions of symmetric and transitive closure for R. This will actually be the most difficult part of figuring out these proofs. Once we've found the right relations, verifying that they satisfy the definitions of symmetric and transitive closure will be somewhat long, but not difficult.

The case of the symmetric closure is the easier of the two. Suppose R is a relation on a set A, and we want to find the symmetric closure of R. Looking at the earlier examples, it appears that all we need to do is to add to R all ordered pairs (x, y) such that $(y, x) \in R$. In other words, it looks like the symmetric closure of R will be $R \cup R^{-1}$.

Theorem 4.5.5. *Suppose R is a relation on a set A. Then R has a symmetric closure.*

Proof. Let $S = R \cup R^{-1}$. We will show that S is the symmetric closure of R.

Clearly $R \subseteq S$, so the first clause of the definition of symmetric closure is satisfied. For the second, suppose $(x, y) \in S$. Then by the definition of S, either $(x, y) \in R$ or $(x, y) \in R^{-1}$. If $(x, y) \in R$, then $(y, x) \in R^{-1} \subseteq S$. If $(x, y) \in R^{-1}$ then $(y, x) \in R \subseteq S$. Thus, we can conclude that if $(x, y) \in S$ then $(y, x) \in S$. Since (x, y) was arbitrary, this shows that S is symmetric.

Finally, for the third clause in the definition of symmetric closure, suppose $R \subseteq T \subseteq A \times A$ and T is symmetric. Suppose $(x, y) \in S$. As before, this means that either $(x, y) \in R$ or $(x, y) \in R^{-1}$, so we consider these two possibilities separately. If $(x, y) \in R$ then $(x, y) \in T$, since $R \subseteq T$. If $(x, y) \in R^{-1}$ then $(y, x) \in R$, so since $R \subseteq T$, $(y, x) \in T$. But then since T is symmetric, it follows that $(x, y) \in T$. Since (x, y) was arbitrary, we have shown that $S \subseteq T$, as required. □

Commentary. The overall form of this proof is quite similar to the form of the proof of Theorem 4.5.2, but some of the details are a little trickier. We start by specifying a relation S that will be the symmetric closure of R, and then we prove the three statements in the definition of symmetric closure one at a time. The logical form of the definition of symmetric suggests that to prove that S is symmetric we should start with an arbitrary ordered pair $(x, y) \in S$ and prove $(y, x) \in S$. Because we defined S to be $R \cup R^{-1}$, the assumption that $(x, y) \in S$ means $(x, y) \in R \lor (x, y) \in R^{-1}$, and since this is a disjunction it suggests the use of proof by cases.

As in the proof of Theorem 4.5.2, to prove the third statement in the definition of symmetric closure we let T be an arbitrary relation on A such that $R \subseteq T$ and T is symmetric, and we prove $S \subseteq T$. To prove $S \subseteq T$ we let (x, y) be an arbitrary element of S and prove $(x, y) \in T$. As before, the assumption that $(x, y) \in S$ is a disjunction, so it leads to a proof by cases.

Before moving on to the proof that all relations have transitive closures, we would like to describe an alternative method of proving Theorem 4.5.5. If R is a relation on a set A, then we know that the symmetric closure of R, if it exists, must be the smallest element of the family $\mathcal{F} = \{T \subseteq A \times A \mid R \subseteq T$ and T is symmetric$\}$. Now according to exercise 20 in the last section, the smallest element of a set is also always the greatest lower bound of the set, and by Theorem 4.4.11, the g.l.b. of a nonempty family of sets $\mathcal{F}$ is always $\cap\mathcal{F}$. Thus, if the symmetric closure of R exists, then it must be equal to $\cap\mathcal{F}$. This suggests an alternative approach to proving Theorem 4.5.5. We could define the family $\mathcal{F}$, prove that $\mathcal{F} \neq \varnothing$, and then let $S = \cap\mathcal{F}$ (instead of $R \cup R^{-1}$) and prove that S satisfies the definition of symmetric closure. Recall that we only defined $\cap\mathcal{F}$ for $\mathcal{F} \neq \varnothing$, so we must check that $\mathcal{F} \neq \varnothing$ before we can form $\cap\mathcal{F}$. You are asked in exercise 10 to work out the details of this alternative proof.

There are also two ways we could prove that any relation R on a set A has a transitive closure. One possibility would be to try to form the transitive closure by starting with R and then adding extra ordered pairs to try to create a transitive relation, as we did in the earlier examples. Although this can be done, a careful treatment of the details of this proof would require the method of mathematical induction, which we have not yet discussed. We will present this proof in Chapter 6 after we've discussed mathematical induction. For now, we'll use the alternative method suggested by the last paragraph of letting $\mathcal{F} = \{T \subseteq A \times A \mid R \subseteq T$ and T is transitive$\}$ and then showing that $\mathcal{F} \neq \varnothing$ and $\cap\mathcal{F}$ is the transitive closure of R.

Theorem 4.5.6. *Suppose R is a relation on a set A. Then R has a transitive closure.*

Proof. Let $\mathcal{F} = \{T \subseteq A \times A \mid R \subseteq T$ and T is transitive$\}$. First of all, you should be able to check that $R \subseteq A \times A$ and $A \times A$ is a transitive relation on A, so $A \times A \in \mathcal{F}$, and therefore $\mathcal{F} \neq \varnothing$. Thus, we can let $S = \cap\mathcal{F}$. We will show that S is the transitive closure of R.

To prove the first clause in the definition of transitive closure, suppose $(x, y) \in R$. Let T be an arbitrary element of $\mathcal{F}$. Then by the definition of $\mathcal{F}$, $R \subseteq T$, so $(x, y) \in T$. Since T was arbitrary, this shows that $\forall T \in \mathcal{F}((x, y) \in T)$, so $(x, y) \in \cap\mathcal{F} = S$. Thus, $R \subseteq S$.

For the second clause, suppose $(x, y) \in S$ and $(y, z) \in S$, and again let T be an arbitrary element of $\mathcal{F}$. Then since $(x, y) \in S = \cap\mathcal{F}$, $(x, y) \in T$, and similarly $(y, z) \in T$. But since $T \in \mathcal{F}$, T is transitive, so it follows that $(x, z) \in T$. Since T was arbitrary, we can conclude that $\forall T \in \mathcal{F}((x, z) \in T)$, so $(x, z) \in \cap\mathcal{F} = S$. Thus, we have shown that if $(x, y) \in S$ and $(y, z) \in S$ then $(x, z) \in S$, so S is transitive.

Finally, for the third clause, suppose T is a relation on A, $R \subseteq T$, and T is transitive. Then $T \in \mathcal{F}$, and by exercise 9 of Section 3.3, it follows that $S = \cap\mathcal{F} \subseteq T$. □

Commentary. Once again we start by defining S, but this time the definition $S = \cap\mathcal{F}$ doesn't make sense unless we know $\mathcal{F} \neq \varnothing$, so we must prove this first. Because $\mathcal{F} \neq \varnothing$ means $\exists Q(Q \in \mathcal{F})$, we prove it by giving an example of an element of $\mathcal{F}$. The example is $A \times A$, so we must prove $A \times A \in \mathcal{F}$, and according to the definition of $\mathcal{F}$ this means $A \times A \subseteq A \times A$, $R \subseteq A \times A$, and $A \times A$ is transitive. The statement in the proof that "you should be able to check" that these statements are true really does mean that you should do the checking. In particular, you should verify that $A \times A$ is transitive by assuming that $(x, y) \in A \times A$ and $(y, z) \in A \times A$ and then proving that $(x, z) \in A \times A$.

As in Theorems 4.5.2 and 4.5.5, we must now prove the three statements in the definition of transitive closure. To prove the first statement, $R \subseteq S$, we let (x, y) be an arbitrary element of R and prove $(x, y) \in S$. Since $S = \cap\mathcal{F}$, the goal $(x, y) \in S$ means $\forall T \in \mathcal{F}((x, y) \in T)$, so to prove it we let T be an arbitrary element of $\mathcal{F}$ and prove $(x, y) \in T$. To prove that S is transitive we assume $(x, y) \in S$ and $(y, z) \in S$, and prove $(x, z) \in S$. Once again, by the definition of S this goal means $\forall T \in \mathcal{F}((x, z) \in T)$, so we let T be an arbitrary element of $\mathcal{F}$ and prove $(x, z) \in T$.

Exercises

*1. Find the reflexive, symmetric, and transitive closures of the following relations.
   (a) $A = \{a, b, c\}$, $R = \{(a, a), (a, b), (b, c), (c, b)\}$.
   (b) $R = \{(x, y) \in \mathbb{R} \times \mathbb{R} \mid x < y\}$.
   (c) D_r, as defined in part 3 of Example 4.3.1, for any positive real number r.

2. Find the reflexive, symmetric, and transitive closures of the relations in exercise 4 of Section 4.3.

*3. Suppose R is a relation on A. R is called *asymmetric* if $\forall x \in A \forall y \in A((x, y) \in R \rightarrow (y, x) \notin R)$.
   (a) Show that if R is asymmetric then R is antisymmetric.
   (b) Show that if R is a strict partial order, then R is asymmetric. Note that it follows by part (a) that it is also antisymmetric.

4. Suppose R is a strict partial order on A. Let S be the reflexive closure of R.
   (a) Show that S is a partial order on A.

(b) Show that if R is a strict total order, then S is a total order.

*5. Suppose R is a relation on A. Let $S = R \setminus i_A$.
   (a) Prove that S is the largest element of the set $\{T \subseteq A \times A \mid T \subseteq R$ and T is irreflexive$\}$.
   (b) Prove that if R is a partial order on A, then S is a strict partial order on A.

6. Let P be the set of all people, and let $R = \{(p, q) \in P \times P \mid$ the person p is a parent of the person $q\}$.
   (a) Let S be the transitive closure of R. Describe the relation S.
   (b) Describe the relation $S \circ S^{-1}$.

*7. Suppose R is a relation on A.
   (a) Prove that R is reflexive iff R is its own reflexive closure.
   (b) Do similar theorems hold for symmetry and transitivity? Justify your answers with proofs or counterexamples.

8. Suppose R is a relation on A, and let S be the symmetric closure of R. Prove that $\text{Dom}(S) = \text{Ran}(S) = \text{Dom}(R) \cup \text{Ran}(R)$.

*9. Suppose R is a relation on A, and let S be the transitive closure of R. Prove that $\text{Dom}(S) = \text{Dom}(R)$ and $\text{Ran}(S) = \text{Ran}(R)$.

10. Suppose R is a relation on A. Let $\mathcal{F} = \{T \subseteq A \times A \mid R \subseteq T$ and T is symmetric$\}$. Complete the alternative proof of Theorem 4.5.5 suggested in the text as follows:
   (a) Prove that $\mathcal{F} \neq \varnothing$.
   (b) Let $S = \cap\mathcal{F}$. Prove that S is the symmetric closure of R.

11. Suppose R_1 and R_2 are relations on A and $R_1 \subseteq R_2$.
   (a) Let S_1 and S_2 be the reflexive closures of R_1 and R_2 respectively. Prove that $S_1 \subseteq S_2$.
   (b) Do similar theorems hold for the symmetric and transitive closures? Justify your answers with proofs or counterexamples.

*12. Suppose R_1 and R_2 are relations on A, and let $R = R_1 \cup R_2$.
   (a) Let S_1, S_2, and S be the reflexive closures of R_1, R_2, and R respectively. Prove that $S_1 \cup S_2 = S$.
   (b) Let S_1, S_2, and S be the symmetric closures of R_1, R_2, and R respectively. Prove that $S_1 \cup S_2 = S$.
   (c) Let S_1, S_2, and S be the transitive closures of R_1, R_2, and R respectively. Prove that $S_1 \cup S_2 \subseteq S$, and give an example to show that it may happen that $S_1 \cup S_2 \neq S$.

13. Suppose R_1 and R_2 are relations on A, and let $R = R_1 \cap R_2$.
   (a) Let S_1, S_2, and S be the reflexive closures of R_1, R_2, and R respectively. What is the relationship between $S_1 \cap S_2$ and S? Justify your conclusions with proofs or counterexamples.

(b) Let S_1, S_2, and S be the symmetric closures of R_1, R_2, and R respectively. What is the relationship between $S_1 \cap S_2$ and S? Justify your conclusions with proofs or counterexamples.

(c) Let S_1, S_2, and S be the transitive closures of R_1, R_2, and R respectively. What is the relationship between $S_1 \cap S_2$ and S? Justify your conclusions with proofs or counterexamples.

14. Find an example of relations R_1 and R_2 on some set A such that, if we let $R = R_1 \setminus R_2$ and we let S_1, S_2, and S be the transitive closures of R_1, R_2, and R respectively, then $S_1 \setminus S_2 \not\subseteq S$ and $S \not\subseteq S_1 \setminus S_2$.

*15. Suppose R is a relation on A. The *reflexive symmetric closure* of R is the smallest set $S \subseteq A \times A$ such that $R \subseteq S$, S is reflexive, and S is symmetric, if there is such a smallest set. Prove that every relation has a reflexive symmetric closure.

16. Suppose R is a relation on A, and let S be the reflexive closure of R.
    (a) Prove that if R is symmetric, then so is S.
    (b) Prove that if R is transitive, then so is S.

17. Suppose R is a relation on A, and let S be the transitive closure of R. Prove that if R is symmetric, then so is S. (Hint: Assume that R is symmetric. Prove that $R \subseteq S^{-1}$ and S^{-1} is transitive. What can you conclude about S and S^{-1}?)

*18. Suppose R is a relation on A. The *symmetric transitive closure* of R is the smallest set $S \subseteq A \times A$ such that $R \subseteq S$, S is symmetric, and S is transitive, if there is such a smallest set.

    Let Q be the symmetric closure of R, and let S be the transitive closure of Q. Also, let Q' be the transitive closure of R, and let S' be the symmetric closure of Q'.
    (a) Prove that S is the symmetric transitive closure of R. (Hint: Use exercise 17.)
    (b) Prove that $S' \subseteq S$.
    (c) Must it be the case that $S' = S$? Justify your answer with either a proof or a counterexample.

19. Consider the following putative theorem:

    Theorem? *Suppose R is a reflexive, antisymmetric relation on A. Let S be the transitive closure of R. Then S is a partial order on A.*

    Is the following proof correct? If so, what proof strategies does it use? If not, can it be fixed? Is the theorem correct?

    Proof. R is already reflexive and antisymmetric. To form the relation S we add more ordered pairs to make it transitive as well. Thus, S is reflexive, antisymmetric, and transitive, so it is a partial order. □

*20. A bus company is trying to decide what bus routes to run among the cities in the set $C = \{$San Francisco, Chicago, Dallas, New York, Washington D.C.$\}$. Their routes will be represented by a relation B on C, as in the example in the text. The company wants to make sure you can get from any city in C to any other city in C, so they want to make sure that the transitive closure of B is $C \times C$. Let $\mathcal{F} = \{B \subseteq C \times C \mid$ the transitive closure of B is $C \times C\}$. However, they don't want to run any bus routes unnecessarily, so they want the relation B to be a minimal element of $\mathcal{F}$. (As usual, we mean minimal with respect to the subset ordering on $\mathcal{F}$. You will have to work out what this means, according to the definition of *minimal* in Section 4.4.)
   (a) Find a minimal element of $\mathcal{F}$.
   (b) Does $\mathcal{F}$ have a smallest element?

4.6. Equivalence Relations

We saw in Example 4.3.3 that the identity relation i_A on any set A is always reflexive, symmetric, and transitive. Relations with this combination of properties come up often in mathematics, and they have some important properties that we will investigate in this section. These relations are called *equivalence relations*.

Definition 4.6.1. Suppose R is a relation on a set A. Then R is called an *equivalence relation on* A (or just an *equivalence relation* if A is clear from context) if it is reflexive, symmetric, and transitive.

As we observed earlier, the identity relation i_A on a set A is an equivalence relation. For another example, let T be the set of all triangles, and let C be the relation of congruence of triangles. In other words, $C = \{(s, t) \in T \times T \mid$ the triangle s is congruent to the triangle $t\}$. (Recall that a triangle is congruent to another if it can be moved without distorting it so that it coincides with the other.) Clearly every triangle is congruent to itself, so C is reflexive. Also, if triangle s is congruent to triangle t, then t is congruent to s, so C is symmetric; and if r is congruent to s and s is congruent to t, then r is congruent to t, so C is transitive. Thus, C is an equivalence relation on T.

As another example, let P be the set of all people, and let $B = \{(p, q) \in P \times P \mid$ the person p has the same birthday as the person $q\}$. (By "same birthday" we mean same month and day, but not necessarily the same year.) Everyone has the same birthday as himself, so B is reflexive. If p has the same birthday as q, then q has the same birthday as p, so B is symmetric. And if p has the same birthday as q and q has the same birthday as r, then p has the same birthday as r, so B is transitive. Therefore B is an equivalence relation.

It may be instructive to look at the relation B more closely. We can think of this relation as splitting the set P of all people into 366 categories, one for each possible birthday. (Remember, some people were born on February 29th!) An ordered pair of people will be an element of B if the people come from the same category, but will not be an element of B if the people come from different categories. We could think of these categories as forming a family of subsets of P, which we could write as an indexed family as follows. First of all, let D be the set of all possible birthdays. In other words, $D = \{$Jan. 1, Jan. 2, Jan. 3, ..., Dec. 30, Dec. 31$\}$. Now for each $d \in D$, let $P_d = \{p \in P \mid$ the person p was born on the day $d\}$. Then the family $\mathcal{F} = \{P_d \mid d \in D\}$ is an indexed family of subsets of P. The elements of $\mathcal{F}$ are called *equivalence classes* for the relation B, and every person is an element of exactly one of these equivalence classes. The relation B consists of those pairs $(p, q) \in P \times P$ such that the people p and q are in the same equivalence class. In other words,

$$\begin{aligned} B &= \{(p, q) \in P \times P \mid \exists d \in D(p \in P_d \text{ and } q \in P_d)\} \\ &= \{(p, q) \in P \times P \mid \exists d \in D((p, q) \in P_d \times P_d)\} \\ &= \bigcup_{d \in D} (P_d \times P_d). \end{aligned}$$

We will call the family $\mathcal{F}$ a *partition* of P because it breaks the set P into disjoint pieces. It turns out that every equivalence relation on a set A determines a partition of A, whose elements are the equivalence classes for the equivalence relation. But before we can work out the details of why this is true, we must define the terms *partition* and *equivalence class* more carefully.

Definition 4.6.2. Suppose A is a set and $\mathcal{F} \subseteq \mathscr{P}(A)$. We will say that $\mathcal{F}$ is *pairwise disjoint* if every pair of distinct elements of $\mathcal{F}$ are disjoint, or in other words $\forall X \in \mathcal{F} \forall Y \in \mathcal{F}(X \neq Y \rightarrow X \cap Y = \varnothing)$. (This concept was discussed in exercise 5 of Section 3.6.) $\mathcal{F}$ is called a *partition* of A if it has the following properties:

1. $\cup\mathcal{F} = A$.
2. $\mathcal{F}$ is pairwise disjoint.
3. $\forall X \in \mathcal{F}(X \neq \varnothing)$.

For example, suppose $A = \{1, 2, 3, 4\}$ and $\mathcal{F} = \{\{2\}, \{1, 3\}, \{4\}\}$. Then $\cup\mathcal{F} = \{2\} \cup \{1, 3\} \cup \{4\} = \{1, 2, 3, 4\} = A$, so $\mathcal{F}$ satisfies the first clause in the definition of partition. Also, no two sets in $\mathcal{F}$ have any elements in common, so $\mathcal{F}$ is pairwise disjoint, and clearly all the sets in $\mathcal{F}$ are nonempty. Thus,

$\mathcal{F}$ is a partition of A. On the other hand, the family $\mathcal{G} = \{\{1, 2\}, \{1, 3\}, \{4\}\}$ is not pairwise disjoint, because $\{1, 2\} \cap \{1, 3\} = \{1\} \neq \varnothing$, so it is not a partition of A. The family $\mathcal{H} = \{\varnothing, \{2\}, \{1, 3\}, \{4\}\}$ is also not a partition of A, because it fails on the third requirement in the definition.

Definition 4.6.3. Suppose R is an equivalence relation on a set A, and $x \in A$. Then the *equivalence class of x with respect to R* is the set

$$[x]_R = \{y \in A \mid yRx\}.$$

If R is clear from context, then we just write $[x]$ instead of $[x]_R$. The set of all equivalence classes of elements of A is called *A modulo R*, and is denoted A/R. Thus,

$$A/R = \{[x]_R \mid x \in A\} = \{X \subseteq A \mid \exists x \in A(X = [x]_R)\}.$$

In the case of the same-birthday relation B, if p is any person, then according to Definition 4.6.3,

$$\begin{aligned}[p]_B &= \{q \in P \mid qBp\} \\ &= \{q \in P \mid \text{the person } q \text{ has the same birthday as the person } p\}.\end{aligned}$$

For example, if John was born on August 10, then

$$\begin{aligned}[\text{John}]_B &= \{q \in P \mid \text{ the person } q \text{ has the same birthday as John}\} \\ &= \{q \in P \mid \text{ the person } q \text{ was born on August 10}\}.\end{aligned}$$

In the notation we introduced earlier, this is just the set P_d, for $d =$ August 10. In fact, it should be clear now that for any person p, if we let d be p's birthday, then $[p]_B = P_d$. This is in agreement with our earlier statement that the sets P_d are the equivalence classes for the equivalence relation B. According to Definition 4.6.3, the set of all of these equivalence classes is called P modulo B:

$$P/B = \{[p]_B \mid p \in P\} = \{P_d \mid d \in D\}.$$

You are asked to give a more careful proof of this equation in exercise 5. As we observed before, this family is a partition of P.

Let's consider one more example. Let S be the relation on $\mathbb{R}$ defined as follows:

$$S = \{(x, y) \in \mathbb{R} \times \mathbb{R} \mid x - y \in \mathbb{Z}\}.$$

For example, $(5.73, 2.73) \in S$ and $(-1.27, 2.73) \in S$, since $5.73 - 2.73 = 3 \in \mathbb{Z}$ and $-1.27 - 2.73 = -4 \in \mathbb{Z}$, but $(1.27, 2.73) \notin S$, since $1.27 - 2.73 = -1.46 \notin \mathbb{Z}$. Clearly for any $x \in \mathbb{R}$, $x - x = 0 \in \mathbb{Z}$, so $(x, x) \in S$, and therefore

S is reflexive. To see that S is symmetric, suppose $(x, y) \in S$. By the definition of S, this means that $x - y \in \mathbb{Z}$. But then $y - x = -(x - y) \in \mathbb{Z}$ too, since the negative of any integer is also an integer, so $(y, x) \in S$. Because (x, y) was an arbitrary element of S, this shows that S is symmetric. Finally, to see that S is transitive, suppose that $(x, y) \in S$ and $(y, z) \in S$. Then $x - y \in \mathbb{Z}$ and $y - z \in \mathbb{Z}$. Because the sum of any two integers is an integer, it follows that $x - z = (x - y) + (y - z) \in \mathbb{Z}$, so $(x, z) \in S$, as required. Thus, S is an equivalence relation on $\mathbb{R}$.

What do the equivalence classes for this equivalence relation look like? We have already observed that $(5.73, 2.73) \in S$ and $(-1.27, 2.73) \in S$, so $5.73 \in [2.73]$ and $-1.27 \in [2.73]$. In fact, it is not hard to see what the other elements of this equivalence class will be:

$$[2.73] = \{\ldots, -1.27, -0.27, 0.73, 1.73, 2.73, 3.73, 4.73, 5.73, \ldots\}.$$

In other words, the equivalence class contains all positive real numbers of the form "__.73" and all negative real numbers of the form "–__.27." In general, for any real number x, the equivalence class of x will contain all real numbers that differ from x by an integer amount:

$$[x] = \{\ldots, x - 3, x - 2, x - 1, x, x + 1, x + 2, x + 3, \ldots\}.$$

Here are a few facts about these equivalence classes that you might try to prove to yourself. As you can see in the last equation, x is always an element of $[x]$. If we choose any number $x \in [2.73]$, then $[x]$ will be exactly the same as $[2.73]$. For example, taking $x = 4.73$ we find that

$$[4.73] = \{\ldots, -1.27, -0.27, 0.73, 1.73, 2.73, 3.73, \\ 4.73, 5.73, \ldots\} = [2.73].$$

Thus, $[4.73]$ and $[2.73]$ are just two different names for the same set. But if we choose $x \notin [2.73]$, then $[x]$ will be different from $[2.73]$. For example,

$$[1.3] = \{\ldots, -1.7, -0.7, 0.3, 1.3, 2.3, 3.3, 4.3, \ldots\}.$$

In fact, you can see from these equations that $[1.3]$ and $[2.73]$ have no elements in common. In other words, $[1.3]$ is actually *disjoint* from $[2.73]$. In general, for any two real numbers x and y, the equivalence classes $[x]$ and $[y]$ are either identical or disjoint. Each equivalence class has many different names, but different equivalence classes are disjoint. Because $[x]$ always contains x as an element, every equivalence class is nonempty, and every real number x is in exactly one equivalence class, namely $[x]$. In other words, the set of all of the equivalence classes, $\mathbb{R}/S$, is a partition of $\mathbb{R}$. This is another illustration of the fact that the equivalence classes determined by an equivalence relation always form a partition.

Theorem 4.6.4. *Suppose R is an equivalence relation on a set A. Then A/R is a partition of A.*

The proof of Theorem 4.6.4 will be easier to understand if we first prove a few facts about equivalence classes. Facts that are proven primarily for the purpose of using them to prove a theorem are usually called *lemmas.*

Lemma 4.6.5. *Suppose R is an equivalence relation on A. Then:*

1. *For every $x \in A$, $x \in [x]$.*
2. *For every $x \in A$ and $y \in A$, $y \in [x]$ iff $[y] = [x]$.*

Proof.

1. Let $x \in A$ be arbitrary. Since R is reflexive, xRx. Therefore, by the definition of equivalence class, $x \in [x]$.
2. ($\rightarrow$) Suppose $y \in [x]$. Then by the definition of equivalence class, yRx. Now suppose $z \in [y]$. Then zRy. Since zRy and yRx, by transitivity of R we can conclude that zRx, so $z \in [x]$. Since z was arbitrary, this shows that $[y] \subseteq [x]$.

   Now suppose $z \in [x]$, so zRx. We already know yRx, and since R is symmetric we can conclude that xRy. Applying transitivity to zRx and xRy, we can conclude that zRy, so $z \in [y]$. Therefore $[x] \subseteq [y]$, so $[x] = [y]$.

   ($\leftarrow$) Suppose $[y] = [x]$. By part 1 we know that $y \in [y]$, so since $[y] = [x]$, it follows that $y \in [x]$.

□

Commentary

1. According to the definition of equivalence classes, $x \in [x]$ means xRx. This is what leads us to apply the fact that R is reflexive.
2. Of course, the iff form of the goal leads us to prove both directions separately. For the $\rightarrow$ direction, the goal is $[y] = [x]$, and, since $[y]$ and $[x]$ are sets, we can prove this by proving $[y] \subseteq [x]$ and $[x] \subseteq [y]$. We prove each of these statements by the usual method of taking an arbitrary element of one set and proving that it is in the other. Throughout the proof we use the definition of equivalence classes repeatedly, as we did in the proof of statement 1.

Proof of Theorem 4.6.4. To prove that A/R is a partition of A, we must prove the three properties in Definition 4.6.2. For the first, we must show that $\cup(A/R) = A$, or in other words that $\cup_{x \in A}[x] = A$. Now every equivalence class in A/R is a subset of A, so it should be clear that their union is also a subset of A. Thus, $\cup(A/R) \subseteq A$, so all we need to show to finish the proof is that $A \subseteq \cup(A/R)$. To prove this, suppose $x \in A$. Then by Lemma 4.6.5, $x \in [x]$, and of course $[x] \in A/R$, so $x \in \cup(A/R)$. Thus, $\cup(A/R) = A$.

To see that A/R is pairwise disjoint, suppose that X and Y are two elements of A/R, and $X \cap Y \neq \varnothing$. By definition of A/R, X and Y are equivalence classes, so we must have $X = [x]$ and $Y = [y]$ for some $x, y \in A$. Since $X \cap Y \neq \varnothing$, we can choose some z such that $z \in X \cap Y = [x] \cap [y]$. Now by Lemma 4.6.5, since $z \in [x]$ and $z \in [y]$, it follows that $[x] = [z] = [y]$. Thus, $X = Y$. This shows that if $X \neq Y$ then $X \cap Y = \varnothing$, so A/R is pairwise disjoint.

Finally, for the last clause of the definition of partition, suppose $X \in A/R$. As before, this means that $X = [x]$ for some $x \in A$. Now by Lemma 4.6.5, $x \in [x] = X$, so $X \neq \varnothing$, as required. □

Commentary. We have given an intuitive reason why $\cup(A/R) \subseteq A$, but if you're not sure why this is correct, you should write out a formal proof. (You might also want to look at exercise 16 in Section 3.3.) The proof that $A \subseteq \cup(A/R)$ is straightforward.

The definition of pairwise disjoint suggests that to prove that A/R is pairwise disjoint we should let X and Y be arbitrary elements of A/R and then prove $X \neq Y \rightarrow X \cap Y = \varnothing$. Recall that the statement that a set is empty is really a negative statement, so both the antecedent and the consequent of this conditional are negative. This suggests that it will probably be easier to prove the contrapositive, so we assume $X \cap Y \neq \varnothing$ and prove $X = Y$. The givens $X \in A/R$, $Y \in A/R$, and $X \cap Y \neq \varnothing$ are all existential statements, so we use them to introduce the variables x, y, and z. Lemma 4.6.5 now takes care of the proof that $X = Y$ as well as the proof of the final clause in the definition of partition.

Theorem 4.6.4 shows that if R is an equivalence relation on A then A/R is a partition of A. In fact, it turns out that *every* partition of A arises in this way.

Theorem 4.6.6. *Suppose A is a set and $\mathcal{F}$ is a partition of A. Then there is an equivalence relation R on A such that $A/R = \mathcal{F}$.*

Before proving this theorem, it might be worthwhile to discuss the strategy for the proof briefly. Because the conclusion of the theorem is an existential statement, we should try to find an equivalence relation R such that $A/R = \mathcal{F}$.

Clearly for different choices of $\mathcal{F}$ we will need to choose R differently, so the definition of R should depend on $\mathcal{F}$ in some way. Looking back at the same-birthday example may help you see how to proceed. Recall that in that example the equivalence relation B consisted of all pairs of people (p, q) such that p and q were in the same set in the partition $\{P_d \mid d \in D\}$. In fact, we found that we could also express this by saying that $B = \cup_{d \in D}(P_d \times P_d)$. This suggests that in the proof of Theorem 4.6.6 we should let R be the set of all pairs $(x, y) \in A \times A$ such that x and y are in the same set in the partition $\mathcal{F}$. An alternative way to write this would be $R = \cup_{X \in \mathcal{F}}(X \times X)$.

For example, consider again the example of a partition given after Definition 4.6.2. In that example we had $A = \{1, 2, 3, 4\}$ and $\mathcal{F} = \{\{2\}, \{1, 3\}, \{4\}\}$. Now let's define a relation R on A as suggested in the last paragraph. This gives us:

$$\begin{aligned} R &= \bigcup_{X \in \mathcal{F}} (X \times X) \\ &= (\{2\} \times \{2\}) \cup (\{1, 3\} \times \{1, 3\}) \cup (\{4\} \times \{4\}) \\ &= \{(2, 2)\} \cup \{(1, 1), (1, 3), (3, 1), (3, 3)\} \cup \{(4, 4)\} \\ &= \{(2, 2), (1, 1), (1, 3), (3, 1), (3, 3), (4, 4)\}. \end{aligned}$$

The directed graph for this relation is shown in Figure 1. We will let you check that R is an equivalence relation and that the equivalence classes are

$$[2] = \{2\}, \quad [1] = [3] = \{1, 3\}, \quad [4] = \{4\}.$$

Thus, the set of all equivalence classes is $A/R = \{\{2\}, \{1, 3\}, \{4\}\}$, which is precisely the same as the partition $\mathcal{F}$ we started with.

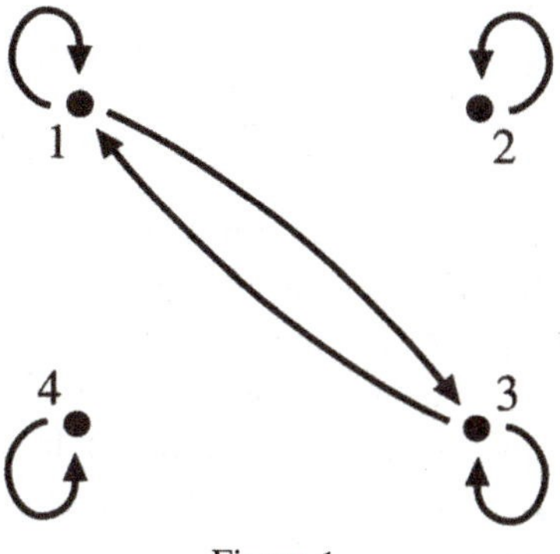

Figure 1

Of course, the reasoning that led us to the formula $R = \cup_{X \in \mathcal{F}}(X \times X)$ will not be part of the proof of Theorem 4.6.6. When we write the proof, we can simply define R in this way and then verify that it is an equivalence relation on A and that $A/R = \mathcal{F}$. It may make the proof easier to follow if we once again prove some lemmas first.

Lemma 4.6.7. *Suppose A is a set and $\mathcal{F}$ is a partition of A. Let $R = \bigcup_{X\in\mathcal{F}}(X \times X)$. Then R is an equivalence relation on A. We will call R the* equivalence relation determined by $\mathcal{F}$.

Proof. We'll prove that R is reflexive and leave the rest for you to do in exercise 7. Let x be an arbitrary element of A. Since $\mathcal{F}$ is a partition of A, $\bigcup\mathcal{F} = A$, so $x \in \bigcup\mathcal{F}$. Thus, we can choose some $X \in \mathcal{F}$ such that $x \in X$. But then $(x, x) \in X \times X$, so $(x, x) \in \bigcup_{X\in\mathcal{F}}(X \times X) = R$. Therefore, R is reflexive. □

Commentary. After letting x be an arbitrary element of A, we must prove $(x, x) \in R$. Because $R = \bigcup_{X\in\mathcal{F}}(X \times X)$, this means we must prove $\exists X \in \mathcal{F}((x, x) \in X \times X)$, or in other words $\exists X \in \mathcal{F}(x \in X)$. But this just means $x \in \bigcup\mathcal{F}$, so this suggests using the first clause in the definition of partition, which says that $\bigcup\mathcal{F} = A$.

Lemma 4.6.8. *Suppose A is a set and $\mathcal{F}$ is a partition of A. Let R be the equivalence relation determined by $\mathcal{F}$. Suppose $X \in \mathcal{F}$ and $x \in X$. Then $[x]_R = X$.*

Proof. Suppose $y \in [x]_R$. Then $(y, x) \in R$, so by the definition of R there must be some $Y \in \mathcal{F}$ such that $(y, x) \in Y \times Y$, and therefore $y \in Y$ and $x \in Y$. Since $x \in X$ and $x \in Y$, $X \cap Y \neq \emptyset$, and since $\mathcal{F}$ is pairwise disjoint it follows that $X = Y$. Thus, since $y \in Y$, $y \in X$. Since y was an arbitrary element of $[x]_R$, we can conclude that $[x]_R \subseteq X$.

Now suppose $y \in X$. Then $(y, x) \in X \times X$, so $(y, x) \in R$ and therefore $y \in [x]_R$. Thus $X \subseteq [x]_R$, so $[x]_R = X$. □

Commentary. To prove $[x]_R = X$ we prove $[x]_R \subseteq X$ and $X \subseteq [x]_R$. For the first we start with an arbitrary $y \in [x]_R$ and prove $y \in X$. Writing out the definition of $[x]_R$ we get $(y, x) \in R$, and since R was defined to be $\bigcup_{Y\in\mathcal{F}}(Y \times Y)$, this means $\exists Y \in \mathcal{F}((y, x) \in Y \times Y)$. Of course, since this is an existential statement we immediately introduce the new variable Y by existential instantiation. Since this gives us $y \in Y$ and our goal is $y \in X$, it is not surprising that the proof is completed by proving $Y = X$.

The proof that $X \subseteq [x]_R$ also uses the definitions of $[x]_R$ and R, but is more straightforward.

Proof of Theorem 4.6.6. Let $R = \bigcup_{X\in\mathcal{F}}(X \times X)$. We have already seen that R is an equivalence relation, so we need only check that $A/R = \mathcal{F}$. To see this, suppose $X \in A/R$. This means that $X = [x]$ for some $x \in A$. Since $\mathcal{F}$ is a partition, we know that $\bigcup\mathcal{F} = A$, so $x \in \bigcup\mathcal{F}$, and therefore we can choose some

$Y \in \mathcal{F}$ such that $x \in Y$. But then by Lemma 4.6.8, $[x] = Y$. Thus $X = Y \in \mathcal{F}$, so $A/R \subseteq \mathcal{F}$.

Now suppose $X \in \mathcal{F}$. Then since $\mathcal{F}$ is a partition, $X \neq \varnothing$, so we can choose some $x \in X$. Therefore by Lemma 4.6.8, $X = [x] \in A/R$, so $\mathcal{F} \subseteq A/R$. Thus, $A/R = \mathcal{F}$. □

Commentary. We prove that $A/R = \mathcal{F}$ by proving that $A/R \subseteq \mathcal{F}$ and $\mathcal{F} \subseteq A/R$. For the first, we take an arbitrary $X \in A/R$ and prove that $X \in \mathcal{F}$. Because $X \in A/R$ means $\exists x \in A(X = [x])$, we immediately introduce the new variable x to stand for an element of A such that $X = [x]$. The proof that $X \in \mathcal{F}$ now proceeds by the slightly roundabout route of finding a set $Y \in \mathcal{F}$ such that $X = Y$. This is motivated by Lemma 4.6.8, which suggests a way of showing that an element of $\mathcal{F}$ is equal to $[x] = X$. The proof that $\mathcal{F} \subseteq A/R$ also relies on Lemma 4.6.8.

We have seen how an equivalence relation R on a set A can be used to define a partition A/R of A and also how a partition $\mathcal{F}$ of A can be used to define an equivalence relation $\cup_{X \in \mathcal{F}}(X \times X)$ on A. The proof of Theorem 4.6.6 demonstrates an interesting relationship between these operations. If you start with a partition $\mathcal{F}$ of A, use $\mathcal{F}$ to define the equivalence relation $R = \cup_{X \in \mathcal{F}}(X \times X)$ and then use R to define a partition A/R, then you end up back where you started. In other words, the final partition A/R is the same as the original partition $\mathcal{F}$. You might wonder if the same idea would work in the other order. In other words, suppose you start with an equivalence relation R on A, use R to define a partition $\mathcal{F} = A/R$, and then use $\mathcal{F}$ to define an equivalence relation $S = \cup_{X \in \mathcal{F}}(X \times X)$. Would the final equivalence relation S be the same as the original equivalence relation R? You are asked in exercise 9 to show that the answer is yes.

We end this section by considering a few more examples of equivalence relations. A very useful family of equivalence relations is given by the next definition.

Definition 4.6.9. Suppose m is a positive integer. For any integers x and y, we will say that x is *congruent* to y *modulo* m if $\exists k \in \mathbb{Z}(x - y = km)$. In other words, x is congruent to y modulo m iff $m \mid (x - y)$. We will use the notation $x \equiv y \pmod{m}$ to mean that x is congruent to y modulo m.

For example, $12 \equiv 27 \pmod{5}$, since $12 - 27 = -15 = (-3) \cdot 5$. Now it turns out that for every positive integer m, the relation $C_m = \{(x, y) \in \mathbb{Z} \times \mathbb{Z} \mid x \equiv y \pmod{m}\}$ is an equivalence relation on $\mathbb{Z}$. We will check transitivity for C_m and let you check reflexivity and symmetry in exercise 10.

To see that C_m is transitive, suppose that $x \equiv y \pmod{m}$ and $y \equiv z$ (mod m). Then $m \mid (x - y)$ and $m \mid (y - z)$, so by exercise 18(a) in Section 3.3, $m \mid [(x - y) + (y - z)]$. But $(x - y) + (y - z) = x - z$, so it follows that $m \mid (x - z)$, and therefore $x \equiv z \pmod{m}$. For more about these equivalence relations, see exercise 10.

Equivalence relations often come up when we want to group together elements of a set that have something in common. For example, if you've studied vectors in a previous math course or perhaps in a physics course, then you may have been told that vectors can be thought of as arrows. But you were probably also told that different arrows that point in the same direction and have the same length must be thought of as representing the same vector. Here's a more lucid explanation of the relationship between vectors and arrows. Let A be the set of all arrows, and let $R = \{(x, y) \in A \times A \mid$ the arrows x and y point in the same direction and have the same length$\}$. We will let you check for yourself that R is an equivalence relation on A. Each equivalence class consists of arrows that all have the same length and point in the same direction. We can now think of vectors as being represented, not by arrows, but by equivalence classes of arrows.

Students who are familiar with computer programming may be interested in our next example. Suppose we let P be the set of all computer programs, and for any computer programs p and q we say that p and q are *equivalent* if they always produce the same output when given the same input. Let $R = \{(p, q) \in P \times P \mid$ the programs p and q are equivalent$\}$. It is not hard to check that R is an equivalence relation on P. The equivalence classes group together programs that produce the same output when given the same input.

Exercises

*1. Find all partitions of the set $A = \{1, 2, 3\}$.

2. Find all equivalence relations on the set $A = \{1, 2, 3\}$.

*3. Let W = the set of all words in the English language. Which of the following relations on W are equivalence relations? For those that are equivalence relations, what are the equivalence classes?
   (a) $R = \{(x, y) \in W \times W \mid$ the words x and y start with the same letter$\}$.
   (b) $S = \{(x, y) \in W \times W \mid$ the words x and y have at least one letter in common$\}$.
   (c) $T = \{(x, y) \in W \times W \mid$ the words x and y have the same number of letters$\}$.

4. Which of the following relations on $\mathbb{R}$ are equivalence relations? For those that are equivalence relations, what are the equivalence classes?
   (a) $R = \{(x, y) \in \mathbb{R} \times \mathbb{R} \mid x - y \in \mathbb{N}\}$.
   (b) $S = \{(x, y) \in \mathbb{R} \times \mathbb{R} \mid x - y \in \mathbb{Q}\}$.
   (c) $T = \{(x, y) \in \mathbb{R} \times \mathbb{R} \mid \exists n \in \mathbb{Z}(y = x \cdot 10^n)\}$.

*5. In the discussion of the same-birthday equivalence relation B, we claimed that $P/B = \{P_d \mid d \in D\}$. Give a careful proof of this claim. You will find when you work out the proof that there is an assumption you must make about people's birthdays (a very reasonable assumption) to make the proof work. What is this assumption?

6. Let T be the set of all triangles, and let $S = \{(s, t) \in T \times T \mid$ the triangles s and t are similar$\}$. (Recall that two triangles are similar if the angles of one triangle are equal to the corresponding angles of the other.) Verify that S is an equivalence relation.

7. Complete the proof of Lemma 4.6.7.

8. Suppose R and S are equivalence relations on A and $A/R = A/S$. Prove that $R = S$.

*9. Suppose R is an equivalence relation on A. Let $\mathcal{F} = A/R$, and let S be the equivalence relation determined by $\mathcal{F}$. In other words, $S = \cup_{X \in \mathcal{F}}(X \times X)$. Prove that $S = R$.

10. Let C_m be the congruence mod m relation defined in the text, for a positive integer m.
    (a) Complete the proof that C_m is an equivalence relation on $\mathbb{Z}$ by showing that it is reflexive and symmetric.
    (b) Find all the equivalence classes for C_2 and C_3. How many equivalence classes are there in each case? In general, how many equivalence classes do you think there are for C_m?

11. Prove that for every integer n, either $n^2 \equiv 0 \pmod 4$ or $n^2 \equiv 1 \pmod 4$.

*12. Suppose m is a positive integer. Prove that for all integers a, b, c, and d, if $a \equiv c \pmod m$ and $b \equiv d \pmod m$ then $a + b \equiv c + d \pmod m$ and $ab \equiv cd \pmod m$.

13. Suppose R is an equivalence relation on A and $B \subseteq A$. Let $S = R \cap (B \times B)$.
    (a) Prove that S is an equivalence relation on B.
    (b) Prove that for all $x \in B$, $[x]_S = [x]_R \cap B$.

14. Suppose $B \subseteq A$, and define a relation R on $\mathscr{P}(A)$ as follows:

$$R = \{(X, Y) \in \mathscr{P}(A) \times \mathscr{P}(A) \mid (X \Delta Y) \subseteq B\}.$$

    (a) Prove that R is an equivalence relation on $\mathscr{P}(A)$.

(b) Prove that for every $X \in \mathscr{P}(A)$ there is exactly one $Y \in [X]_R$ such that $Y \cap B = \varnothing$.

*15. Suppose $\mathcal{F}$ is a partition of A, $\mathcal{G}$ is a partition of B, and A and B are disjoint. Prove that $\mathcal{F} \cup \mathcal{G}$ is a partition of $A \cup B$.

16. Suppose R is an equivalence relation on A, S is an equivalence relation on B, and A and B are disjoint.
    (a) Prove that $R \cup S$ is an equivalence relation on $A \cup B$.
    (b) Prove that for all $x \in A$, $[x]_{R\cup S} = [x]_R$, and for all $y \in B$, $[y]_{R\cup S} = [y]_S$.
    (c) Prove that $(A \cup B)/(R \cup S) = (A/R) \cup (B/S)$.

17. Suppose $\mathcal{F}$ and $\mathcal{G}$ are partitions of a set A. We define a new family of sets $\mathcal{F} \cdot \mathcal{G}$ as follows:

$$\mathcal{F} \cdot \mathcal{G} = \{Z \in \mathscr{P}(A) \mid Z \neq \varnothing \text{ and } \exists X \in \mathcal{F} \exists Y \in \mathcal{G}(Z = X \cap Y)\}.$$

Prove that $\mathcal{F} \cdot \mathcal{G}$ is a partition of A.

18 Let $\mathcal{F} = \{\mathbb{R}^-, \mathbb{R}^+, \{0\}\}$ and $\mathcal{G} = \{\mathbb{Z}, \mathbb{R} \setminus \mathbb{Z}\}$, and note that both $\mathcal{F}$ and $\mathcal{G}$ are partitions of $\mathbb{R}$. List the elements of $\mathcal{F} \cdot \mathcal{G}$. (See exercise 17 for the meaning of the notation used here.)

*19. Suppose R and S are equivalence relations on a set A. Let $T = R \cap S$.
    (a) Prove that T is an equivalence relation on A.
    (b) Prove that for all $x \in A$, $[x]_T = [x]_R \cap [x]_S$.
    (c) Prove that $A/T = (A/R) \cdot (A/S)$. (See exercise 17 for the meaning of the notation used here.)

20. Suppose $\mathcal{F}$ is a partition of A and $\mathcal{G}$ is a partition of B. We define a new family of sets $\mathcal{F} \otimes \mathcal{G}$ as follows:

$$\mathcal{F} \otimes \mathcal{G} = \{Z \in \mathscr{P}(A \times B) \mid \exists X \in \mathcal{F} \exists Y \in \mathcal{G}(Z = X \times Y)\}.$$

Prove that $\mathcal{F} \otimes \mathcal{G}$ is a partition of $A \times B$.

*21. Let $\mathcal{F} = \{\mathbb{R}^-, \mathbb{R}^+, \{0\}\}$, which is a partition of $\mathbb{R}$. List the elements of $\mathcal{F} \otimes \mathcal{F}$, and describe them geometrically as subsets of the xy-plane. (See exercise 20 for the meaning of the notation used here.)

22. Suppose R is an equivalence relation on A and S is an equivalence relation on B. Define a relation T on $A \times B$ as follows:

$$T = \{((a, b), (a', b')) \in (A \times B) \times (A \times B) \mid aRa' \text{ and } bSb'\}.$$

    (a) Prove that T is an equivalence relation on $A \times B$.
    (b) Prove that if $a \in A$ and $b \in B$ then $[(a, b)]_T = [a]_R \times [b]_S$.
    (c) Prove that $(A \times B)/T = (A/R) \otimes (B/S)$. (See exercise 20 for the meaning of the notation used here.)

*23. Suppose R and S are relations on a set A, and S is an equivalence relation. We will say that R is *compatible* with S if for all x, y, x', and y' in A, if xSx' and ySy' then xRy iff $x'Ry'$.
   (a) Prove that if R is compatible with S, then there is a unique relation T on A/S such that for all x and y in A, $[x]_S T[y]_S$ iff xRy.
   (b) Suppose T is a relation on A/S and for all x and y in A, $[x]_S T[y]_S$ iff xRy. Prove that R is compatible with S.

24. Suppose R is a relation on A and R is reflexive and transitive. (Such a relation is called a *preorder* on A.) Let $S = R \cap R^{-1}$.
   (a) Prove that S is an equivalence relation on A.
   (b) Prove that there is a unique relation T on A/S such that for all x and y in A, $[x]_S T[y]_S$ iff xRy. (Hint: Use exercise 23.)
   (c) Prove that T is a partial order on A/S, where T is the relation from part (b).

25. Let $I = \{1, 2, \ldots, 100\}$, $A = \mathscr{P}(I)$, and $R = \{(X, Y) \in A \times A \mid Y$ has at least as many elements as $X\}$.
   (a) Prove that R is a preorder on A. (See exercise 24 for the definition of *preorder*.)
   (b) Let S and T be defined as in exercise 24. Describe the elements of A/S and the partial order T. How many elements does A/S have? Is T a total order?

26. Suppose A is a set. If $\mathcal{F}$ and $\mathcal{G}$ are partitions of A, then we'll say that $\mathcal{F}$ *refines* $\mathcal{G}$ if $\forall X \in \mathcal{F} \exists Y \in \mathcal{G}(X \subseteq Y)$. Let P be the set of all partitions of A, and let $R = \{(\mathcal{F}, \mathcal{G}) \in P \times P \mid \mathcal{F}$ refines $\mathcal{G}\}$.
   (a) Prove that R is a partial order on P.
   (b) Suppose that S and T are equivalence relations on A. Let $\mathcal{F} = A/S$ and $\mathcal{G} = A/T$. Prove that $S \subseteq T$ iff $\mathcal{F}$ refines $\mathcal{G}$.
   (c) Suppose $\mathcal{F}$ and $\mathcal{G}$ are partitions of A. Prove that $\mathcal{F} \cdot \mathcal{G}$ is the greatest lower bound of the set $\{\mathcal{F}, \mathcal{G}\}$ in the partial order R. (See exercise 17 for the meaning of the notation used here.)

5

Functions

5.1. Functions

Suppose P is the set of all people, and let $H = \{(p, n) \in P \times \mathbb{N} \mid \text{the person } p \text{ has } n \text{ children}\}$. Then H is a relation from P to $\mathbb{N}$, and it has the following important property. For every $p \in P$, there is *exactly one* $n \in \mathbb{N}$ such that $(p, n) \in H$. Mathematicians express this by saying that H is a *function* from P to $\mathbb{N}$.

Definition 5.1.1. Suppose F is a relation from A to B. Then F is called a *function from A to B* if for every $a \in A$ there is exactly one $b \in B$ such that $(a, b) \in F$. In other words, to say that F is a function from A to B means:

$$\forall a \in A \exists! b \in B((a, b) \in F).$$

To indicate that F is a function from A to B, we will write $F : A \to B$.

Example 5.1.2.

1. Let $A = \{1, 2, 3\}$, $B = \{4, 5, 6\}$, and $F = \{(1, 5), (2, 4), (3, 5)\}$. Is F a function from A to B?
2. Let $A = \{1, 2, 3\}$, $B = \{4, 5, 6\}$, and $G = \{(1, 5), (2, 4), (1, 6)\}$. Is G a function from A to B?
3. Let C be the set of all cities, N the set of all countries, and let $L = \{(c, n) \in C \times N \mid \text{the city } c \text{ is in the country } n\}$. Is L a function from C to N?
4. Let P be the set of all people, and let $C = \{(p, q) \in P \times P \mid \text{the person } p \text{ is a parent of the person } q\}$. Is C a function from P to P?
5. Let P be the set of all people, and let $D = \{(p, x) \in P \times \mathscr{P}(P) \mid x = \text{the set of all children of } p\}$. Is D a function from P to $\mathscr{P}(P)$?

6. Let A be any set. Recall that $i_A = \{(a, a) \mid a \in A\}$ is called the identity relation on A. Is it a function from A to A?
7. Let $f = \{(x, y) \in \mathbb{R} \times \mathbb{R} \mid y = x^2\}$. Is f a function from $\mathbb{R}$ to $\mathbb{R}$?

Solutions

1. Yes. Note that 1 is paired with 5 in the relation F, but it is not paired with any other element of B. Similarly, 2 is paired only with 4, and 3 with 5. In other words, each element of A appears as the first coordinate of exactly one ordered pair in F. Therefore F is a function from A to B. Note that the definition of function does *not* require that each element of B be paired with exactly one element of A. Thus, it doesn't matter that 5 occurs as the second coordinate of two different pairs in F and that 6 doesn't occur in any ordered pairs at all.
2. No. G fails to be a function from A to B for two reasons. First of all, 3 isn't paired with any element of B in the relation G, which violates the requirement that every element of A must be paired with some element of B. Second, 1 is paired with two different elements of B, 5 and 6, which violates the requirement that each element of A be paired with *only one* element of B.
3. If we make the reasonable assumption that every city is in exactly one country, then L is a function from C to N.
4. Because some people have no children and some people have more than one child, C is not a function from P to P.
5. Yes. D is a function from P to $\mathscr{P}(P)$. Each person p is paired with exactly one set $x \subseteq P$, namely the set of all children of p. Note that in the relation D, a person p is paired with the set consisting of all of p's children, *not* with the children themselves. Even if p does not have exactly one child, it is still true that there is exactly one set that contains precisely the children of p and nothing else.
6. Yes. Each $a \in A$ is paired in the relation i_A with exactly one element of A, namely a itself. In other words, $(a, a) \in i_A$, but for every $a' \neq a$, $(a, a') \notin i_A$. Thus, we can call i_A the identity *function* on A.
7. Yes. For each real number x there is exactly one value of y, namely $y = x^2$, such that $(x, y) \in f$.

Suppose $f : A \to B$. If $a \in A$, then we know that there is exactly one $b \in B$ such that $(a, b) \in f$. This unique b is called "the value of f at a," or "the image of a under f," or "the result of applying f to a," or just "f of a," and it is written $f(a)$. In other words, for every $a \in A$ and $b \in B$, $b = f(a)$ iff $(a, b) \in f$.

For example, for the function $F = \{(1, 5), (2, 4), (3, 5)\}$ in part 1 of the last example, we could say that $F(1) = 5$, since $(1, 5) \in F$. Similarly, $F(2) = 4$ and $F(3) = 5$. If L is the function in part 3 and c is any city, then $L(c)$ would be the unique country n such that $(c, n) \in L$. In other words, $L(c) =$ the country in which c is located. For example, $L(\text{Paris}) = \text{France}$. For the function D in part 5, we could say that for any person p, $D(p) =$ the set of all children of p. If A is any set and $a \in A$, then $(a, a) \in i_A$, so $i_A(a) = a$. And if f is the function in part 7, then for every real number x, $f(x) = x^2$.

A function f from a set A to another set B is often specified by giving a rule that can be used to determine $f(a)$ for any $a \in A$. For example, if A is the set of all people and $B = \mathbb{R}^+$, then we could define a function f from A to B by the rule that for every $a \in A$, $f(a) = a$'s height in inches. Although this definition doesn't say explicitly which ordered pairs are elements of f, we can determine this by using our rule that for all $a \in A$ and $b \in B$, $(a, b) \in f$ iff $b = f(a)$. Thus,

$$\begin{aligned} f &= \{(a, b) \in A \times B \mid b = f(a)\} \\ &= \{(a, b) \in A \times B \mid b = a\text{'s height in inches}\}. \end{aligned}$$

For example, if Joe Smith is 68 inches tall, then $(\text{Joe Smith}, 68) \in f$ and $f(\text{Joe Smith}) = 68$.

It is often useful to think of a function f from A to B as representing a rule that associates, with each $a \in A$, some corresponding object $b = f(a) \in B$. However, it is important to remember that although a function can be defined by giving such a rule, it need not be defined in this way. Any subset of $A \times B$ that satisfies the requirements given in Definition 5.1.1 is a function from A to B.

Example 5.1.3. Here are some more examples of functions defined by rules.

1. Suppose every student is assigned an academic advisor who is a professor. Let S be the set of students and P the set of professors. Then we can define a function f from S to P by the rule that for every student s, $f(s) =$ the advisor of s. In other words,

$$\begin{aligned} f &= \{(s, p) \in S \times P \mid p = f(s)\} \\ &= \{(s, p) \in S \times P \mid \text{the professor } p \text{ is the academic advisor} \\ &\qquad \text{of the student } s\}. \end{aligned}$$

2. We can define a function g from $\mathbb{Z}$ to $\mathbb{R}$ by the rule that for every $x \in \mathbb{Z}$, $g(x) = 2x + 3$. Then

$$\begin{aligned} g &= \{(x, y) \in \mathbb{Z} \times \mathbb{R} \mid y = g(x)\} \\ &= \{(x, y) \in \mathbb{Z} \times \mathbb{R} \mid y = 2x + 3\} \\ &= \{\ldots, (-2, -1), (-1, 1), (0, 3), (1, 5), (2, 7), \ldots\}. \end{aligned}$$

3. Let h be the function from $\mathbb{R}$ to $\mathbb{R}$ defined by the rule that for every $x \in \mathbb{R}$, $h(x) = 2x + 3$. Note that the formula for $h(x)$ is the same as the formula for $g(x)$ in part 2. However, h and g are *not* the same function. You can see this by noting that, for example, $(\pi, 2\pi + 3) \in h$ but $(\pi, 2\pi + 3) \notin g$, since $\pi \notin \mathbb{Z}$. (For more on the relationship between g and h, see exercise 7.)

Notice that when a function f from A to B is specified by giving a rule for finding $f(a)$, the rule must determine the value of $f(a)$ for *every* $a \in A$. Sometimes when mathematicians are stating such a rule they don't say explicitly that the rule applies to all $a \in A$. For example, a mathematician might say "let f be the function from $\mathbb{R}$ to $\mathbb{R}$ defined by the formula $f(x) = x^2 + 7$." It is understood in this case that the equation $f(x) = x^2 + 7$ applies to all $x \in \mathbb{R}$ even though it hasn't been said explicitly. This means that you can plug in any real number for x in this equation, and the resulting equation will be true. For example, you can conclude that $f(3) = 3^2 + 7 = 16$. Similarly, if w is a real number, then you can write $f(w) = w^2 + 7$, or even $f(2w - 3) = (2w - 3)^2 + 7 = 4w^2 - 12w + 16$.

Because a function f from A to B is completely determined by the rule for finding $f(a)$, two functions that are defined by equivalent rules must be equal. More precisely, we have the following theorem:

Theorem 5.1.4. *Suppose f and g are functions from A to B. If $\forall a \in A(f(a) = g(a))$, then $f = g$.*

Proof. Suppose $\forall a \in A(f(a) = g(a))$, and let (a, b) be an arbitrary element of f. Then $b = f(a)$. But by our assumption $f(a) = g(a)$, so $b = g(a)$ and therefore $(a, b) \in g$. Thus, $f \subseteq g$. A similar argument shows $g \subseteq f$, so $f = g$. □

Commentary. Because f and g are sets, we prove $f = g$ by proving $f \subseteq g$ and $g \subseteq f$. Each of these goals is proven by showing that an arbitrary element of one set must be an element of the other. Note that, now that we have proven Theorem 5.1.4, we have another method for proving that two functions f and g from a set A to another set B are equal. In the future, to prove $f = g$ we will usually prove $\forall a \in A(f(a) = g(a))$ and then apply Theorem 5.1.4.

Because functions are just relations of a special kind, the concepts introduced in Chapter 4 for relations can be applied to functions as well. For example, suppose $f : A \to B$. Then f is a relation from A to B, so it makes sense to talk about the domain of f, which is a subset of A, and the range of f, which is a subset of B. According to the definition of function, every element of A must appear as the first coordinate of some (in fact, exactly one) ordered pair in f, so the domain of f must actually be all of A. But the range of f need not be all of B. The elements of the range of f will be the second coordinates of all the ordered pairs in f, and the second coordinate of an ordered pair in f is what we have called the image of its first coordinate. Thus, the range of f could also be described as the set of all images of elements of A under f:

$$\text{Ran}(f) = \{f(a) \mid a \in A\}.$$

For example, for the function f defined in part 1 of Example 5.1.3, $\text{Ran}(f) = \{f(s) \mid s \in S\}$ = the set of all advisors of students.

We can draw diagrams of functions in exactly the same way we drew diagrams for relations in Chapter 4. If $f : A \to B$, then as before, every ordered pair $(a, b) \in f$ would be represented in the diagram by an edge connecting a to b. By the definition of function, every $a \in A$ occurs as the first coordinate of exactly one ordered pair in f, and the second coordinate of this ordered pair is $f(a)$. Thus, for every $a \in A$ there will be exactly one edge coming from a, and it will connect a to $f(a)$. For example, Figure 1 shows what the diagram for the function L defined in part 3 of Example 5.1.2 would look like.

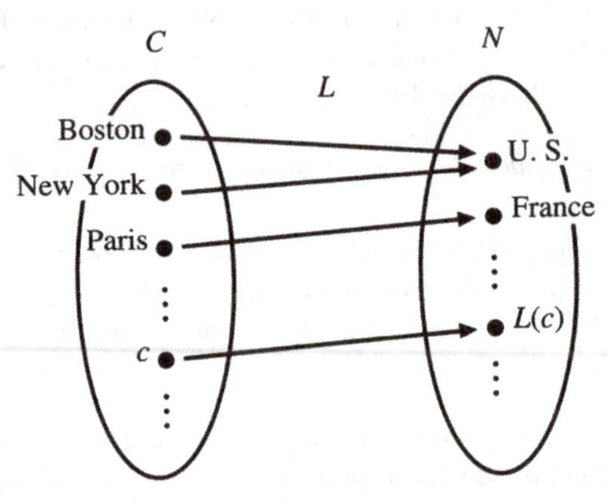

Figure 1

The definition of composition of relations can also be applied to functions. If $f : A \to B$ and $g : B \to C$, then f is a relation from A to B and g is a relation from B to C, so it makes sense to talk about $g \circ f$, which will be a relation from A to C. In fact, it turns out that $g \circ f$ is a function from A to C, as the next theorem shows.

Theorem 5.1.5. *Suppose $f : A \to B$ and $g : B \to C$. Then $g \circ f : A \to C$, and for any $a \in A$, the value of $g \circ f$ at a is given by the formula $(g \circ f)(a) = g(f(a))$.*

Scratch work

Before proving this theorem, it might be helpful to discuss the scratch work for the proof. According to the definition of function, to show that $g \circ f : A \to C$ we must prove that $\forall a \in A \exists! c \in C((a, c) \in g \circ f)$, so we will start out by letting a be an arbitrary element of A and then try to prove that $\exists! c \in C((a, c) \in g \circ f)$. As we saw in Section 3.6, we can prove this statement by proving existence and uniqueness separately. To prove existence, we should try to find a $c \in C$ such that $(a, c) \in g \circ f$. For uniqueness, we should assume that $(a, c_1) \in g \circ f$ and $(a, c_2) \in g \circ f$, and then try to prove that $c_1 = c_2$.

Proof. Let a be an arbitrary element of A. We must show that there is a unique $c \in C$ such that $(a, c) \in g \circ f$.

Existence: Let $b = f(a) \in B$. Let $c = g(b) \in C$. Then $(a, b) \in f$ and $(b, c) \in g$, so by the definition of composition of relations, $(a, c) \in g \circ f$. Thus, $\exists c \in C((a, c) \in g \circ f)$.

Uniqueness: Suppose $(a, c_1) \in g \circ f$ and $(a, c_2) \in g \circ f$. Then by the definition of composition, we can choose $b_1 \in B$ such that $(a, b_1) \in f$ and $(b_1, c_1) \in g$, and we can also choose $b_2 \in B$ such that $(a, b_2) \in f$ and $(b_2, c_2) \in g$. Since f is a function, there can be only one $b \in B$ such that $(a, b) \in f$. Thus, since (a, b_1) and (a, b_2) are both elements of f, it follows that $b_1 = b_2$. But now applying the same reasoning to g, since $(b_1, c_1) \in g$ and $(b_1, c_2) = (b_2, c_2) \in g$, it follows that $c_1 = c_2$, as required.

This completes the proof that $g \circ f$ is a function from A to C. Finally, to derive the formula for $(g \circ f)(a)$, note that we showed in the *existence* half of the proof that for any $a \in A$, if we let $b = f(a)$ and $c = g(b)$, then $(a, c) \in g \circ f$. Thus,

$$(g \circ f)(a) = c = g(b) = g(f(a)).$$ □

When we first introduced the idea of the composition of two relations in Chapter 4, we pointed out that the notation was somewhat peculiar and promised to explain the reason for the notation in this chapter. We can now provide this explanation. The reason for the notation we've used for composition of relations is that it leads to the convenient formula $(g \circ f)(x) = g(f(x))$

derived in Theorem 5.1.5. Note that because functions are just relations of a special kind, everything we have proven about composition of relations applies to composition of functions. In particular, by Theorem 4.2.5, we know that composition of functions is associative.

Example 5.1.6. Here are some examples of compositions of functions.

1. Let C and N be the sets of all cities and countries, respectively, and let $L : C \to N$ be the function defined in part 3 of Example 5.1.2. Thus, for every city c, $L(c) =$ the country in which c is located. Let B be the set of all buildings located in cities, and define $F : B \to C$ by the formula $F(b) =$ the city in which the building b is located. Then $L \circ F : B \to N$. For example, F(Eiffel Tower) = Paris, so according to the formula derived in Theorem 5.1.5,

$$(L \circ F)(\text{Eiffel Tower}) = L(F(\text{Eiffel Tower})) \\ = L(\text{Paris}) = \text{France}.$$

   In general, for every building $b \in B$,

$$(L \circ F)(b) = L(F(b)) = L(\text{the city in which } b \text{ is located}) \\ = \text{the country in which } b \text{ is located}.$$

   A diagram of this function is shown in Figure 2.

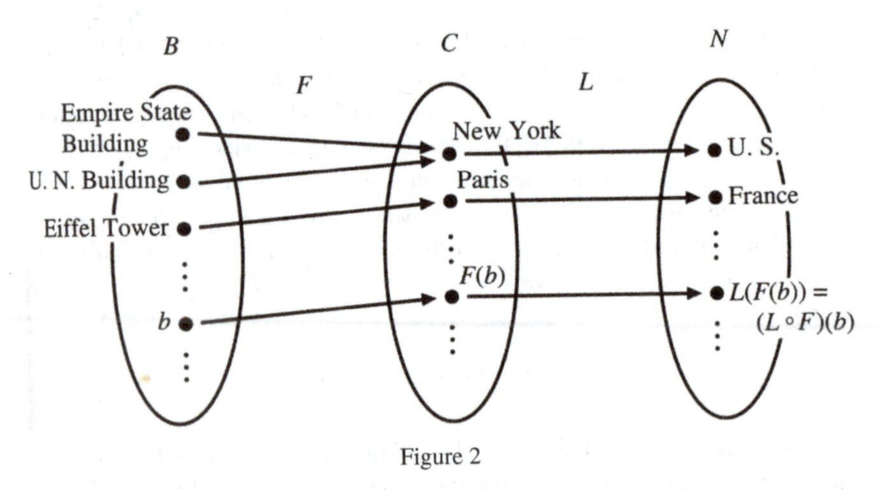

Figure 2

2. Let $g : \mathbb{Z} \to \mathbb{R}$ be the function from part 2 of Example 5.1.3, which was defined by the formula $g(x) = 2x + 3$. Let $f : \mathbb{Z} \to \mathbb{Z}$ be defined by the formula $f(n) = n^2 - 3n + 1$. Then $g \circ f : \mathbb{Z} \to \mathbb{R}$. For example,

$f(2) = 2^2 - 3 \cdot 2 + 1 = -1$, so $(g \circ f)(2) = g(f(2)) = g(-1) = 1$. In general, for every $n \in \mathbb{Z}$,

$$(g \circ f)(n) = g(f(n)) = g(n^2 - 3n + 1) = 2(n^2 - 3n + 1) + 3$$
$$= 2n^2 - 6n + 5.$$

Exercises

*1. (a) Let $A = \{1, 2, 3\}$, $B = \{4\}$, and $f = \{(1, 4), (2, 4), (3, 4)\}$. Is f a function from A to B?

(b) Let $A = \{1\}$, $B = \{2, 3, 4\}$, and $f = \{(1, 2), (1, 3), (1, 4)\}$. Is f a function from A to B?

(c) Let C be the set of all cars registered in your state, and let S be the set of all finite sequences of letters and digits. Let $L = \{(c, s) \in C \times S \mid$ the license plate number of the car c is $s\}$. Is L a function from C to S?

2. (a) Let f be the relation represented by the following graph. Is f a function from A to B?

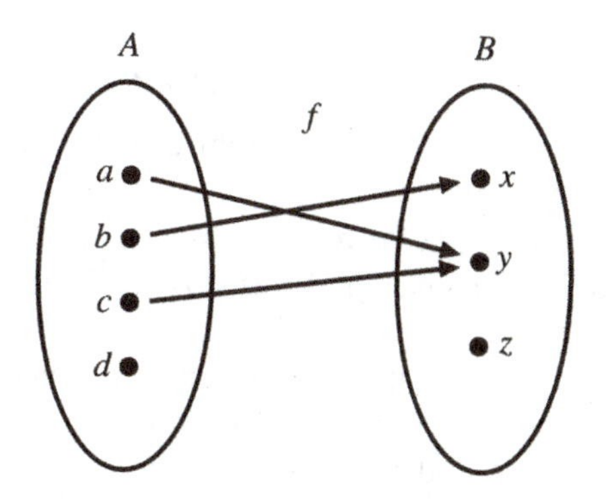

(b) Let W be the set of all words of English, and let A be the set of all letters of the alphabet. Let $f = \{(w, a) \in W \times A \mid$ the letter a occurs in the word $w\}$, and let $g = \{(w, a) \in W \times A \mid$ the letter a is the first letter of the word $w\}$. Is f a function from W to A? How about g?

(c) John, Mary, Susan, and Fred go out to dinner and sit at a round table. Let $P = \{\text{John, Mary, Susan, Fred}\}$, and let $R = \{(p, q) \in P \times P \mid$ the person p is sitting immediately to the right of the person $q\}$. Is R a function from P to P?

*3. (a) Let $A = \{a, b, c\}$, $B = \{a, b\}$, and $f = \{(a, b), (b, b), (c, a)\}$. Then $f : A \to B$. What are $f(a)$, $f(b)$, and $f(c)$?

(b) Let $f : \mathbb{R} \to \mathbb{R}$ be the function defined by the formula $f(x) = x^2 - 2x$. What is $f(2)$?

(c) Let $f = \{(x, n) \in \mathbb{R} \times \mathbb{Z} \mid n \leq x < n + 1\}$. Then $f : \mathbb{R} \to \mathbb{Z}$. What is $f(\pi)$? What is $f(-\pi)$?

4. (a) Let N be the set of all countries and C the set of all cities. Let $H : N \to C$ be the function defined by the rule that for every country n, $H(n) =$ the capital of the country n. What is $H(\text{Italy})$?

   (b) Let $A = \{1, 2, 3\}$ and $B = \mathscr{P}(A)$. Let $F : B \to B$ be the function defined by the formula $F(X) = A \setminus X$. What is $F(\{1, 3\})$?

   (c) Let $f : \mathbb{R} \to \mathbb{R} \times \mathbb{R}$ be the function defined by the formula $f(x) = (x + 1, x - 1)$. What is $f(2)$?

*5. Let L be the function defined in part 3 of Example 5.1.2 and let H be the function defined in exercise 4(a). Describe $L \circ H$ and $H \circ L$.

6. Let f and g be functions from $\mathbb{R}$ to $\mathbb{R}$ defined by the following formulas:

$$f(x) = \frac{1}{x^2 + 2}, \qquad g(x) = 2x - 1.$$

Find formulas for $(f \circ g)(x)$ and $(g \circ f)(x)$.

*7. Suppose $f : A \to B$ and $C \subseteq A$. The set $f \cap (C \times B)$, which is a relation from C to B, is called the *restriction* of f to C, and is sometimes denoted $f \restriction C$. In other words,

$$f \restriction C = f \cap (C \times B).$$

   (a) Prove that $f \restriction C$ is a function from C to B and that for all $c \in C$, $f(c) = (f \restriction C)(c)$.

   (b) Suppose $g : C \to B$. Prove that $g = f \restriction C$ iff $g \subseteq f$.

   (c) Let g and h be the functions defined in parts 2 and 3 of Example 5.1.3. Show that $g = h \restriction \mathbb{Z}$.

8. Suppose A is a set. Show that i_A is the only relation on A that is both an equivalence relation on A and also a function from A to A.

9. Suppose $f : A \to C$ and $g : B \to C$.

   (a) Prove that if A and B are disjoint, then $f \cup g : A \cup B \to C$.

   (b) More generally, prove that $f \cup g : A \cup B \to C$ iff $f \restriction (A \cap B) = g \restriction (A \cap B)$. (See exercise 7 for the meaning of the notation used here.)

*10. Suppose R is a relation from A to B, S is a relation from B to C, $\text{Ran}(R) = \text{Dom}(S) = B$, and $S \circ R : A \to C$.

   (a) Prove that $S : B \to C$.

   (b) Give an example to show that it need not be the case that $R : A \to B$.

11. Suppose $f : A \to B$ and S is a relation on B. Define a relation R on A as follows:

$$R = \{(x, y) \in A \times A \mid (f(x), f(y)) \in S\}.$$

(a) Prove that if S is reflexive, then so is R.
(b) Prove that if S is symmetric, then so is R.
(c) Prove that if S is transitive, then so is R.

*12. Suppose $f : A \to B$ and R is a relation on A. Define a relation S on B as follows:

$$S = \{(x, y) \in B \times B \mid \exists u \in A \exists v \in A(f(u) = x \wedge f(v) = y \wedge (u, v) \in R)\}.$$

Justify your answers to the following questions with either proofs or counterexamples.
(a) If R is reflexive, must it be the case that S is reflexive?
(b) If R is symmetric, must it be the case that S is symmetric?
(c) If R is transitive, must it be the case that S is transitive?

13. Suppose A and B are sets, and let $\mathcal{F} = \{f \mid f : A \to B\}$. Also, suppose R is a relation on B, and define a relation S on $\mathcal{F}$ as follows:

$$S = \{(f, g) \in \mathcal{F} \times \mathcal{F} \mid \forall x \in A((f(x), g(x)) \in R)\}.$$

Justify your answers to the following questions with either proofs or counterexamples.
(a) If R is reflexive, must it be the case that S is reflexive?
(b) If R is symmetric, must it be the case that S is symmetric?
(c) If R is transitive, must it be the case that S is transitive?

14. Suppose A is a nonempty set and $f : A \to A$.
(a) Suppose there is some $a \in A$ such that $\forall x \in A(f(x) = a)$. (In this case, f is called a *constant* function.) Prove that for all $g : A \to A$, $f \circ g = f$.
(b) Suppose that for all $g : A \to A$, $f \circ g = f$. Prove that f is a constant function. (Hint: What happens if g is a constant function?)

15. Let $\mathcal{F} = \{f \mid f : \mathbb{R} \to \mathbb{R}\}$. Let $R = \{(f, g) \in \mathcal{F} \times \mathcal{F} \mid \exists a \in \mathbb{R} \forall x > a(f(x) = g(x))\}$.
(a) Let $f : \mathbb{R} \to \mathbb{R}$ and $g : \mathbb{R} \to \mathbb{R}$ be the functions defined by the formulas $f(x) = |x|$ and $g(x) = x$. Show that $(f, g) \in R$.
(b) Prove that R is an equivalence relation.

*16. Let $\mathcal{F} = \{f \mid f : \mathbb{Z}^+ \to \mathbb{R}\}$. For $g \in \mathcal{F}$, we define the set $O(g)$ as follows:

$$O(g) = \{f \in \mathcal{F} \mid \exists a \in \mathbb{Z}^+ \exists c \in \mathbb{R}^+ \forall x > a(|f(x)| \leq c|g(x)|)\}.$$

(If $f \in O(g)$, then mathematicians say that "f is big-oh of g".)
(a) Let $f : \mathbb{Z}^+ \to \mathbb{R}$ and $g : \mathbb{Z}^+ \to \mathbb{R}$ be defined by the formulas $f(x) = 7x + 3$ and $g(x) = x^2$. Prove that $f \in O(g)$, but $g \notin O(f)$.

(b) Let $S = \{(f, g) \in \mathcal{F} \times \mathcal{F} \mid f \in O(g)\}$. Prove that S is a preorder, but not a partial order. (See exercise 24 of Section 4.6 for the definition of *preorder*.)

(c) Suppose $f_1 \in O(g)$ and $f_2 \in O(g)$, and s and t are real numbers. Define a function $f : \mathbb{Z}^+ \to \mathbb{R}$ by the formula $f(x) = sf_1(x) + tf_2(x)$. Prove that $f \in O(g)$. (Hint: You may find the triangle inequality helpful. See exercise 12(c) of Section 3.5.)

17. (a) Suppose $g : A \to B$ and let $R = \{(x, y) \in A \times A \mid g(x) = g(y)\}$. Show that R is an equivalence relation on A.

    (b) Suppose R is an equivalence relation on A and let $g : A \to A/R$ be the function defined by the formula $g(x) = [x]_R$. Show that $R = \{(x, y) \in A \times A \mid g(x) = g(y)\}$.

*18. Suppose $f : A \to B$ and R is an equivalence relation on A. We will say that f is *compatible* with R if $\forall x \in A \forall y \in A(xRy \to f(x) = f(y))$. (You might want to compare this exercise to exercise 23 of Section 4.6.)

    (a) Suppose f is compatible with R. Prove that there is a unique function $h : A/R \to B$ such that for all $x \in A$, $h([x]_R) = f(x)$.

    (b) Suppose $h : A/R \to B$ and for all $x \in A$, $h([x]_R) = f(x)$. Prove that f is compatible with R.

19. Let $R = \{(x, y) \in \mathbb{Z} \times \mathbb{Z} \mid x \equiv y \pmod 5\}$. Recall that we saw in Section 4.6 that R is an equivalence relation on $\mathbb{Z}$.

    (a) Show that there is a unique function $h : \mathbb{Z}/R \to \mathbb{Z}/R$ such that for every integer x, $h([x]_R) = [x^2]_R$. (Hint: Use exercise 18.)

    (b) Show that there is no function $h : \mathbb{Z}/R \to \mathbb{Z}/R$ such that for every integer x, $h([x]_R) = [2^x]_R$.

5.2. One-to-one and Onto

In the last section we saw that the composition of two functions is again a function. What about inverses of functions? If $f : A \to B$, then f is a relation from A to B, so f^{-1} is a relation from B to A. Is it a function from B to A? We'll answer this question in the next section. As we will see, the answer hinges on the following two properties of functions.

Definition 5.2.1. Suppose $f : A \to B$. We will say that f is *one-to-one* if

$$\neg\exists a_1 \in A \exists a_2 \in A(f(a_1) = f(a_2) \wedge a_1 \neq a_2).$$

We say that f is *onto* if

$$\forall b \in B \exists a \in A(f(a) = b).$$

One-to-one functions are sometimes also called *injections*, and onto functions are sometimes called *surjections*.

Note that our definition of one-to-one starts with the negation symbol $\neg$. In other words, to say that f is one-to-one means that a certain situation does *not* occur. The situation that must not occur is that there are two different elements of the domain of f, a_1 and a_2, such that $f(a_1) = f(a_2)$. This situation is illustrated in Figure 1(a). Thus, the function in Figure 1(a) is not one-to-one. Figure 1(b) shows a function that is one-to-one.

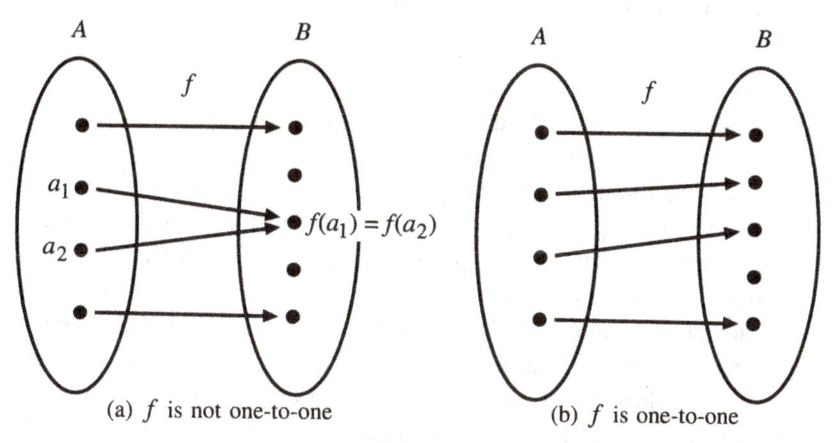

(a) f is not one-to-one (b) f is one-to-one

Figure 1

If $f : A \to B$, then to say that f is onto means that every element of B is the image under f of some element of A. In other words, in the diagram of f, every element of B has an edge pointing to it. Neither of the functions in Figure 1 is onto, because in both cases there are elements of B without edges pointing to them. Figure 2 shows two functions that are onto.

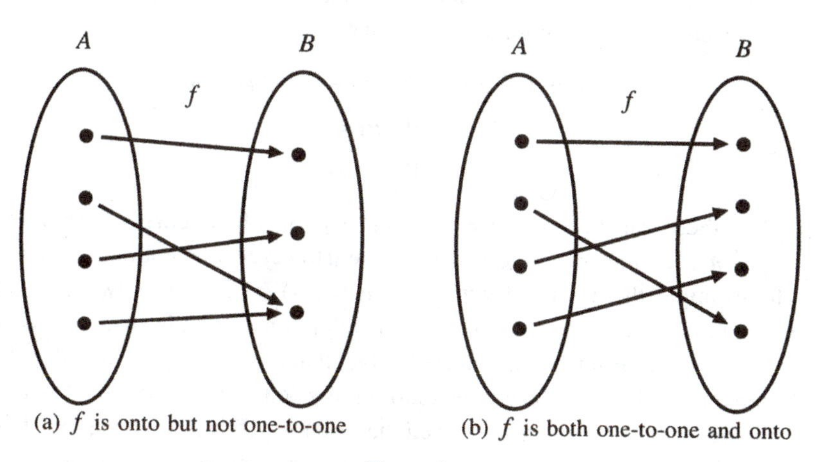

(a) f is onto but not one-to-one (b) f is both one-to-one and onto

Figure 2

Example 5.2.2. Are the following functions one-to-one? Are they onto?

1. The function F from part 1 of Example 5.1.2.
2. The function L from part 3 of Example 5.1.2.
3. The identity function i_A, for any set A.
4. The function g from part 2 of Example 5.1.3.
5. The function h from part 3 of Example 5.1.3.

Solutions

1. F is not one-to-one because $F(1) = 5 = F(3)$. It is also not onto, because $6 \in B$ but there is no $a \in A$ such that $F(a) = 6$.
2. L is not one-to-one because there are many pairs of different cities c_1 and c_2 for which $L(c_1) = L(c_2)$. For example, $L(\text{Chicago}) = \text{United States} = L(\text{Seattle})$. To say that L is onto means that $\forall n \in N \exists c \in C(L(c) = n)$, or in other words, for every country n there is a city c such that the city c is located in the country n. This is probably true, since it is unlikely that there is a country that contains no cities at all. Thus, L is probably onto.
3. To decide whether i_A is one-to-one we must determine whether there are two elements a_1 and a_2 of A such that $i_A(a_1) = i_A(a_2)$ and $a_1 \neq a_2$. But as we saw in Section 5.1, for every $a \in A, i_A(a) = a$, so $i_A(a_1) = i_A(a_2)$ means $a_1 = a_2$. Thus, there cannot be elements a_1 and a_2 of A such that $i_A(a_1) = i_A(a_2)$ and $a_1 \neq a_2$, so i_A is one-to-one.

   To say that i_A is onto means that for every $a \in A, a = i_A(b)$ for some $b \in A$. This is clearly true because, in fact, $a = i_A(a)$. Thus i_A is also onto.
4. As in solution 3, to decide whether g is one-to-one, we must determine whether there are integers n_1 and n_2 such that $g(n_1) = g(n_2)$ and $n_1 \neq n_2$. According to the definition of g, we have

$$\begin{aligned} g(n_1) = g(n_2) &\text{ iff } 2n_1 + 3 = 2n_2 + 3 \\ &\text{ iff } 2n_1 = 2n_2 \\ &\text{ iff } n_1 = n_2. \end{aligned}$$

   Thus there can be no integers n_1 and n_2 for which $g(n_1) = g(n_2)$ and $n_1 \neq n_2$. In other words, g is one-to-one. However, g is not onto because, for example, there is no integer n for which $g(n) = 0$. To see why, suppose n is an integer and $g(n) = 0$. Then by the definition of g we have $2n + 3 = 0$, so $n = -3/2$. But this contradicts the fact that n is an integer. Note that the domain of g is $\mathbb{Z}$, so for g to be onto it must be the case that for every real number y there is an *integer* n such that $g(n) = y$. Since we have seen that there is no integer n such that $g(n) = 0$, we can conclude that g is not onto.

5. This function is both one-to-one and onto. The verification that h is one-to-one is very similar to the verification in solution 4 that g is one-to-one, and it is left to the reader. To see that h is onto, we must show that $\forall y \in \mathbb{R} \exists x \in \mathbb{R}(h(x) = y)$. Here is a brief proof of this statements. Let y be an arbitrary real number. Let $x = (y - 3)/2$. Then $g(x) = 2x + 3 = 2 \cdot ((y - 3)/2) + 3 = y - 3 + 3 = y$. Thus, $\forall y \in \mathbb{R} \exists x \in \mathbb{R}(h(x) = y)$, so h is onto.

Although the definition of one-to-one is easiest to understand when it is stated as a negative statement, as in Definition 5.2.1, we know from Chapter 3 that the definition will be easier to use in proofs if we reexpress it as an equivalent positive statement. The following theorem shows how to do this. It also gives a useful equivalence for the definition of onto.

Theorem 5.2.3. *Suppose* $f : A \to B$.

1. f *is one-to-one iff* $\forall a_1 \in A \forall a_2 \in A(f(a_1) = f(a_2) \to a_1 = a_2)$.
2. f *is onto iff* $\text{Ran}(f) = B$.

Proof

1. We use the rules from Chapters 1 and 2 for reexpressing negative statements as positive ones.

$$\begin{aligned} f \text{ is one-to-one iff } & \neg\exists a_1 \in A \exists a_2 \in A(f(a_1) = f(a_2) \wedge a_1 \neq a_2) \\ \text{iff } & \forall a_1 \in A \forall a_2 \in A \neg(f(a_1) = f(a_2) \wedge a_1 \neq a_2) \\ \text{iff } & \forall a_1 \in A \forall a_2 \in A(f(a_1) \neq f(a_2) \vee a_1 = a_2) \\ \text{iff } & \forall a_1 \in A \forall a_2 \in A(f(a_1) = f(a_2) \to a_1 = a_2). \end{aligned}$$

2. First we relate the definition of onto to the definition of range.

$$\begin{aligned} f \text{ is onto iff } & \forall b \in B \exists a \in A(f(a) = b) \\ \text{iff } & \forall b \in B \exists a \in A((a, b) \in f) \\ \text{iff } & \forall b \in B(b \in \text{Ran}(f)) \\ \text{iff } & B \subseteq \text{Ran}(f). \end{aligned}$$

Now we are ready to prove part 2 of the theorem.

($\to$) Suppose f is onto. By the equivalence just derived we have $B \subseteq \text{Ran}(f)$, and by the definition of range we have $\text{Ran}(f) \subseteq B$. Thus, it follows that $\text{Ran}(f) = B$.

($\leftarrow$) Suppose $\text{Ran}(f) = B$. Then certainly $B \subseteq \text{Ran}(f)$, so by the equivalence, f is onto. □

Commentary. It is often most efficient to write the proof of an iff statement as a string of equivalences, if this can be done. In the case of statement 1 this is easy, using rules of logic. For statement 2 this strategy doesn't quite work, but it does give us an equivalence that turns out to be useful in the proof.

Example 5.2.4. Let $A = \mathbb{R} \setminus \{-1\}$, and define $f : A \to \mathbb{R}$ by the formula

$$f(a) = \frac{2a}{a+1}.$$

Prove that f is one-to-one but not onto.

Scratch work

By part 1 of Theorem 5.2.3, we can prove that f is one-to-one by proving the equivalent statement $\forall a_1 \in A \forall a_2 \in A(f(a_1) = f(a_2) \to a_1 = a_2)$. Thus, we let a_1 and a_2 be arbitrary elements of A, assume $f(a_1) = f(a_2)$, and then prove $a_1 = a_2$. This is the strategy that is almost always used when proving that a function is one-to-one. The remaining details of the proof involve only simple algebra and are given later.

To show that f is not onto we must prove $\neg\forall x \in \mathbb{R}\exists a \in A(f(a) = x)$. Re-expressing this as a positive statement, we see that we must prove $\exists x \in \mathbb{R}\forall a \in A(f(a) \neq x)$, so we should try to find a particular real number x such that $\forall a \in A(f(a) \neq x)$. Unfortunately, it is not at all clear what value we should use for x. We'll use a somewhat unusual procedure to overcome this difficulty. Instead of trying to prove that f is not onto, let's try to prove that it *is* onto! Of course, we're expecting that this proof won't work, but maybe seeing *why* it won't work will help us figure out what value of x to use in the proof that f is *not* onto.

To prove that f is onto we would have to prove $\forall x \in \mathbb{R}\exists a \in A(f(a) = x)$, so we should let x be an arbitrary real number and try to find some $a \in A$ such that $f(a) = x$. Filling in the definition of f, we see that we must find $a \in A$ such that

$$\frac{2a}{a+1} = x.$$

To find this value of a, we simply solve the equation for a:

$$\frac{2a}{a+1} = x \Rightarrow 2a = ax + x \Rightarrow a(2-x) = x \Rightarrow a = \frac{x}{2-x}.$$

Aha! The last step in this derivation wouldn't work if $x = 2$, because then we would be dividing by 0. This is the only value of x that seems to cause trouble when we try to find a value of a for which $f(a) = x$. Perhaps $x = 2$ is the value to use in the proof that f is *not* onto.

Let's return now to the proof that f is not onto. If we let $x = 2$, then to complete the proof we must show that $\forall a \in A(f(a) \neq 2)$. We'll do this by letting a be an arbitrary element of A, assuming $f(a) = 2$, and then trying to derive a contradiction. The remaining details of the proof are not hard.

Solution

Proof To see that f is one-to-one, let a_1 and a_2 be arbitrary elements of A and assume $f(a_1) = f(a_2)$. Applying the definition of f, it follows that $\frac{2a_1}{a_1+1} = \frac{2a_2}{a_2+1}$. Thus, $2a_1(a_2 + 1) = 2a_2(a_1 + 1)$. Multiplying out both sides gives us $2a_1a_2 + 2a_1 = 2a_1a_2 + 2a_2$, so $2a_1 = 2a_2$ and therefore $a_1 = a_2$.

To show that f is not onto we will prove that $\forall a \in A(f(a) \neq 2)$. Suppose $a \in A$ and $f(a) = 2$. Applying the definition of f, we get $\frac{2a}{a+1} = 2$. Thus, $2a = 2a + 2$, which is clearly impossible. Thus, $2 \notin \text{Ran}(f)$, so $\text{Ran}(f) \neq \mathbb{R}$ and therefore f is not onto. □

As we saw in the preceding example, when proving that a function f is one-to-one it is usually easiest to prove the equivalent statement $\forall a_1 \in A \forall a_2 \in A(f(a_1) = f(a_2) \rightarrow a_1 = a_2)$ given in part 1 of Theorem 5.2.3. Of course, this is just an example of the fact that it is generally easier to prove a positive statement than a negative one. This equivalence is also often used in proofs in which we are *given* that a function is one-to-one, as you will see in the proof of part 1 of the following theorem.

Theorem 5.2.5. *Suppose $f : A \rightarrow B$ and $g : B \rightarrow C$. As we saw in Theorem 5.1.5, it follows that $g \circ f : A \rightarrow C$.*

1. *If f and g are both one-to-one, then so is $g \circ f$.*
2. *If f and g are both onto, then so is $g \circ f$.*

Proof

1. Suppose f and g are both one-to-one. Let a_1 and a_2 be arbitrary elements of A and suppose that $(g \circ f)(a_1) = (g \circ f)(a_2)$. By Theorem 5.1.5 this means that $g(f(a_1)) = g(f(a_2))$. Since g is one-to-one it follows that $f(a_1) = f(a_2)$, and similarly since f is one-to-one we can then conclude that $a_1 = a_2$. Thus, $g \circ f$ is one-to-one.
2. Suppose f and g are both onto, and let c be an arbitrary element of C. Since g is onto, we can find some $b \in B$ such that $g(b) = c$. Similarly, since f is onto, there is some $a \in A$ such that $f(a) = b$. Then $(g \circ f)(a) = g(f(a)) = g(b) = c$. Thus, $g \circ f$ is onto. □

Commentary

1. As in Example 5.2.4, we prove that $g \circ f$ is one-to-one by proving that $\forall a_1 \in A \forall a_2 \in A((g \circ f)(a_1) = (g \circ f)(a_2) \to a_1 = a_2)$. Thus, we let a_1 and a_2 be arbitrary elements of A, assume that $(g \circ f)(a_1) = (g \circ f)(a_2)$, which means $g(f(a_1)) = g(f(a_2))$, and then prove that $a_1 = a_2$. The next sentence of the proof says that the assumption that g is one-to-one is being used, but it might not be clear *how* it is being used. To understand this step, let's write out what it means to say that g is one-to-one. As we observed before, rather than using the original definition, which is a negative statement, we are probably better off using the equivalent positive statement $\forall b_1 \in B \forall b_2 \in B(g(b_1) = g(b_2) \to b_1 = b_2)$. The natural way to use a given of this form is to plug something in for b_1 and b_2. Plugging in $f(a_1)$ and $f(a_2)$, we get $g(f(a_1)) = g(f(a_2)) \to f(a_1) = f(a_2)$, and since we know $g(f(a_1)) = g(f(a_2))$, it follows by modus ponens that $f(a_1) = f(a_2)$. None of this was explained in the proof; readers of the proof are expected to work it out for themselves. Make sure you understand how, using similar reasoning, you can get from $f(a_1) = f(a_2)$ to $a_1 = a_2$ by applying the fact that f is one-to-one.
2. After the assumption that f and g are both onto, the form of the rest of the proof is entirely guided by the logical form of the goal of proving that $g \circ f$ is onto. Because this means $\forall c \in C \exists a \in A((g \circ f)(a) = c)$, we let c be an arbitrary element of C and then find some $a \in A$ for which we can prove $(g \circ f)(a) = c$.

Functions that are both one-to-one and onto are particularly important in mathematics. Such functions are sometimes called *one-to-one correspondences* or *bijections*. Figure 2(b) shows an example of a one-to-one correspondence. Notice in the figure that both A and B have four elements. In fact, you should be able to convince yourself that if there is a one-to-one correspondence between two finite sets, then the sets must have the same number of elements. This is one of the reasons why one-to-one correspondences are so important. We will discuss one-to-one correspondences between infinite sets in Chapter 7.

Here's another example of a one-to-one correspondence. Suppose A is the set of all members of the audience at a sold-out concert and S is the set of all seats in the concert hall. Let $f : A \to S$ be the function defined by the rule

$$f(a) = \text{the seat in which } a \text{ is sitting.}$$

Because different people would not be sitting in the same seat, f is one-to-one. Because the concert is sold out, every seat is taken, so f is onto. Thus, f is a one-to-one correspondence. Even without counting people or seats, we can tell that the number of people in the audience must be the same as the number of seats in the concert hall.

Exercises

1. Which of the functions in exercise 1 of Section 5.1 are one-to-one? Which are onto?

*2. Which of the functions in exercise 2 of Section 5.1 are one-to-one? Which are onto?

3. Which of the functions in exercise 3 of Section 5.1 are one-to-one? Which are onto?

4. Which of the functions in exercise 4 of Section 5.1 are one-to-one? Which are onto?

*5. Let $A = \mathbb{R} \setminus \{1\}$, and let $f : A \to A$ be defined as follows:

$$f(x) = \frac{x+1}{x-1}.$$

(a) Show that f is one-to-one and onto.
(b) Show that $f \circ f = i_A$.

6. Let $A = \mathscr{P}(\mathbb{R})$. Define $f : \mathbb{R} \to A$ by the formula $f(x) = \{y \in \mathbb{R} \mid y^2 < x\}$.
(a) Find $f(2)$.
(b) Is f one-to-one? Is it onto?

*7. Let $A = \mathscr{P}(\mathbb{R})$ and $B = \mathscr{P}(A)$. Define $f : B \to A$ by the formula $f(\mathcal{F}) = \cup\mathcal{F}$.
(a) Find $f(\{\{1, 2\}, \{3, 4\}\})$.
(b) Is f one-to-one? Is it onto?

8. Suppose $f : A \to B$ and $g : B \to C$.
(a) Prove that if $g \circ f$ is onto then g is onto.
(b) Prove that if $g \circ f$ is one-to-one then f is one-to-one.

9. Suppose $f : A \to B$ and $g : B \to C$.
(a) Prove that if f is onto and g is not one-to-one, then $g \circ f$ is not one-to-one.
(b) Prove that if f is not onto and g is one-to-one, then $g \circ f$ is not onto.

*10. Suppose $f : A \to B$ and $C \subseteq A$. In exercise 7 of Section 5.1 we defined $f \restriction C$ (the restriction of f to C), and you showed that $f \restriction C : C \to B$.
(a) Prove that if f is one-to-one, then so is $f \restriction C$.
(b) Prove that if $f \restriction C$ is onto, then so is f.

(c) Give examples to show that the converses of parts (a) and (b) are not true.

11. Suppose $f : A \to B$, and there is some $b \in B$ such that $\forall x \in A(f(x) = b)$. (Thus, f is a *constant* function.)
    (a) Prove that if A has more than one element then f is not one-to-one.
    (b) Prove that if B has more than one element then f is not onto.
12. Suppose $f : A \to C$, $g : B \to C$, and A and B are disjoint. In exercise 9(a) of Section 5.1 you proved that $f \cup g : A \cup B \to C$. Now suppose in addition that f and g are both one-to-one. Prove that $f \cup g$ is one-to-one iff $\text{Ran}(f)$ and $\text{Ran}(g)$ are disjoint.
13. Suppose R is a relation from A to B, S is a relation from B to C, $\text{Ran}(R) = \text{Dom}(S) = B$, and $S \circ R : A \to C$. In exercise 10(a) of Section 5.1 you proved that $S : B \to C$. Now prove that if S is one-to-one then $R : A \to B$.

*14. Suppose $f : A \to B$ and R is a relation on A. As in exercise 12 of Section 5.1, define a relation S on B as follows:

$$S = \{(x, y) \in B \times B \mid \exists u \in A \exists v \in A(f(u) = x \wedge f(v) = y \wedge (u, v) \in R\}.$$

   (a) Prove that if R is reflexive and f is onto then S is reflexive.
   (b) Prove that if R is transitive and f is one-to-one then S is transitive.

15. Suppose R is an equivalence relation on A, and let $g : A \to A/R$ be defined by the formula $g(x) = [x]_R$, as in exercise 17 in Section 5.1.
    (a) Show that g is onto.
    (b) Show that g is one-to-one iff $R = i_A$.
16. Suppose $f : A \to B$, R is an equivalence relation on A, and f is compatible with R. (See exercise 18 of Section 5.1 for the definition of *compatible*.) In exercise 18(a) of Section 5.1 you proved that there is a unique function $h : A/R \to B$ such that for all $x \in A$, $h([x]_R) = f(x)$. Now prove that h is one-to-one iff $\forall x \in A \forall y \in A(f(x) = f(y) \to xRy)$.

*17. Suppose A, B, and C are sets and $f : A \to B$.
   (a) Prove that if f is onto, $g : B \to C$, $h : B \to C$, and $g \circ f = h \circ f$, then $g = h$.
   (b) Suppose that C has at least two elements, and for all functions g and h from B to C, if $g \circ f = h \circ f$ then $g = h$. Prove that f is onto.

18. Suppose A, B, and C are sets and $f : B \to C$.
    (a) Prove that if f is one-to-one, $g : A \to B$, $h : A \to B$, and $f \circ g = f \circ h$, then $g = h$.
    (b) Suppose that $A \neq \varnothing$, and for all functions g and h from A to B, if $f \circ g = f \circ h$ then $g = h$. Prove that f is one-to-one.

19. Let $\mathcal{F} = \{f \mid f : \mathbb{R} \to \mathbb{R}\}$, and define a relation R on $\mathcal{F}$ as follows:

$$R = \{(f, g) \in \mathcal{F} \times \mathcal{F} \mid \exists h \in \mathcal{F}(f = h \circ g)\}.$$

   (a) Let f, g, and h be the functions from $\mathbb{R}$ to $\mathbb{R}$ defined by the formulas $f(x) = x^2 + 1$, $g(x) = x^3 + 1$, and $h(x) = x^4 + 1$. Prove that hRf, but it is not the case that gRf.
   (b) Prove that R is a preorder. (See exercise 24 of Section 4.6 for the definition of *preorder*.)
   (c) Prove that for all $f \in \mathcal{F}$, $fRi_{\mathbb{R}}$.
   (d) Prove that for all $f \in \mathcal{F}$, $i_{\mathbb{R}}Rf$ iff f is one-to-one. (Hint for right-to-left direction: Suppose f is one-to-one. Let $A = \text{Ran}(f)$, and let $h = f^{-1} \cup ((\mathbb{R} \setminus A) \times \{0\})$. Now prove that $h : \mathbb{R} \to \mathbb{R}$ and $i_{\mathbb{R}} = h \circ f$.)
   (e) Suppose that $g \in \mathcal{F}$ is a constant function; in other words, there is some real number c such that $\forall x \in \mathbb{R}(g(x) = c)$. Prove that for all $f \in \mathcal{F}$, gRf. (Hint: See exercise 14 of Section 5.1.)
   (f) Suppose that $g \in \mathcal{F}$ is a constant function. Prove that for all $f \in \mathcal{F}$, fRg iff f is a constant function.
   (g) As in exercise 24 of Section 4.6, if we let $S = R \cap R^{-1}$, then S is an equivalence relation on $\mathcal{F}$. Also, there is a unique relation T on $\mathcal{F}/S$ such that for all f and g in $\mathcal{F}$, $[f]_S T [g]_S$ iff fRg, and T is a partial order on $\mathcal{F}/S$. Prove that the set of all one-to-one functions from $\mathbb{R}$ to $\mathbb{R}$ is the largest element of $\mathcal{F}/S$ in the partial order T, and the set of all constant functions from $\mathbb{R}$ to $\mathbb{R}$ is the smallest element.

5.3. Inverses of Functions

We are now ready to return to the question of whether the inverse of a function from A to B is always a function from B to A. Consider again the function F from part 1 of Example 5.1.2. Recall that in that example we had $A = \{1, 2, 3\}$, $B = \{4, 5, 6\}$, and $F = \{(1, 5), (2, 4), (3, 5)\}$. As we saw in Example 5.1.2, F is a function from A to B. According to the definition of the inverse of a relation, $F^{-1} = \{(5, 1), (4, 2), (5, 3)\}$, which is clearly a relation from B to A. But F^{-1} fails to be a function from B to A for two reasons. First of all, $6 \in B$, but 6 isn't paired with any element of A in the relation F^{-1}. Second, 5 is paired with two different elements of A, 1 and 3. Thus, this example shows that the inverse of a function from A to B is not always a function from B to A.

You may have noticed that the reasons why F^{-1} isn't a function from B to A are related to the reasons why F is neither one-to-one nor onto, which were discussed in part 1 of Example 5.2.2. This suggests the following theorem.

Theorem 5.3.1. *Suppose* $f : A \to B$. *If f is one-to-one and onto, then* $f^{-1} : B \to A$.

Proof. Suppose f is one-to-one and onto, and let b be an arbitrary element of B. To show that f^{-1} is a function from B to A, we must prove that $\exists! a \in A((b, a) \in f^{-1})$, so we prove existence and uniqueness separately.

Existence: Since f is onto, there is some $a \in A$ such that $f(a) = b$. Thus, $(a, b) \in f$, so $(b, a) \in f^{-1}$.

Uniqueness: Suppose $(b, a_1) \in f^{-1}$ and $(b, a_2) \in f^{-1}$ for some $a_1, a_2 \in A$. Then $(a_1, b) \in f$ and $(a_2, b) \in f$, so $f(a_1) = b = f(a_2)$. Since f is one-to-one, it follows that $a_1 = a_2$. □

Commentary. The form of the proof is guided by the logical form of the statement that $f^{-1} : B \to A$. Because this means $\forall b \in B \exists! a \in A((b, a) \in f^{-1})$, we let b be an arbitrary element of B and then prove existence and uniqueness for the required $a \in A$ separately. Note that the assumption that f is onto is the key to the existence half of the proof, and the assumption that f is one-to-one is the key to the uniqueness half.

Suppose f is any function from a set A to a set B. Theorem 5.3.1 says that a sufficient condition for f^{-1} to be a function from B to A is that f be one-to-one and onto. Is it also a necessary condition? In other words, is the converse of Theorem 5.3.1 true? (If you don't remember what the words *sufficient*, *necessary*, and *converse* mean, you should review Section 1.5!) We will show in Theorem 5.3.4 that the answer to this question is yes. In other words, if f^{-1} is a function from B to A, then f must be one-to-one and onto.

If $f^{-1} : B \to A$ then, by the definition of function, for every $b \in B$ there is exactly one $a \in A$ such that $(b, a) \in f^{-1}$, and

$$\begin{aligned} f^{-1}(b) &= \text{the unique } a \in A \text{ such that } (b, a) \in f^{-1} \\ &= \text{the unique } a \in A \text{ such that } (a, b) \in f \\ &= \text{the unique } a \in A \text{ such that } f(a) = b. \end{aligned}$$

This gives another useful way to think about f^{-1}. If f^{-1} is a function from B to A, then it is the function that assigns, to each $b \in B$, the unique $a \in A$ such that $f(a) = b$. The assumption in Theorem 5.3.1 that f is one-to-one and onto guarantees that there is exactly one such a.

As an example, consider again the function f that assigns, to each person in the audience at a sold-out concert, the seat in which that person is sitting. As we saw at the end of the last section, f is a one-to-one, onto function from the set A of all members of the audience to the set S of all seats in the concert hall.

Thus, f^{-1} must be a function from S to A, and for each $s \in S$,

$$\begin{aligned} f^{-1}(s) &= \text{the unique } a \in A \text{ such that } f(a) = s \\ &= \text{the unique person } a \text{ such that the seat in which } a \text{ is sitting is } s \\ &= \text{the person who is sitting in the seat } s. \end{aligned}$$

In other words, the function f assigns to each person the seat in which that person is sitting, and the function f^{-1} assigns to each seat the person sitting in that seat.

Because $f : A \to S$ and $f^{-1} : S \to A$, it follows by Theorem 5.1.5 that $f^{-1} \circ f : A \to A$ and $f \circ f^{-1} : S \to S$. What are these functions? To figure out what the first function is, let's let a be an arbitrary element of A and compute $(f^{-1} \circ f)(a)$.

$$\begin{aligned} (f^{-1} \circ f)(a) &= f^{-1}(f(a)) \\ &= f^{-1}(\text{the seat in which } a \text{ is sitting}) \\ &= \text{the person sitting in the seat in which } a \text{ is sitting} \\ &= a. \end{aligned}$$

But recall that for every $a \in A$, $i_A(a) = a$. Thus, we have shown that $\forall a \in A((f^{-1} \circ f)(a) = i_A(a))$, so by Theorem 5.1.4, $f^{-1} \circ f = i_A$. Similarly, you should be able to check that $f \circ f^{-1} = i_S$.

When mathematicians find an unusual phenomenon like this in an example, they always wonder whether it's just a coincidence or if it's part of a more general pattern. In other words, can we prove a theorem that says that what happened in this example will happen in other examples too? In this case, it turns out that we can.

Theorem 5.3.2. *Suppose f is a function from A to B, and suppose that f^{-1} is a function from B to A. Then $f^{-1} \circ f = i_A$ and $f \circ f^{-1} = i_B$.*

Proof Let a be an arbitrary element of A. Let $b = f(a) \in B$. Then $(a, b) \in f$, so $(b, a) \in f^{-1}$ and therefore $f^{-1}(b) = a$. Thus,

$$(f^{-1} \circ f)(a) = f^{-1}(f(a)) = f^{-1}(b) = a = i_A(a).$$

Since a was arbitrary, we have shown that $\forall a \in A((f^{-1} \circ f)(a) = i_A(a))$, so $f^{-1} \circ f = i_A$. The proof of the second half of the theorem is similar and is left as an exercise (see exercise 8). □

Commentary. To prove that two functions are equal, we usually apply Theorem 5.1.4. Thus, since $f^{-1} \circ f$ and i_A are both functions from A to A, to prove that they are equal we prove that $\forall a \in A((f^{-1} \circ f)(a) = i_A(a))$.

Theorem 5.3.2 says that if $f : A \to B$ and $f^{-1} : B \to A$, then each function undoes the effect of the other. For any $a \in A$, applying the function f gives us $f(a) \in B$. According to Theorem 5.3.2, $f^{-1}(f(a)) = (f^{-1} \circ f)(a) = i_A(a) = a$. Thus, applying f^{-1} to $f(a)$ undoes the effect of applying f, giving us back the original element a. Similarly, for any $b \in B$, applying f^{-1} we get $f^{-1}(b) \in A$, and we can undo the effect of applying f^{-1} by applying f, since $f(f^{-1}(b)) = b$.

For example, let $f : \mathbb{R} \to \mathbb{R}$ be defined by the formula $f(x) = 2x$. You should be able to check that f is one-to-one and onto, so $f^{-1} : \mathbb{R} \to \mathbb{R}$, and for any $x \in \mathbb{R}$,

$$f^{-1}(x) = \text{the unique } y \text{ such that } f(y) = x.$$

Because $f^{-1}(x)$ is the unique solution for y in the equation $f(y) = x$, we can find a formula for $f^{-1}(x)$ by solving this equation for y. Filling in the definition of f in the equation gives us $2y = x$, so $y = x/2$. Thus, for every $x \in \mathbb{R}$, $f^{-1}(x) = x/2$. Notice that applying f to any number doubles the number and applying f^{-1} halves the number, and each of these operations undoes the effect of the other. In other words, if you double a number and then halve the result, you get back the number you started with. Similarly, halving any number and then doubling the result gives you back the original number.

Are there other circumstances in which the composition of two functions is equal to the identity function? Investigation of this question leads to the following theorem.

Theorem 5.3.3. *Suppose $f : A \to B$.*

1. *If there is a function $g : B \to A$ such that $g \circ f = i_A$ then f is one-to-one.*
2. *If there is a function $g : B \to A$ such that $f \circ g = i_B$ then f is onto.*

Proof

1. Suppose $g : B \to A$ and $g \circ f = i_A$. Let a_1 and a_2 be arbitrary elements of A, and suppose that $f(a_1) = f(a_2)$. Applying g to both sides of this equation we get $g(f(a_1)) = g(f(a_2))$. But $g(f(a_1)) = (g \circ f)(a_1) = i_A(a_1) = a_1$, and similarly, $g(f(a_2)) = a_2$. Thus, we can conclude that $a_1 = a_2$, and therefore f is one-to-one.
2. See exercise 9. □

Commentary. The assumption that there is a $g : B \to A$ such that $g \circ f = i_A$ is an existential statement, so we immediately imagine that a particular function g has been chosen. The proof that f is one-to-one follows the usual pattern for such proofs, based on Theorem 5.2.3.

We have come full circle. In Theorem 5.3.1 we found that if f is a one-to-one, onto function from A to B, then f^{-1} is a function from B to A. From this conclusion it follows, as we showed in Theorem 5.3.2, that the composition of f with its inverse must be the identity function. And in Theorem 5.3.3 we found that when the composition of two functions is the identity function, we are led back to the properties one-to-one and onto! Thus, combining Theorems 5.3.1–5.3.3, we get the following theorem.

Theorem 5.3.4. *Suppose $f : A \to B$. Then the following statements are equivalent.*

1. *f is one-to-one and onto.*
2. *$f^{-1} : B \to A$.*
3. *There is a function $g : B \to A$ such that $g \circ f = i_A$ and $f \circ g = i_B$.*

Proof. 1 → 2. This is precisely what Theorem 5.3.1 says.
2 → 3. Suppose $f^{-1} : B \to A$. Let $g = f^{-1}$ and apply Theorem 5.3.2.
3 → 1. Apply Theorem 5.3.3. □

Commentary. As we saw in Section 3.6, the easiest way to prove that several statements are equivalent is to prove a circle of implications. In this case we have proven the circle 1 → 2 → 3 → 1. Note that the proofs of these implications are quite sketchy. You should make sure you know how to fill in all the details.

For example, let f and g be functions from $\mathbb{R}$ to $\mathbb{R}$ defined by the following formulas:

$$f(x) = \frac{x+7}{5}, \qquad g(x) = 5x - 7\,.$$

Then for any real number x,

$$(g \circ f)(x) = g(f(x)) = g\left(\frac{x+7}{5}\right) = 5 \cdot \frac{x+7}{5} - 7 = x + 7 - 7 = x.$$

Thus, $g \circ f = i_{\mathbb{R}}$. A similar computation shows that $f \circ g = i_{\mathbb{R}}$. Thus, it follows from Theorem 5.3.4 that f must be one-to-one and onto, and f^{-1} must also be a function from $\mathbb{R}$ to $\mathbb{R}$. What is f^{-1}? Of course, a logical guess would be that $f^{-1} = g$, but this doesn't actually follow from the theorems we've proven. You could check it directly by solving for $f^{-1}(x)$, using the fact that $f^{-1}(x)$ must be the unique solution for y in the equation $f(y) = x$. However, there is no need to check. The next theorem shows that f^{-1} must be equal to g.

Theorem 5.3.5. *Suppose $f : A \to B$, $g : B \to A$, $g \circ f = i_A$, and $f \circ g = i_B$. Then $g = f^{-1}$.*

Proof. By Theorem 5.3.4, $f^{-1} : B \to A$. Therefore, by Theorem 5.3.2, $f^{-1} \circ f = i_A$. Thus,

$$\begin{aligned} g &= i_A \circ g && \text{(exercise 9 of Section 4.3)} \\ &= (f^{-1} \circ f) \circ g && \\ &= f^{-1} \circ (f \circ g) && \text{(Theorem 4.2.5)} \\ &= f^{-1} \circ i_B && \\ &= f^{-1} && \text{(exercise 9 of Section 4.3).} \end{aligned}$$

□

Commentary. This proof gets the desired conclusion quickly by clever use of previous theorems and exercises. For a more direct but somewhat longer proof, see exercise 10.

Example 5.3.6. In each part, determine whether or not f is one-to-one and onto. If it is, find f^{-1}.

1. Let $A = \mathbb{R} \setminus \{0\}$ and $B = \mathbb{R} \setminus \{2\}$, and define $f : A \to B$ by the formula

$$f(x) = \frac{1}{x} + 2.$$

   (Note that for all $x \in A$, $1/x$ is defined and nonzero, so $f(x) \neq 2$ and therefore $f(x) \in B$.)
2. Let $A = \mathbb{R}$ and $B = \{x \in \mathbb{R} \mid x \geq 0\}$, and define $f : A \to B$ by the formula

$$f(x) = x^2.$$

Solutions

1. You can check directly that f is one-to-one and onto, but we won't bother to check. Instead, we'll simply try to find a function $g : B \to A$ such that $g \circ f = i_A$ and $f \circ g = i_B$. We know by Theorems 5.3.4 and 5.3.5 that if we find such a g, then we can conclude that f is one-to-one and onto and $g = f^{-1}$.

Because we're hoping to have $g = f^{-1}$, we know that for any $x \in B = \mathbb{R} \setminus \{2\}$, $g(x)$ must be the unique $y \in A$ such that $f(y) = x$. Thus, to find a formula for $g(x)$, we solve for y in the equation $f(y) = x$. Filling in the definition of f, we see that the equation we must solve is

$$\frac{1}{y} + 2 = x.$$

Solving this equation we get

$$\frac{1}{y} + 2 = x \quad \Rightarrow \quad \frac{1}{y} = x - 2 \quad \Rightarrow \quad y = \frac{1}{x-2}.$$

Thus, we define $g : B \to A$ by the formula

$$g(x) = \frac{1}{x-2}.$$

Let's check that g has the required properties. For any $x \in A$, we have

$$g(f(x)) = g\left(\frac{1}{x} + 2\right) = \frac{1}{1/x + 2 - 2} = \frac{1}{1/x} = x.$$

Thus, $g \circ f = i_A$. Similarly, for any $x \in B$,

$$f(g(x)) = f\left(\frac{1}{x-2}\right) = \frac{1}{1/(x-2)} + 2 = x - 2 + 2 = x,$$

so $f \circ g = i_B$. Therefore, as we observed earlier, f must be one-to-one and onto, and $g = f^{-1}$.

2. Imitating the solution to part 1, let's try to find a function $g : B \to A$ such that $g \circ f = i_A$ and $f \circ g = i_B$. Because applying f to a number squares the number and we want g to undo the effect of f, a reasonable guess would be to let $g(x) = \sqrt{x}$. Let's see if this works.

   For any $x \in B$ we have

$$f(g(x)) = f(\sqrt{x}) = (\sqrt{x})^2 = x,$$

so $f \circ g = i_B$. But for $x \in A$ we have

$$g(f(x)) = g(x^2) = \sqrt{x^2},$$

and this is *not* always equal to x. For example, $g(f(-3)) = \sqrt{(-3)^2} = \sqrt{9} = 3 \neq -3$. Thus, $g \circ f \neq i_A$. This example illustrates that you must check *both* $f \circ g = i_B$ *and* $g \circ f = i_A$. It is possible for one to work but not the other.

What went wrong? We know that if f^{-1} is a function from B to A, then for any $x \in B$, $f^{-1}(x)$ must be the unique solution for y in the equation $f(y) = x$. Applying the definition of f gives us $y^2 = x$, so $y = \pm\sqrt{x}$. Thus, there is not a *unique* solution for y in the equation $f(y) = x$, there are two solutions. For example, when $x = 9$ we get $y = \pm 3$. In other words, $f(3) = f(-3) = 9$. But this means that f is not one-to-one! Thus, f^{-1} is not a function from B to A.

Functions that undo each other come up often in mathematics. For example, if you are familiar with logarithms, then you will recognize the formulas $10^{\log x} = x$ and $\log 10^x = x$. (We are using base 10 logarithms here.) We can rephrase these formulas in the language of this section by defining functions $f : \mathbb{R} \to \mathbb{R}^+$ and $g : \mathbb{R}^+ \to \mathbb{R}$ as follows:

$$f(x) = 10^x, \qquad g(x) = \log x.$$

Then for any $x \in \mathbb{R}$ we have $g(f(x)) = \log 10^x = x$, and for any $x \in \mathbb{R}^+$, $f(g(x)) = 10^{\log x} = x$. Thus, $g \circ f = i_{\mathbb{R}}$ and $f \circ g = i_{\mathbb{R}^+}$, so $g = f^{-1}$. In other words, the logarithm function is the inverse of the "raise 10 to the power" function.

We saw another example of functions that undo each other in Section 4.6. Suppose A is any set, let $\mathcal{E}$ be the set of all equivalence relations on A, and let $\mathcal{P}$ be the set of all partitions of A. Define a function $f : \mathcal{E} \to \mathcal{P}$ by the formula $f(R) = A / R$, and define another function $g : \mathcal{P} \to \mathcal{E}$ by the formula

$$\begin{aligned} g(\mathcal{F}) &= \text{the equivalence relation determined by } \mathcal{F} \\ &= \bigcup_{X \in \mathcal{F}} (X \times X). \end{aligned}$$

You should verify that the proof of Theorem 4.6.6 shows that $f \circ g = i_{\mathcal{P}}$, and exercise 9 in Section 4.6 shows that $g \circ f = i_{\mathcal{E}}$. Thus, f is one-to-one and onto, and $g = f^{-1}$. One interesting consequence of this is that if A has a finite number of elements, then we can say that the number of equivalence relations on A is exactly the same as the number of partitions of A, even though we don't know what this number is.

Exercises

*1. Let R be the function defined in exercise 2(c) of Section 5.1. In exercise 2 of Section 5.2, you showed that R is one-to-one and onto, so $R^{-1} : P \to P$. If $p \in P$, what is $R^{-1}(p)$?

2. Let F be the function defined in exercise 4(b) of Section 5.1. In exercise 4 of Section 5.2, you showed that F is one-to-one and onto, so $F^{-1} : B \to B$. If $X \in B$, what is $F^{-1}(X)$?

*3. Let $f : \mathbb{R} \to \mathbb{R}$ be defined by the formula
$$f(x) = \frac{2x+5}{3}.$$
Show that f is one-to-one and onto, and find a formula for $f^{-1}(x)$. (You may want to imitate the method used in the example after Theorem 5.3.2, or in Example 5.3.6.)

4. Let $f : \mathbb{R} \to \mathbb{R}$ be defined by the formula $f(x) = 2x^3 - 3$. Show that f is one-to-one and onto, and find a formula for $f^{-1}(x)$.

*5. Let $f : \mathbb{R} \to \mathbb{R}^+$ be defined by the formula $f(x) = 10^{2-x}$. Show that f is one-to-one and onto, and find a formula for $f^{-1}(x)$.

6. Let $A = \mathbb{R} \setminus \{2\}$, and let f be the function with domain A defined by the formula
$$f(x) = \frac{3x}{x-2}.$$
   (a) Show that f is a one-to-one, onto function from A to B for some set $B \subseteq \mathbb{R}$. What is the set B?
   (b) Find a formula for $f^{-1}(x)$.

7. In the example after Theorem 5.3.4, we had $f(x) = \frac{x+7}{5}$ and found that $f^{-1}(x) = 5x - 7$. Let f_1 and f_2 be functions from $\mathbb{R}$ to $\mathbb{R}$ defined by the formulas
$$f_1(x) = x + 7, \qquad f_2(x) = \frac{x}{5}.$$
   (a) Show that $f = f_2 \circ f_1$.
   (b) According to part 5 of Theorem 4.2.5, we must have $f^{-1} = (f_2 \circ f_1)^{-1} = (f_1)^{-1} \circ (f_2)^{-1}$. Verify that this is true by computing $(f_1)^{-1} \circ (f_2)^{-1}$ directly.

8. (a) Prove the second half of Theorem 5.3.2 by imitating the proof of the first half.
   (b) Give an alternative proof of the second half of Theorem 5.3.2 by *applying* the first half to f^{-1}.

*9. Prove part 2 of Theorem 5.3.3.

10. Use the following strategy to give an alternative proof of Theorem 5.3.5: Let (b, a) be an arbitrary element of $B \times A$. Assume $(b, a) \in g$ and prove $(b, a) \in f^{-1}$. Then assume $(b, a) \in f^{-1}$ and prove $(b, a) \in g$.

*11. Suppose $f : A \to B$ and $g : B \to A$.
   (a) Prove that if f is one-to-one and $f \circ g = i_B$, then $g = f^{-1}$.
   (b) Prove that if f is onto and $g \circ f = i_A$, then $g = f^{-1}$.
   (c) Prove that if $f \circ g = i_B$ but $g \circ f \neq i_A$, then f is onto but not one-to-one, and g is one-to-one but not onto.

12. Suppose $f : A \to B$ and f is one-to-one. Prove that there is some set $B' \subseteq B$ such that $f^{-1} : B' \to A$.

13. Suppose $f : A \to B$ and f is onto. Let $R = \{(x, y) \in A \times A \mid f(x) = f(y)\}$. By exercise 17(a) of Section 5.1, R is an equivalence relation on A.
   (a) Prove that there is a function $h : A/R \to B$ such that for all $x \in A$, $h([x]_R) = f(x)$. (Hint: See exercise 18 of Section 5.1.)
   (b) Prove that h is one-to-one and onto. (Hint: See exercise 16 of Section 5.2.)
   (c) It follows from part (b) that $h^{-1} : B \to A/R$. Prove that for all $b \in B$, $h^{-1}(b) = \{x \in A \mid f(x) = b\}$.
   (d) Suppose $g : B \to A$. Prove that $f \circ g = i_B$ iff $\forall b \in B(g(b) \in h(b))$.

*14. Suppose $f : A \to B$, $g : B \to A$, and $f \circ g = i_B$. Let $A' = \text{Ran}(g) \subseteq A$.
   (a) Prove that for all $x \in A'$, $(g \circ f)(x) = x$.
   (b) Prove that $f \restriction A'$ is a one-to-one, onto function from A' to B and $g = (f \restriction A')^{-1}$. (See exercise 7 of Section 5.1 for the meaning of the notation used here.)

15. Let $B = \{x \in \mathbb{R} \mid x \geq 0\}$. Let $f : \mathbb{R} \to B$ and $g : B \to \mathbb{R}$ be defined by the formulas $f(x) = x^2$ and $g(x) = \sqrt{x}$. As we saw in part 2 of Example 5.3.6, $g \neq f^{-1}$. Show that $g = (f \restriction B)^{-1}$. (Hint: See exercise 14.)

*16. Let $f : \mathbb{R} \to \mathbb{R}$ be defined by the formula $f(x) = 4x - x^2$. Let $B = \text{Ran}(f)$.
   (a) Find B.
   (b) Find a set $A \subseteq \mathbb{R}$ such that $f \restriction A$ is a one-to-one, onto function from A to B, and find a formula for $(f \restriction A)^{-1}$. (Hint: See exercise 14.)

17. Let A be any set. Let $\mathcal{P}$ be the set of all partial orders on A, and let $\mathcal{S}$ be the set of all strict partial orders on A. In exercises 4 and 5 of Section 4.5 you showed that if $R \in \mathcal{P}$ then $R \setminus i_A \in \mathcal{S}$, and if $R \in \mathcal{S}$ then $R \cup i_A \in \mathcal{P}$. (Recall that we showed in the proof of Theorem 4.5.2 that $R \cup i_A$ is the reflexive closure of R.) Let $f : \mathcal{P} \to \mathcal{S}$ and $g : \mathcal{S} \to \mathcal{P}$ be defined by the formulas

$$f(R) = R \setminus i_A, \qquad g(R) = R \cup i_A.$$

Show that f is one-to-one and onto, and $g = f^{-1}$.

18. Suppose A is a set, and let $\mathcal{F} = \{f \mid f : A \to A\}$ and $\mathcal{P} = \{f \in \mathcal{F} \mid f$ is one-to-one and onto$\}$. Define a relation R on $\mathcal{F}$ as follows:

$$R = \{(f, g) \in \mathcal{F} \times \mathcal{F} \mid \exists h \in \mathcal{P}(f = h^{-1} \circ g \circ h)\}.$$

   (a) Prove that R is an equivalence relation.
   (b) Prove that if $f R g$ then $(f \circ f) R (g \circ g)$.
   (c) For any $f \in \mathcal{F}$ and $a \in A$, if $f(a) = a$ then we say that a is a *fixed point* of f. Prove that if f has a fixed point and $f R g$, then g also has a fixed point.

5.4. Images and Inverse Images: A Research Project

Suppose $f : A \to B$. We have already seen that we can think of f as matching each element of A with exactly one element of B. In this section we will see that f can also be thought of as matching *subsets* of A with subsets of B and vice-versa.

Definition 5.4.1. Suppose $f : A \to B$ and $X \subseteq A$. Then the *image* of X under f is the set $f(X)$ defined as follows:

$$\begin{aligned} f(X) &= \{f(x) \mid x \in X\} \\ &= \{b \in B \mid \exists x \in X(f(x) = b)\}. \end{aligned}$$

(Note that the image of the whole domain A under f is $\{f(a) \mid a \in A\}$, and as we saw in Section 5.1 this is the same as the range of f.)

If $Y \subseteq B$, then the *inverse image* of Y under f is the set $f^{-1}(Y)$ defined as follows:

$$f^{-1}(Y) = \{a \in A \mid f(a) \in Y\}.$$

Note that the function f in Definition 5.4.1 may fail to be one-to-one or onto, and as a result f^{-1} may not be a function from B to A, and for $y \in B$, the notation "$f^{-1}(y)$" may be meaningless. However, even in this case Definition 5.4.1 still assigns a meaning to the notation "$f^{-1}(Y)$" for $Y \subseteq B$. If you find this surprising, look again at the definition of $f^{-1}(Y)$, and notice that it does not treat f^{-1} as a function. The definition refers only to the results of applying f to elements of A, not the results of applying f^{-1} to elements of B.

For example, let L be the function defined in part 3 of Example 5.1.2, which assigns to each city the country in which that city is located. As in Example 5.1.2, let C be the set of all cities and N the set of all countries. If B is the set of all cities with population at least one million, then B is a subset of C, and the image of B under L would be the set

$$\begin{aligned} L(B) &= \{L(b) \mid b \in B\} \\ &= \{n \in N \mid \exists b \in B(L(b) = n)\} \\ &= \{n \in N \mid \text{there is some city with population at least} \\ &\qquad \text{one million that is located in the country } n\}. \end{aligned}$$

Thus, $L(B)$ is the set of all countries that contain a city with population at least one million. Now let A be the subset of N consisting of all countries in Africa. Then the inverse image of A under L is the set

$$\begin{aligned} L^{-1}(A) &= \{c \in C \mid L(c) \in A\} \\ &= \{c \in C \mid \text{the country in which } c \text{ is located is in Africa}\}. \end{aligned}$$

Thus, $L^{-1}(A)$ is the set of all cities in African countries.

Let's do one more example. Let $f : \mathbb{R} \to \mathbb{R}$ be defined by the formula $f(x) = x^2$, and let $X = \{x \in \mathbb{R} \mid 0 \leq x < 2\}$. Then

$$f(X) = \{f(x) \mid x \in X\} = \{x^2 \mid 0 \leq x < 2\}.$$

Thus, $f(X)$ is the set of all squares of real numbers between 0 and 2 (including 0 but not 2). A moment's reflection should convince you that this set is $\{x \in \mathbb{R} \mid 0 \leq x < 4\}$. Now let's let $Y = \{x \in \mathbb{R} \mid 0 \leq x < 4\}$ and compute $f^{-1}(Y)$. According to the definition of inverse image,

$$\begin{aligned} f^{-1}(Y) &= \{x \in \mathbb{R} \mid f(x) \in Y\} \\ &= \{x \in \mathbb{R} \mid 0 \leq f(x) < 4\} \\ &= \{x \in \mathbb{R} \mid 0 \leq x^2 < 4\} \\ &= \{x \in \mathbb{R} \mid -2 < x < 2\}. \end{aligned}$$

By now you have had enough experience writing proofs that you should be ready to put your proof-writing skills to work in answering mathematical questions. Thus, most of this section will be devoted to a research project in which you will discover for yourself the answers to basic mathematical questions about images and inverse images. To get you started, we'll work out the answer to the first question.

Suppose $f : A \to B$, and W and X are subsets of A. A natural question you might ask is whether or not $f(W \cap X)$ must be the same as $f(W) \cap f(X)$. It seems plausible that the answer is yes, so let's see if we can prove it. Thus, our goal will be to prove that $f(W \cap X) = f(W) \cap f(X)$. Because this is an equation between two sets, we proceed by taking an arbitrary element of each set and trying to prove that it is an element of the other.

Suppose first that y is an arbitrary element of $f(W \cap X)$. By the definition of $f(W \cap X)$, this means that $y = f(x)$ for some $x \in W \cap X$. Since $x \in W \cap X$, it follows that $x \in W$ and $x \in X$. But now we have $y = f(x)$ and $x \in W$, so we can conclude that $y \in f(W)$. Similarly, since $y = f(x)$ and $x \in X$, it follows that $y \in f(X)$. Thus, $y \in f(W) \cap f(X)$. This completes the first half of the proof.

Now suppose that $y \in f(W) \cap f(X)$. Then $y \in f(W)$, so there is some $w \in W$ such that $f(w) = y$, and also $y \in f(X)$, so there is some $x \in X$ such that $y = f(x)$. If only we knew that w and x were equal, we could conclude that $w = x \in W \cap X$, so $y = f(x) \in f(W \cap X)$. But the best we can do is to say that $f(w) = y = f(x)$. This should remind you of the definition of one-to-one. If we knew that f was one-to-one, we could conclude from the fact that $f(w) = f(x)$ that $w = x$, and the proof would be done. But without this information we seem to be stuck.

Let's summarize what we've discovered. First of all, the first half of the proof worked fine, so we can certainly say that in general $f(W \cap X) \subseteq f(W) \cap f(X)$. The second half worked *if* we knew that f was one-to-one, so we can also say that if f is one-to-one, then $f(W \cap X) = f(W) \cap f(X)$. But what if f isn't one-to-one? There might be some way of fixing up the proof to show that the equation $f(W \cap X) = f(W) \cap f(X)$ is still true even if f isn't one-to-one. But by now you have probably come to suspect that perhaps $f(W \cap X)$ and $f(W) \cap f(X)$ are not always equal, so maybe we should devote some time to trying to show that the proposed theorem is incorrect. In other words, let's see if we can find a counterexample – an example of a function f and sets W and X for which $f(W \cap X) \neq f(W) \cap f(X)$.

Fortunately, we can do better than just trying examples at random. Of course, we know we'd better use a function that isn't one-to-one, but by examining our attempt at a proof, we can tell more than that. The attempted proof that $f(W \cap X) = f(W) \cap f(X)$ ran into trouble only when W and X contained elements w and x such that $w \neq x$ but $f(w) = f(x)$, so we should choose an example in which this happens. In other words, not only should we make sure f isn't one-to-one, we should also make sure W and X contain elements that *show* that f isn't one-to-one.

The graph in Figure 1 shows a simple function that isn't one-to-one. Writing it as a set of ordered pairs, we could say $f = \{(1, 4), (2, 5), (3, 5)\}$, and $f : A \to B$, where $A = \{1, 2, 3\}$ and $B = \{4, 5, 6\}$. The two elements of A that show that f is not one-to-one are 2 and 3, so these should be elements of W and X, respectively. Why not just try letting $W = \{2\}$ and $X = \{3\}$? With these choices

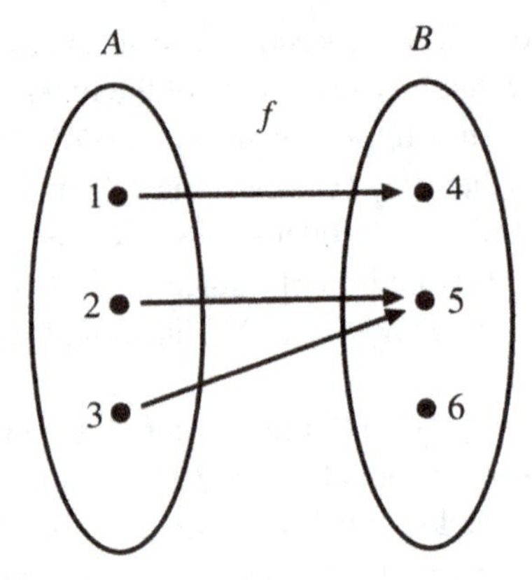

Figure 1

we get $f(W) = \{f(2)\} = \{5\}$ and $f(X) = \{f(3)\} = \{5\}$, so $f(W) \cap f(X) = \{5\} \cap \{5\} = \{5\}$. But $f(W \cap X) = f(\varnothing) = \varnothing$, so $f(W \cap X) \neq f(W) \cap f(X)$. (If you're not sure why $f(\varnothing) = \varnothing$, work it out using Definition 5.4.1!) If you want to see an example in which $W \cap X \neq \varnothing$, try $W = \{1, 2\}$ and $X = \{1, 3\}$.

This example shows that it would be incorrect to state a theorem saying that $f(W \cap X)$ and $f(W) \cap f(X)$ are always equal. But our proof shows that the following theorem is correct:

Theorem 5.4.2. *Suppose $f : A \to B$, and W and X are subsets of A. Then $f(W \cap X) \subseteq f(W) \cap f(X)$. Furthermore, if f is one-to-one, then $f(W \cap X) = f(W) \cap f(X)$.*

Now, here are some questions for you to try to answer. In each case, try to figure out as much as you can. Justify your answers with proofs and counterexamples.

Suppose $f : A \to B$.

1. Suppose W and X are subsets of A.
   (a) Will it always be true that $f(W \cup X) = f(W) \cup f(X)$?
   (b) Will it always be true that $f(W \setminus X) = f(W) \setminus f(X)$?
   (c) Will it always be true that $W \subseteq X \leftrightarrow f(W) \subseteq f(X)$?
2. Suppose that Y and Z are subsets of B.
   (a) Will it always be true that $f^{-1}(Y \cap Z) = f^{-1}(Y) \cap f^{-1}(Z)$?
   (b) Will it always be true that $f^{-1}(Y \cup Z) = f^{-1}(Y) \cup f^{-1}(Z)$?
   (c) Will it always be true that $f^{-1}(Y \setminus Z) = f^{-1}(Y) \setminus f^{-1}(Z)$?
   (d) Will it always be true that $Y \subseteq Z \leftrightarrow f^{-1}(Y) \subseteq f^{-1}(Z)$?

3. Suppose $X \subseteq A$. Will it always be true that $f^{-1}(f(X)) = X$?
4. Suppose $Y \subseteq B$. Will it always be true that $f(f^{-1}(Y)) = Y$?
5. Suppose $g : B \to C$. Can you prove any interesting theorems about images and inverse images of sets under $g \circ f$?

Note: An observant reader may have noticed an ambiguity in our notation for images and inverse images. If $f : A \to B$ and $Y \subseteq B$, then we have used the notation $f^{-1}(Y)$ to stand for the inverse image of Y under f. But if f is one-to-one and onto, then, as we saw in Section 5.3, f^{-1} is a function from B to A. Thus, $f^{-1}(Y)$ could also be interpreted as the image of Y under the function f^{-1}. Fortunately, this ambiguity is harmless, as the next problem shows.

6. Suppose $f : A \to B$, f is one-to-one and onto, and $Y \subseteq B$. Show that the inverse image of Y under f and the image of Y under f^{-1} are equal. (Hint: First write out the definitions of the two sets carefully!)

6

Mathematical Induction

6.1. Proof by Mathematical Induction

In Chapter 3 we studied proof techniques that could be used in reasoning about any mathematical topic. In this chapter we'll discuss one more proof technique, called *mathematical induction*, that is designed for proving statements about what is perhaps the most fundamental of all mathematical structures, the natural numbers. Recall that the set of all natural numbers is $\mathbb{N} = \{0, 1, 2, 3, \ldots\}$.

Suppose you want to prove that every natural number has some property P. In other words, you want to show that $0, 1, 2, \ldots$ all have the property P. Of course, there are infinitely many numbers in this list, so you can't check one-by-one that they all have property P. The key idea behind mathematical induction is that to list all the natural numbers all you have to do is start with 0 and repeatedly add 1. Thus, you can show that every natural number has the property P by showing that 0 has property P, and that whenever you add 1 to a number that has property P, the resulting number also has property P. This would guarantee that, as you go through the list of all natural numbers, starting with 0 and repeatedly adding 1, every number you encounter must have property P. In other words, all natural numbers have property P. Here, then, is how the method of mathematical induction works.

To prove a goal of the form $\forall n \in \mathbb{N} P(n)$:

First prove $P(0)$, and then prove $\forall n \in \mathbb{N}(P(n) \rightarrow P(n+1))$. The first of these proofs is sometimes called the *base case* and the second the *induction step*.

Form of the final proof:

Base case: [Proof of $P(0)$ goes here.]

Induction step: [Proof of $\forall n \in \mathbb{N}(P(n) \rightarrow P(n+1))$ goes here.]

We'll say more about the justification of the method of mathematical induction later, but first let's look at an example of a proof that uses mathematical induction. The following list of calculations suggests a surprising pattern:

$$2^0 = 1 = 2^1 - 1$$
$$2^0 + 2^1 = 1 + 2 = 3 = 2^2 - 1$$
$$2^0 + 2^1 + 2^2 = 1 + 2 + 4 = 7 = 2^3 - 1$$
$$2^0 + 2^1 + 2^2 + 2^3 = 1 + 2 + 4 + 8 = 15 = 2^4 - 1$$

The general pattern appears to be:

$$2^0 + 2^1 + \cdots + 2^n = 2^{n+1} - 1.$$

Will this pattern hold for all values of n? Let's see if we can prove it.

Example 6.1.1. Prove that for every natural number n, $2^0 + 2^1 + \cdots + 2^n = 2^{n+1} - 1$.

Scratch work

Our goal is to prove the statement $\forall n \in \mathbb{N} P(n)$, where $P(n)$ is the statement $2^0 + 2^1 + \cdots + 2^n = 2^{n+1} - 1$. According to our strategy, we can do this by proving two other statements, $P(0)$ and $\forall n \in \mathbb{N}(P(n) \rightarrow P(n+1))$.

Plugging in 0 for n, we see that $P(0)$ is simply the statement $2^0 = 2^1 - 1$, the first statement in our list of calculations. The proof of this is easy – just do the arithmetic to verify that both sides are equal to 1. Often the base case of an induction proof is very easy, and the only hard work in figuring out the proof is in carrying out the induction step.

For the induction step, we must prove $\forall n \in \mathbb{N}(P(n) \rightarrow P(n+1))$. Of course, all of the proof techniques discussed in Chapter 3 can be used in mathematical induction proofs, so we can do this by letting n be an arbitrary natural number, assuming that $P(n)$ is true, and then proving that $P(n+1)$ is true. In other words, we'll let n be an arbitrary natural number, assume that $2^0 + 2^1 + \cdots + 2^n = 2^{n+1} - 1$, and then prove that $2^0 + 2^1 + \cdots + 2^{n+1} = 2^{n+2} - 1$. This gives us the following givens and goal:

Givens	*Goal*
$n \in \mathbb{N}$	$2^0 + 2^1 + \cdots + 2^{n+1} = 2^{n+2} - 1$
$2^0 + 2^1 + \cdots + 2^n = 2^{n+1} - 1$	

Clearly the second given is similar to the goal. Is there some way to start with the second given and derive the goal using algebraic steps? The key to the

proof is to recognize that the left side of the equation in the goal is exactly the same as the left side of the second given, but with the extra term 2^{n+1} added on. So let's try adding 2^{n+1} to both sides of the second given.
This gives us

$$(2^0 + 2^1 + \cdots + 2^n) + 2^{n+1} = (2^{n+1} - 1) + 2^{n+1},$$

or in other words,

$$2^0 + 2^1 + \cdots + 2^{n+1} = 2 \cdot 2^{n+1} - 1 = 2^{n+2} - 1.$$

This is the goal, so we are done!

Solution

Theorem. *For every natural number* n, $2^0 + 2^1 + \cdots + 2^n = 2^{n+1} - 1$.
Proof. We use mathematical induction.

Base case: Setting $n = 0$, we get $2^0 = 1 = 2^1 - 1$ as required.

Induction step: Let n be an arbitrary natural number and suppose that $2^0 + 2^1 + \cdots + 2^n = 2^{n+1} - 1$. Then

$$\begin{aligned} 2^0 + 2^1 + \cdots + 2^{n+1} &= (2^0 + 2^1 + \cdots + 2^n) + 2^{n+1} \\ &= (2^{n+1} - 1) + 2^{n+1} \\ &= 2 \cdot 2^{n+1} - 1 \\ &= 2^{n+2} - 1. \end{aligned}$$

□

Does the proof in Example 6.1.1 convince you that the equation $2^0 + 2^1 + \cdots + 2^n = 2^{n+1} - 1$, which we called $P(n)$ in our scratch work, is true for all natural numbers n? Well, certainly $P(0)$ is true, since we checked that explicitly in the base case of the proof. In the induction step we showed that $\forall n \in \mathbb{N}(P(n) \to P(n+1))$, so we know that for every natural number n, $P(n) \to P(n+1)$. For example, plugging in $n = 0$ we can conclude that $P(0) \to P(1)$. But now we know that both $P(0)$ and $P(0) \to P(1)$ are true, so applying modus ponens we can conclude that $P(1)$ is true too. Similarly, plugging in $n = 1$ in the induction step we get $P(1) \to P(2)$, so applying modus ponens to the statements $P(1)$ and $P(1) \to P(2)$ we can conclude that $P(2)$ is true. Setting $n = 2$ in the induction step we get $P(2) \to P(3)$, so by modus ponens, $P(3)$ is true. Continuing in this way, you should be able to see that by repeatedly applying the induction step you can show that $P(n)$ must be true for every natural number n. In other words, the proof really does show that $\forall n \in \mathbb{N} P(n)$.

As we saw in the last example, the hardest part of a proof by mathematical induction is usually the induction step, in which you must prove the statement $\forall n \in \mathbb{N}(P(n) \rightarrow P(n+1))$. It is usually best to do this by letting n be an arbitrary natural number, assuming $P(n)$ is true, and then proving that $P(n+1)$ is true. The assumption that $P(n)$ is true is sometimes called the *inductive hypothesis*, and the key to the proof is usually to work out some relationship between the inductive hypothesis $P(n)$ and the goal $P(n+1)$.

Here's another example of a proof by mathematical induction.

Example 6.1.2. Prove that $\forall n \in \mathbb{N}(3 \mid (n^3 - n))$.

Scratch work

As usual, the base case is easy to check. The details are given in the following proof. For the induction step, we let n be an arbitrary natural number and assume that $3 \mid (n^3 - n)$, and we must prove that $3 \mid ((n+1)^3 - (n+1))$. Filling in the definition of *divides*, we can sum up our situation as follows:

Givens	*Goal*
$n \in \mathbb{N}$	$\exists j \in \mathbb{Z}(3j = (n+1)^3 - (n+1))$
$\exists k \in \mathbb{Z}(3k = n^3 - n)$	

The second given is the inductive hypothesis, and we need to figure out how it can be used to establish the goal.

According to our techniques for dealing with existential quantifiers in proofs, the best thing to do first is to use the second given and let k stand for a particular integer such that $3k = n^3 - n$. To complete the proof we'll need to find an integer j (which will probably be related to k in some way) such that $3j = (n+1)^3 - (n+1)$. We expand the right side of this equation, looking for some way to relate it to the given equation $3k = n^3 - n$:

$$\begin{aligned}(n+1)^3 - (n+1) &= n^3 + 3n^2 + 3n + 1 - n - 1 \\ &= (n^3 - n) + 3n^2 + 3n \\ &= 3k + 3n^2 + 3n \\ &= 3(k + n^2 + n).\end{aligned}$$

It should now be clear that we can complete the proof by letting $j = k + n^2 + n$. As in similar earlier proofs, we don't bother to mention j in the proof.

Solution

Theorem. *For every natural number n, $3 \mid (n^3 - n)$.*

Proof. We use mathematical induction.

Base case: If $n = 0$, then $n^3 - n = 0 = 3 \cdot 0$, so $3 \mid (n^3 - n)$.

Induction step: Let n be an arbitrary natural number and suppose $3 \mid (n^3 - n)$. Then we can choose an integer k such that $3k = n^3 - n$. Thus,

$$\begin{aligned}(n+1)^3 - (n+1) &= n^3 + 3n^2 + 3n + 1 - n - 1 \\ &= (n^3 - n) + 3n^2 + 3n \\ &= 3k + 3n^2 + 3n \\ &= 3(k + n^2 + n).\end{aligned}$$

Therefore $3 \mid ((n+1)^3 - (n+1))$, as required. □

Once you understand why mathematical induction works, you should be able to understand proofs that involve small variations on the method of induction. The next example illustrates such a variation. In this example we'll try to figure out which is larger, n^2 or 2^n. Let's try out a few values of n:

n	n^2	2^n	Which is larger?
0	0	1	2^n
1	1	2	2^n
2	4	4	tie
3	9	8	n^2
4	16	16	tie
5	25	32	2^n
6	36	64	2^n

It's a close race at first, but starting with $n = 5$, it looks like 2^n is taking a decisive lead over n^2. Can we prove that it will stay ahead for larger values of n?

Example 6.1.3. Prove that $\forall n \geq 5(2^n > n^2)$.

Scratch work

We are only interested in proving the inequality $2^n > n^2$ for $n \geq 5$. Thus, it would make no sense to use $n = 0$ in the base case of our induction proof. We'll take $n = 5$ as the base case for our induction rather than $n = 0$. Once we've checked that the inequality holds when $n = 5$, the induction step will show that the inequality must continue to hold if we repeatedly add 1 to n. Thus, it must also hold for $n = 6, 7, 8, \ldots$. In other words, we'll be able to conclude that the inequality holds for all $n \geq 5$.

The base case $n = 5$ has already been checked in the table. For the induction step, we let $n \geq 5$ be arbitrary, assume $2^n > n^2$, and try to prove that $2^{n+1} > (n+1)^2$. How can we relate the inductive hypothesis to the goal? Perhaps the simplest relationship involves the left sides of the two inequalities: $2^{n+1} = 2 \cdot 2^n$. Thus, multiplying both sides of the inductive hypothesis $2^n > n^2$ by 2, we can conclude that $2^{n+1} > 2n^2$. Now compare this inequality to the goal, $2^{n+1} > (n+1)^2$. If we could prove that $2n^2 \geq (n+1)^2$, then the goal would follow easily. So let's forget about the original goal and see if we can prove that $2n^2 \geq (n+1)^2$.

Multiplying out the right side of the new goal we see that we must prove that $2n^2 \geq n^2 + 2n + 1$, or in other words $n^2 \geq 2n + 1$. This isn't hard to prove: Since we've assumed that $n \geq 5$, it follows that $n^2 \geq 5n = 2n + 3n > 2n + 1$.

Solution

Theorem. *For every natural number $n \geq 5$, $2^n > n^2$.*

Proof. By mathematical induction.

Base case: When $n = 5$ we have $2^n = 32 > 25 = n^2$.

Induction step: Let $n \geq 5$ be arbitrary, and suppose that $2^n > n^2$. Then

$$\begin{aligned}
2^{n+1} &= 2 \cdot 2^n \\
&> 2n^2 && \text{(by inductive hypothesis)} \\
&= n^2 + n^2 \\
&\geq n^2 + 5n && \text{(since } n \geq 5\text{)} \\
&= n^2 + 2n + 3n \\
&> n^2 + 2n + 1 = (n+1)^2.
\end{aligned}$$

□

Exercises

*1. Prove that for all $n \in \mathbb{N}$, $0 + 1 + 2 + \cdots + n = n(n+1)/2$.

2. Prove that for all $n \in \mathbb{N}$, $0^2 + 1^2 + 2^2 + \cdots + n^2 = n(n+1)(2n+1)/6$.

*3. Prove that for all $n \in \mathbb{N}$, $0^3 + 1^3 + 2^3 + \cdots + n^3 = [n(n+1)/2]^2$.

4. Find a formula for $1 + 3 + 5 + \cdots + (2n - 1)$, for $n \geq 1$, and prove that your formula is correct. (Hint: First try some particular values of n and look for a pattern.)

5. Prove that for all $n \in \mathbb{N}$, $0 \cdot 1 + 1 \cdot 2 + 2 \cdot 3 + \cdots + n(n+1) = n(n+1)(n+2)/3$.

6. Find a formula for $0 \cdot 1 \cdot 2 + 1 \cdot 2 \cdot 3 + 2 \cdot 3 \cdot 4 + \cdots + n(n+1)(n+2)$, for $n \in \mathbb{N}$, and prove that your formula is correct. (Hint: Compare this exercise to exercises 1 and 5, and try to guess the formula.)

*7. Find a formula for $3^0 + 3^1 + 3^2 + \cdots + 3^n$, for $n \geq 0$, and prove that your formula is correct. (Hint: Try to guess the formula, basing your guess on Example 6.1.1. Then try out some values of n and adjust your guess if necessary.)

8. Prove that for all $n \geq 1$,

$$1 - \frac{1}{2} + \frac{1}{3} - \frac{1}{4} + \cdots + \frac{1}{2n-1} - \frac{1}{2n} = \frac{1}{n+1} + \frac{1}{n+2} + \frac{1}{n+3} + \cdots + \frac{1}{2n}$$

9. (a) Prove that all $n \in \mathbb{N}, 2 \mid (n^2 + n)$.
   (b) Prove that for all $n \in \mathbb{N}, 6 \mid (n^3 - n)$.

*10. Prove that for all $n \in \mathbb{N}, 64 \mid (9^n - 8n - 1)$.

11. Prove that for all $n \in \mathbb{N}, 9 \mid (4^n + 6n - 1)$.

12. Prove that for all integers a and b and all $n \in \mathbb{N}, (a - b) \mid (a^n - b^n)$. (Hint: Let a and b be arbitrary integers and then prove by induction that $\forall n \in \mathbb{N}[(a - b) \mid (a^n - b^n)]$. For the induction step, you must relate $a^{n+1} - b^{n+1}$ to $a^n - b^n$. You might find it useful to start by completing the following equation: $a^{n+1} - b^{n+1} = a(a^n - b^n) + \underline{?}$.)

13. Prove that for all integers a and b and all $n \in \mathbb{N}, (a + b) \mid (a^{2n+1} + b^{2n+1})$.

*14. Prove that for all $n \geq 10, 2^n > n^3$.

15. Prove that for all $n \in \mathbb{N}$, either $n \equiv 0 \pmod 3$ or $n \equiv 1 \pmod 3$ or $n \equiv 2 \pmod 3$. (Recall that this notation was introduced in Definition 4.6.9.)

16. Prove that for all $n \geq 1$, $2 \cdot 2^1 + 3 \cdot 2^2 + 4 \cdot 2^3 + \cdots + (n + 1)2^n = n2^{n+1}$.

17. (a) What's wrong with the following proof that for all $n \in \mathbb{N}$, $1 \cdot 3^0 + 3 \cdot 3^1 + 5 \cdot 3^2 + \cdots + (2n + 1)3^n = n3^{n+1}$?

    Proof. We use mathematical induction. Let n be an arbitrary natural number, and suppose that $1 \cdot 3^0 + 3 \cdot 3^1 + 5 \cdot 3^2 + \cdots + (2n + 1)3^n = n3^{n+1}$. Then

$$\begin{aligned} &1 \cdot 3^0 + 3 \cdot 3^1 + 5 \cdot 3^2 + \cdots + (2n + 1)3^n + (2n + 3)3^{n+1} \\ &= n3^{n+1} + (2n + 3)3^{n+1} \\ &= (3n + 3)3^{n+1} \\ &= (n + 1)3^{n+2}, \end{aligned}$$

    as required. □

   (b) Find a formula for $1 \cdot 3^0 + 3 \cdot 3^1 + 5 \cdot 3^2 + \cdots + (2n + 1)3^n$, and prove that your formula is correct.

18. Suppose a is a real number and $a < 0$. Prove that for all $n \in \mathbb{N}$, if n is even then $a^n > 0$, and if n is odd then $a^n < 0$.

*19. Suppose a and b are real numbers and $0 < a < b$.
   (a) Prove that for all $n \geq 1$, $0 < a^n < b^n$. (Notice that this generalizes Theorem 3.1.2.)
   (b) Prove that for all $n \geq 2$, $0 < \sqrt[n]{a} < \sqrt[n]{b}$.
   (c) Prove that for all $n \geq 1$, $ab^n + ba^n < a^{n+1} + b^{n+1}$.
   (d) Prove that for all $n \geq 2$,

$$\left(\frac{a+b}{2}\right)^n < \frac{a^n + b^n}{2}.$$

6.2. More Examples

We introduced mathematical induction in the last section as a method for proving that all natural numbers have some property. However, the applications of mathematical induction extend far beyond the study of the natural numbers. In this section we'll look at some examples of proofs by mathematical induction that illustrate the wide range of uses of induction.

Example 6.2.1. Suppose R is a partial order on a set A. Prove that every finite, nonempty set $B \subseteq A$ has an R-minimal element.

Scratch work

You might think at first that mathematical induction is not appropriate for this proof, because the goal doesn't seem to have the form $\forall n \in \mathbb{N} P(n)$. In fact, the goal doesn't explicitly mention natural numbers at all! But we can see that natural numbers enter into the problem when we recognize that to say that B is finite and nonempty means that it has n elements, for some $n \in \mathbb{N}$, $n \geq 1$. (We'll give a more careful definition of the number of elements in a finite set in Chapter 7. For the moment, an intuitive understanding of this concept will suffice.) Thus, the goal means $\forall n \geq 1 \forall B \subseteq A(B$ has n elements $\rightarrow B$ has a minimal element). We can now use induction to prove this statement.

In the base case we will have $n = 1$, so we must prove that if B has one element, then it has a minimal element. It is easy to check that in this case the one element of B must be minimal.

For the induction step we let $n \geq 1$ be arbitrary, assume that $\forall B \subseteq A$ (B has n elements $\rightarrow B$ has a minimal element), and try to prove that $\forall B \subseteq A(B$ has $n+1$ elements $\rightarrow B$ has a minimal element). Guided by the form of the goal, we let B be an arbitrary subset of A, assume that B has $n+1$ elements, and try to prove that B has a minimal element.

How can we use the inductive hypothesis to reach our goal? The inductive hypothesis tells us that if we had a subset of A with n elements, then it would have a minimal element. To apply it, we need to find a subset of A with n elements. Our arbitrary set B is a subset of A, and we have assumed that it has $n+1$ elements. Thus, a simple way to produce a subset of A with n elements would be to remove one element from B. It is not clear where this reasoning will lead, but it seems to be the simplest way to make use of the inductive hypothesis. Let's give it a try.

Let b be any element of B, and let $B' = B \setminus \{b\}$. Then B' is a subset of A with n elements, so by the inductive hypothesis, B' has a minimal element. This is an existential statement, so we immediately introduce a new variable, say c, to stand for a minimal element of B'.

Our goal is to prove that B has a minimal element, which is also an existential statement, so we should try to come up with a minimal element of B. We only know about two elements of B at this point, b and c, so we should probably try to prove that one of these is a minimal element of B. Which one? Well, it may depend on whether one of them is smaller than the other according to the partial order R. This suggests that we may need to use proof by cases. In our proof we use the cases bRc and $\neg bRc$. In the first case we prove that b is a minimal element of B, and in the second case we prove that c is a minimal element of B. Note that to say that something is a minimal element of B is a negative statement, so in both cases we use proof by contradiction.

Solution

Theorem. *Suppose R is a partial order on a set A. Then every finite, nonempty set $B \subseteq A$ has an R-minimal element.*

Proof. We will show by induction that for every natural number $n \geq 1$, every subset of A with n elements has a minimal element.

Base case: $n = 1$. Suppose $B \subseteq A$ and B has one element. Then $B = \{b\}$ for some $b \in A$. Clearly $\neg\exists x \in B(x \neq b)$, so certainly $\neg\exists x \in B(xRb \wedge x \neq b)$. Thus, b is minimal.

Induction step: Suppose $n \geq 1$, and suppose that every subset of A with n elements has a minimal element. Now let B be an arbitrary subset of A with $n+1$ elements. Let b be any element of B, and let $B' = B \setminus \{b\}$, a subset of A with n elements. By inductive hypothesis, we can choose a minimal element $c \in B'$.

Case 1. bRc. We claim that b is a minimal element of B. To see why, suppose it isn't. Then we can choose some $x \in B$ such that xRb and $x \neq b$. Since $x \neq b$, $x \in B'$. Also, since xRb and bRc, by transitivity of R it follows that xRc. Thus, since c is a minimal element of B', we must have $x = c$. But then since xRb we have cRb, and we also know bRc, so by antisymmetry of R

it follows that $b = c$. This is clearly impossible, since $c \in B' = B \setminus \{b\}$. Thus, b must be a minimal element of B.

Case 2. $\neg bRc$. We claim in this case that c is a minimal element of B. To see why, suppose it isn't. Then we can choose some $x \in B$ such that xRc and $x \neq c$. Since c is a minimal element of B', we can't have $x \in B'$, so the only other possibility is $x = b$. But then since xRc we must have bRc, which contradicts our assumption that $\neg bRc$. Thus, c is a minimal element of B. □

Note that an infinite subset of a partially ordered set need not have a minimal element, as we saw in part 1 of Example 4.4.5. Thus, the assumption that B is finite was needed in our last theorem. This theorem can be used to prove another interesting fact about partial orders, again using mathematical induction:

Example 6.2.2. Suppose A is a finite set and R is a partial order on A. Prove that R can be extended to a total order on A. In other words, prove that there is a total order T on A such that $R \subseteq T$.

Scratch work

We'll only outline the proof, leaving many details as exercises. The idea is to prove by induction that $\forall n \in \mathbb{N} \forall A \forall R[(A$ has n elements and R is a partial order on $A) \rightarrow \exists T(T$ is a total order on A and $R \subseteq T)]$. The induction step is similar to the induction step of the last example. If R is a partial order on a set A with $n + 1$ elements, then we remove one element, call it a, from A, and apply the inductive hypothesis to the remaining set $A' = A \setminus \{a\}$. This will give us a total order T' on A', and to complete the proof we must somehow turn this into a total order T on A such that $R \subseteq T$. The relation T' already tells us how to compare any two elements of A', but it doesn't tell us how to compare a to the elements of A'. This is what we must decide in order to define T, and the main difficulty in this step of the proof is that we must make this decision in such a way that we end up with $R \subseteq T$. Our resolution of this difficulty in the following proof involves choosing a carefully in the first place. We choose a to be an R-minimal element of A, and then when we define T, we make a smaller in the T ordering than every element of A'. We use the theorem in the last example, with $B = A$, to guarantee that A has an R-minimal element.

Solution

Theorem. *Suppose A is a finite set and R is a partial order on A. Then there is a total order T on A such that $R \subseteq T$.*

Proof. We will show by induction on n that every partial order on a set with n elements can be extended to a total order. Clearly this suffices to prove the theorem.

Base case: $n = 0$. Suppose R is a partial order on A and A has 0 elements. Then clearly $A = R = \varnothing$. It is easy to check that $\varnothing$ is a total order on A, so we are done.

Induction step: Let n be an arbitrary natural number, and suppose that every partial order on a set with n elements can be extended to a total order. Now suppose that A has $n + 1$ elements and R is a partial order on A. By the theorem in the last example, there must be some $a \in A$ such that a is an R-minimal element of A. Let $A' = A \setminus \{a\}$ and let $R' = R \cap (A' \times A')$. You are asked to show in exercise 1 that R' is a partial order on A'. By inductive hypothesis, we can let T' be a total order on A' such that $R' \subseteq T'$. Now let $T = T' \cup (\{a\} \times A)$. You are also asked to show in exercise 1 that T is a total order on A and $R \subseteq T$, as required. □

The theorem in the last example can be extended to apply to partial orders on infinite sets. For a step in this direction, see exercise 17 in Section 7.1.

Example 6.2.3. Prove that for all $n \geq 3$, if n distinct points on a circle are connected in consecutive order with straight lines, then the interior angles of the resulting polygon add up to $(n - 2)180°$.

Solution

Figure 1 shows an example with $n = 4$. We won't give the scratch work separately for this proof.

Theorem. *For all $n \geq 3$, if n distinct points on a circle are connected in consecutive order with straight lines, then the interior angles of the resulting polygon add up to $(n - 2)180°$.*

Proof. We use induction on n.

Base case: Suppose $n = 3$. Then the polygon is a triangle, and it is well known that the interior angles of a triangle add up to 180°.

Induction step: Let n be an arbitrary natural number, $n \geq 3$, and assume the statement is true for n. Now consider the polygon P formed by connecting some $n + 1$ distinct points $A_1, A_2, \ldots, A_{n+1}$ on a circle. If we skip the last point A_{n+1}, then we get a polygon P' with only n vertices, and by inductive hypothesis the interior angles of this polygon add up to $(n - 2)180°$. But now as you can see in Figure 2, the sum of the interior angles of P is equal to the sum of the interior angles of P' plus the sum of the interior angles of the triangle $A_1 A_n A_{n+1}$. Since the sum of the interior angles of the triangle is 180°, we can conclude that the sum of the interior angles of P is

$$(n - 2)180° + 180° = ((n + 1) - 2)180°,$$

as required. □

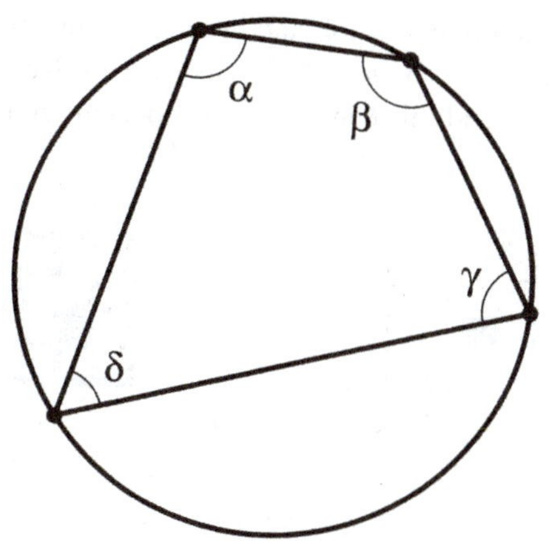

$$\alpha + \beta + \gamma + \delta = (4 - 2)180^\circ = 360^\circ$$

Figure 1

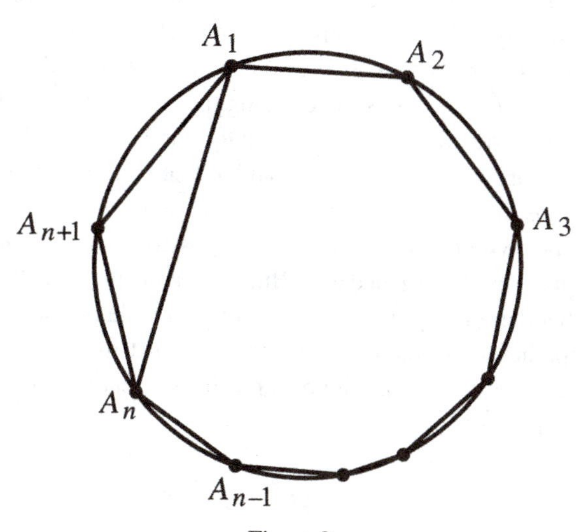

Figure 2

Example 6.2.4. Prove that for any positive integer n, a $2^n \times 2^n$ square grid with any one square removed can be covered with L-shaped tiles that look like this: ⊞

Scratch work

Figure 3 shows an example for the case $n = 2$. In this case $2^n = 4$, so we have a 4×4 grid, and the square that has been removed is shaded. The heavy lines show how the remaining squares can be covered with five L-shaped tiles.

We'll use induction in our proof, and because we're only interested in positive n, the base case will be $n = 1$. In this case we have a 2×2 grid with one square removed, and this can clearly be covered with one L-shaped tile. (Draw a picture!)

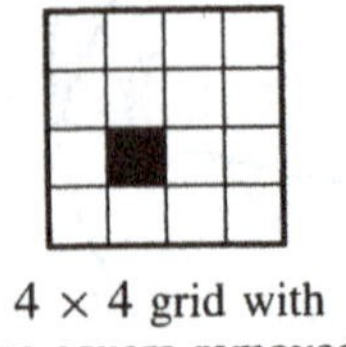

4×4 grid with one square removed.

Grid covered with L-shaped tiles.

Figure 3

For the induction step, we let n be an arbitrary positive integer and assume that a $2^n \times 2^n$ grid with any one square removed can be covered with L-shaped tiles. Now suppose we have a $2^{n+1} \times 2^{n+1}$ grid with one square removed. To use our inductive hypothesis we must somehow relate this to the $2^n \times 2^n$ grid. Since $2^{n+1} = 2^n \cdot 2$, the $2^{n+1} \times 2^{n+1}$ grid is twice as wide and twice as high as the $2^n \times 2^n$ grid. In other words, by dividing the $2^{n+1} \times 2^{n+1}$ grid in half both horizontally and vertically, we can split it into four $2^n \times 2^n$ "subgrids." This is illustrated in Figure 4. The one square that has been removed will be in one of the four subgrids; in Figure 4, it is in the upper right.

The inductive hypothesis tells us that it is possible to cover the upper right subgrid in Figure 4 with L-shaped tiles. But what about the other three subgrids? It turns out that there is a clever way of placing one tile on the grid so that the inductive hypothesis can then be used to show that the remaining subgrids can be covered. See if you can figure it out before reading the answer in the following proof.

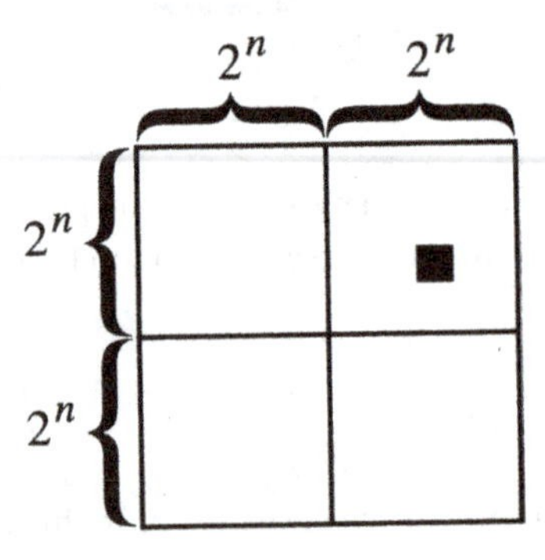

Figure 4

Solution

Theorem. *For any positive integer n, a $2^n \times 2^n$ square grid with any one square removed can be covered with L-shaped tiles.*

Proof. We use induction on n.

Base case: Suppose $n = 1$. Then the grid is a 2×2 grid with one square removed, which can clearly be covered with one L-shaped tile.

Induction step: Let n be an arbitrary positive integer, and suppose that a $2^n \times 2^n$ grid with any one square removed can be covered with L-shaped tiles. Now consider a $2^{n+1} \times 2^{n+1}$ grid with one square removed. Cut the grid in half both vertically and horizontally, splitting it into four $2^n \times 2^n$ subgrids. The one square that has been removed comes from one of these subgrids, so by the inductive hypothesis the rest of this subgrid can be covered with L-shaped tiles. To cover the other three subgrids, first place one L-shaped tile in the center so that it covers one square from each of the three remaining subgrids, as illustrated in Figure 5. The area remaining to be covered now contains every square except one in each of the subgrids, so by applying the inductive hypothesis to each subgrid we can see that this area can be covered with tiles. □

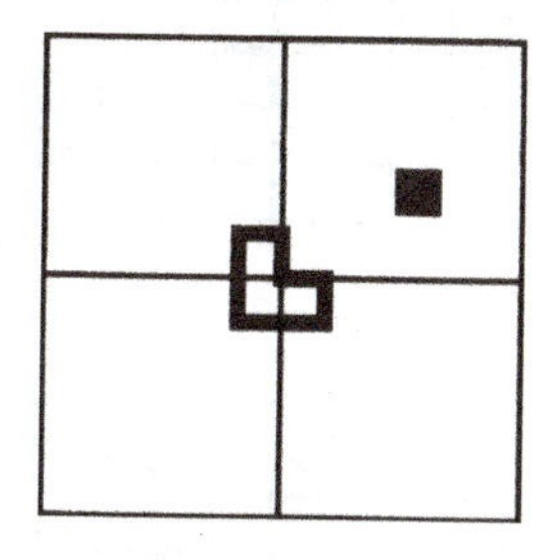

Figure 5

It is interesting to note that this proof can actually be used to figure out how to place tiles on a particular grid. For example, consider the 8×8 grid with one square removed shown in Figure 6.

According to the preceding proof, the first step in covering this grid with tiles is to split it into four 4×4 subgrids and place one tile in the center, covering one square from each subgrid except the upper left. This is illustrated in Figure 7. The area remaining to be covered now consists of four 4×4 subgrids with one square removed from each of them.

How do we cover the remaining 4×4 subgrids? By the same method, of course! For example, let's cover the subgrid in the upper right of Figure 7. We need to cover every square of this subgrid except the lower left corner, which has already been covered. We start by cutting it into four 2×2 subgrids and

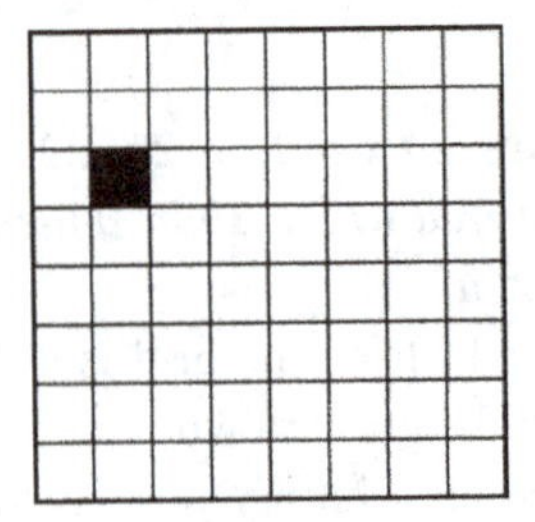

Figure 6

put one tile in the middle, as in Figure 8. The area remaining to be covered now consists of four 2×2 subgrids with one square removed from each. Each of these can be covered with one tile, thus completing the upper right subgrid of Figure 7.

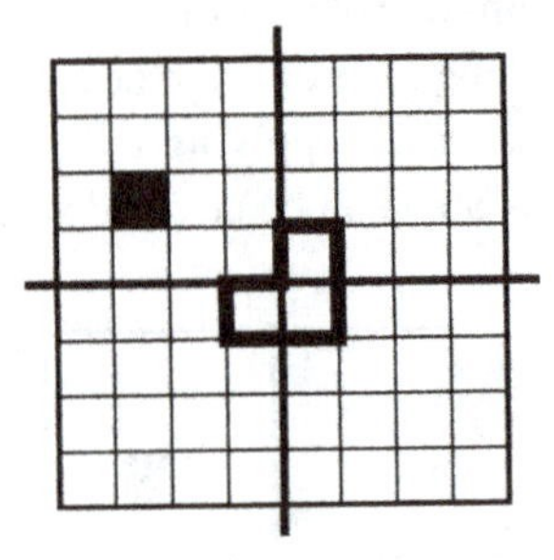

Figure 7

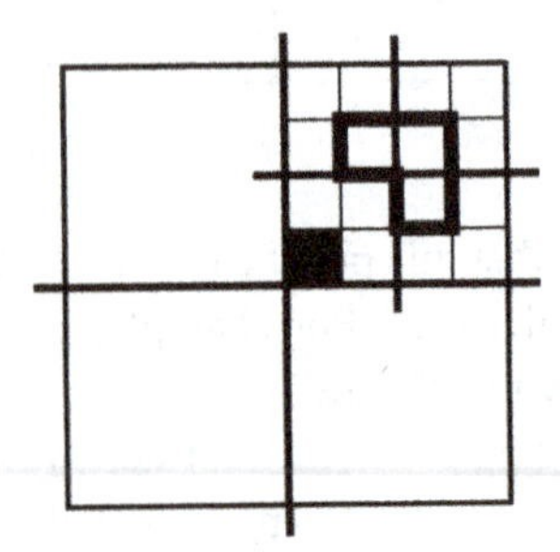

Figure 8

The remaining three quarters of Figure 7 are completed by a similar procedure. The final solution is shown in Figure 9.

The method we used in solving this problem is an example of a *recursive* procedure. We solved the problem for an 8×8 grid by splitting it into four 4×4 grid problems. To solve each of these, we split it into four 2×2 problems, each

Figure 9

of which was easy to solve. If we had started with a larger grid, we might have had to repeat the splitting many times before reaching easy 2×2 problems. Recursion and its relationship to mathematical induction is the subject of our next section.

Exercises

*1. Complete the proof in Example 6.2.2 by doing the following proofs. (We use the same notation here as in the example.)
   (a) Prove that R' is a partial order on A'.
   (b) Prove that T is a total order on A and $R \subseteq T$.

2. Suppose R is a partial order on a set A, $B \subseteq A$, and B is finite. Prove that there is a partial order T on A such that $R \subseteq T$ and $\forall x \in B \forall y \in A(xTy \vee yTx)$. Note that, in particular, if A is finite we can let $B = A$, and the conclusion then means that T is a total order on A. Thus, this gives an alternative approach to the proof of the theorem in Example 6.2.2. (Hint: Use induction on the number of elements in B. For the induction step, assume the conclusion holds for any set $B \subseteq A$ with n elements, and suppose B is a subset of A with $n + 1$ elements. Let b be any element of B and let $B' = B \setminus \{b\}$, a subset of A with n elements. By inductive hypothesis, let T' be a partial order on A such that $R \subseteq T'$ and $\forall x \in B' \forall y \in A(xT'y \vee yT'x)$. Now let $A_1 = \{x \in A \mid xT'b\}$ and $A_2 = A \setminus A_1$, and let $T = T' \cup (A_1 \times A_2)$. Prove that T has all the required properties.)

3. Suppose R is a total order on a set A. Prove that every finite, nonempty set $B \subseteq A$ has an R-smallest element.

*4. (a) Suppose R is a relation on A, and $\forall x \in A \forall y \in A(xRy \vee yRx)$. (Note that this implies that R is reflexive.) Prove that for every finite, nonempty set $B \subseteq A$ there is some $x \in B$ such that $\forall y \in B((x, y) \in R \circ R)$. (Hint: Imitate Example 6.2.1.)

(b) Consider a tournament in which each contestant plays every other contestant exactly once, and one of them wins. We'll say that a contestant x is *excellent* if, for every other contestant y, either x beats y or there is a third contestant z such that x beats z and z beats y. Prove that there is at least one excellent contestant.

5. For each $n \in \mathbb{N}$, let $F_n = 2^{(2^n)} + 1$. (These numbers are called the *Fermat numbers*, after the French mathematician Pierre de Fermat (1601–1665). Fermat showed that F_0, F_1, F_2, F_3, and F_4 are prime, and conjectured that all of the Fermat numbers are prime. However, over 100 years later Euler showed that F_5 is not prime. It is not known if there is any $n > 4$ for which F_n is prime.)

   Prove that for all $n \geq 1$, $F_n = (F_0 \cdot F_1 \cdot F_2 \cdots F_{n-1}) + 2$.

6. Prove that if $n \geq 1$ and $a_1, a_2, \ldots, a_n$ are any real numbers, then $|a_1 + a_2 + \cdots + a_n| \leq |a_1| + |a_2| + \cdots + |a_n|$. (Note that this generalizes the triangle inequality; see exercise 12(c) of Section 3.5.)

7. (a) Prove that if a and b are positive real numbers, then $a/b + b/a \geq 2$. (Hint: Start with the fact that $(a - b)^2 \geq 0$.)

   (b) Suppose that a, b, and c are real numbers and $0 < a \leq b \leq c$. Prove that $b/c + c/a - b/a \geq 1$. (Hint: Start with the fact that $(c - a)(c - b) \geq 0$.)

   (c) Prove that if $n \geq 2$ and $a_1, a_2, \ldots, a_n$ are real numbers such that $0 < a_1 \leq a_2 \leq \ldots \leq a_n$, then $a_1/a_2 + a_2/a_3 + \cdots + a_{n-1}/a_n + a_n/a_1 \geq n$.

*8. If $n \geq 2$ and $a_1, a_2, \ldots, a_n$ is a list of positive real numbers, then the number $(a_1 + a_2 + \cdots + a_n)/n$ is called the *arithmetic mean* of the numbers $a_1, a_2, \ldots, a_n$, and the number $\sqrt[n]{a_1 a_2 \cdots a_n}$ is called their *geometric mean*. In this exercise you will prove the *arithmetic-geometric mean inequality*, which says that the arithmetic mean is always at least as large as the geometric mean.

   (a) Prove that the arithmetic-geometric mean inequality holds for lists of numbers of length 2. In other words, prove that for all positive real numbers a and b, $(a + b)/2 \geq \sqrt{ab}$.

   (b) Prove that the arithmetic-geometric mean inequality holds for any list of numbers whose length is a power of 2. In other words, prove that for all $n \geq 1$, if $a_1, a_2, \ldots, a_{2^n}$ is a list of positive real numbers, then

$$\frac{a_1 + a_2 + \cdots + a_{2^n}}{2^n} \geq \sqrt[2^n]{a_1 a_2 \cdots a_{2^n}}.$$

(c) Suppose that $n_0 \geq 2$ and the arithmetic-geometric mean inequality fails for some list of length n_0. In other words, there are positive real numbers $a_1, a_2, \ldots, a_{n_0}$ such that

$$\frac{a_1 + a_2 + \cdots + a_{n_0}}{n_0} < \sqrt[n_0]{a_1 a_2 \cdots a_{n_0}}.$$

Prove that for all $n \geq n_0$, the arithmetic-geometric mean inequality fails for some list of length n.

(d) Prove that the arithmetic-geometric mean inequality always holds.

9. Prove that if $n \geq 2$ and $a_1, a_2, \ldots, a_n$ is a list of positive real numbers, then

$$\frac{n}{\frac{1}{a_1} + \frac{1}{a_2} + \cdots + \frac{1}{a_n}} \leq \sqrt[n]{a_1 a_2 \cdots a_n}.$$

(Hint: Apply exercise 8. The number on the left side of the inequality above is called the *harmonic mean* of the numbers $a_1, a_2, \ldots, a_n$.)

*10. Prove that for every set A, if A has n elements then $\mathscr{P}(A)$ has 2^n elements.

11. If A is a set, let $\mathscr{P}_2(A)$ be the set of all subsets of A that have exactly two elements. Prove that for every set A, if A has n elements then $\mathscr{P}_2(A)$ has $n(n-1)/2$ elements. (Hint: See the solution for exercise 10.)

12. Suppose n is a positive integer. An equilateral triangle is cut into 4^n congruent equilateral triangles, and one corner is removed. (Figure 10 shows an example in the case $n = 2$.) Show that the remaining area can be covered by trapezoidal tiles like this: ⏢.

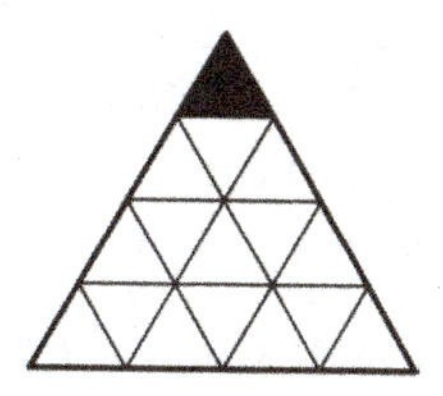

Figure 10

*13. Let n be a positive integer. Suppose n chords are drawn in a circle in such a way that each chord intersects every other, but no three intersect at one point. Prove that the chords cut the circle into $\frac{n^2+n+2}{2}$ regions. (Figure 11 shows an example in the case $n = 4$. Note that there are $\frac{4^2+4+2}{2} = 11$ regions in this figure.)

14. Let n be a positive integer, and suppose that n chords are drawn in a circle, cutting the circle into a number a regions. Prove that the regions can be colored with two colors in such a way that adjacent regions (that

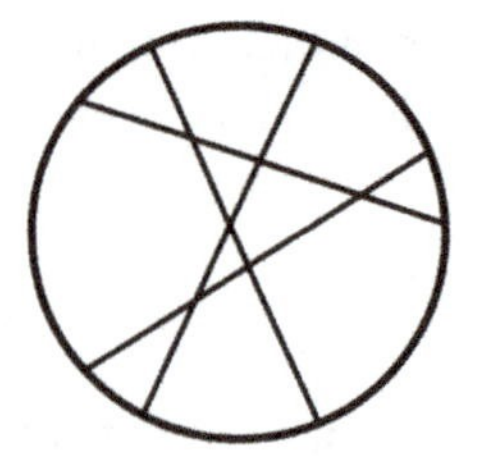

Figure 11

is, regions that share an edge) are different colors. (Figure 12 shows an example in the case $n = 4$.)

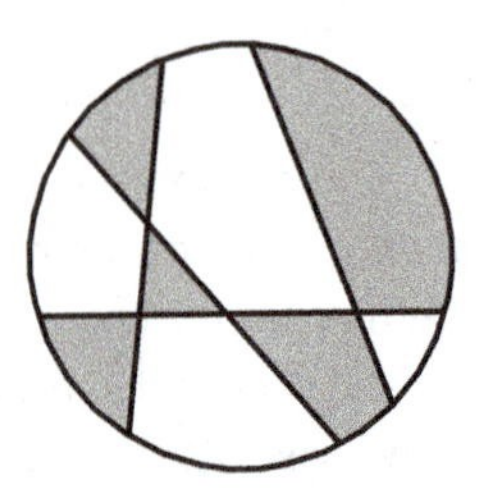

Figure 12

15. What's wrong with the following proof that if $A \subseteq \mathbb{N}$ and $0 \in A$ then $A = \mathbb{N}$?

    Proof. We will prove by induction that $\forall n \in \mathbb{N}(n \in A)$.

    Base case: If $n = 0$, then $n \in A$ by assumption.

    Induction step: Let $n \in \mathbb{N}$ be arbitrary, and suppose that $n \in A$. Since n was arbitrary, it follows that every natural number is an element of A, and therefore in particular $n + 1 \in A$. □

16. Suppose $f : \mathbb{R} \to \mathbb{R}$. What's wrong with the following proof that for every finite, nonempty set $A \subseteq \mathbb{R}$ there is a real number c such that $\forall x \in A(f(x) = c)$?

    Proof. We will prove by induction that for every $n \geq 1$, if A is any subset of $\mathbb{R}$ with n elements then $\exists c \in \mathbb{R} \forall x \in A(f(x) = c)$.

    Base case: $n = 1$. Suppose $A \subseteq \mathbb{R}$ and A has one element. Then $A = \{a\}$, for some $a \in \mathbb{R}$. Let $c = f(a)$. Then clearly $\forall x \in A(f(x) = c)$.

    Induction step: Suppose $n \geq 1$, and for all $A \subseteq \mathbb{R}$, if A has n elements then $\exists c \in \mathbb{R} \forall x \in A(f(x) = c)$. Now suppose $A \subseteq \mathbb{R}$ and A has $n + 1$ elements. Let a_1 be any element of A, and let $A_1 = A \setminus \{a_1\}$. Then A_1 has n elements, so by inductive hypothesis there is some $c_1 \in \mathbb{R}$

such that $\forall x \in A_1(f(x) = c_1)$. If we can show that $f(a_1) = c_1$ then we will be done, since then it will follow that $\forall x \in A(f(x) = c_1)$.

Let a_2 be an element of A that is different from a_1, and let $A_2 = A \setminus \{a_2\}$. Applying the inductive hypothesis again, we can choose a number $c_2 \in \mathbb{R}$ such that $\forall x \in A_2(f(x) = c_2)$. Notice that since $a_1 \neq a_2, a_1 \in A_2$, so $f(a_1) = c_2$. Now let a_3 be an element of A that is different from both a_1 and a_2. Then $a_3 \in A_1$ and $a_3 \in A_2$, so $f(a_3) = c_1$ and $f(a_3) = c_2$. Therefore $c_1 = c_2$, so $f(a_1) = c_1$, as required. □

6.3. Recursion

In Chapter 3 we learned to prove statements of the form $\forall n P(n)$ by letting n be arbitrary and proving $P(n)$. In this chapter we've learned another method for proving such statements, when n ranges over the natural numbers: Prove $P(0)$, and then prove that for any natural number n, if $P(n)$ is true then so is $P(n+1)$. Once we have proven these statements, we can run through all the natural numbers in order and see that P must be true of all of them.

We can use a similar idea to introduce a new way of defining functions. In Chapter 5, we usually defined a function f by saying how to compute $f(n)$ for any n in the domain of f. If the domain of f is the set of all natural numbers, an alternative method to define f would be to say what $f(0)$ is and then, for any natural number n, say how we could compute $f(n+1)$ if we already knew the value of $f(n)$. Such a definition would enable us to run through all the natural numbers in order computing the image of each one under f.

For example, we might use the following equations to define a function f with domain $\mathbb{N}$:

$$f(0) = 1;$$
$$\text{for every } n \in \mathbb{N},\ f(n+1) = (n+1) \cdot f(n).$$

The second equation tells us how to compute $f(n+1)$, but only if we already know the value of $f(n)$. Thus, although we cannot use this equation to tell us directly what the image of any number is under f, we *can* use it to run through all the natural numbers in order and compute their images.

We start with $f(0)$, which we know from the first equation is equal to 1. Plugging in $n = 0$ in the second equation, we see that $f(1) = 1 \cdot f(0) = 1 \cdot 1 = 1$, so we've determined the value of $f(1)$. But now that we know that $f(1) = 1$, we can use the second equation again to compute $f(2)$. Plugging in $n = 1$ in the second equation, we find that $f(2) = 2 \cdot f(1) = 2 \cdot 1 = 2$. Similarly, setting $n = 2$ in the second equation we get $f(3) = 3 \cdot f(2) = 3 \cdot 2 = 6$. Continuing in this way we can compute $f(n)$ for any natural number n. Thus, the two equations really do give us a rule that determines a unique value $f(n)$ for each natural number n, so they define a function f with domain $\mathbb{N}$. Definitions of this kind are called *recursive* definitions.

Sometimes we'll work backwards when using a recursive definition to evaluate a function. For example, suppose we want to compute $f(6)$, where f is the function just defined. According to the second equation in the definition of f, $f(6) = 6 \cdot f(5)$, so to complete the calculation we must compute $f(5)$. Using the second equation again, we find that $f(5) = 5 \cdot f(4)$, so we must compute $f(4)$. Continuing in this way leads to the following calculation:

$$
\begin{aligned}
f(6) &= 6 \cdot f(5) \\
&= 6 \cdot 5 \cdot f(4) \\
&= 6 \cdot 5 \cdot 4 \cdot f(3) \\
&= 6 \cdot 5 \cdot 4 \cdot 3 \cdot f(2) \\
&= 6 \cdot 5 \cdot 4 \cdot 3 \cdot 2 \cdot f(1) \\
&= 6 \cdot 5 \cdot 4 \cdot 3 \cdot 2 \cdot 1 \cdot f(0) \\
&= 6 \cdot 5 \cdot 4 \cdot 3 \cdot 2 \cdot 1 \cdot 1 \\
&= 720.
\end{aligned}
$$

Perhaps now you recognize the function f. For any positive integer n, $f(n) = n \cdot (n-1) \cdot (n-2) \cdots 1$, and $f(0) = 1$. This number is called n *factorial*, denoted $n!$. For example, $6! = 720$. Often, if a function can be written as a formula with an ellipsis (...) in it, then the use of the ellipsis can be avoided by giving a recursive definition for the function. Such a definition is usually easier to work with.

Many familiar functions are most easily defined using recursive definitions. For example, for any number a, we could define a^n with the following recursive definition:

$$
\begin{aligned}
&a^0 = 1; \\
&\text{for every } n \in \mathbb{N},\ a^{n+1} = a^n \cdot a.
\end{aligned}
$$

Using this definition, we would compute a^4 like this:

$$
\begin{aligned}
a^4 &= a^3 \cdot a \\
&= a^2 \cdot a \cdot a \\
&= a^1 \cdot a \cdot a \cdot a \\
&= a^0 \cdot a \cdot a \cdot a \cdot a \\
&= 1 \cdot a \cdot a \cdot a \cdot a.
\end{aligned}
$$

For another example, consider the sum $2^0 + 2^1 + 2^2 + \cdots + 2^n$, which appeared in the first example of this chapter. The ellipsis suggests that we might be able to use a recursive definition. If we let $f(n) = 2^0 + 2^1 + 2^2 + \cdots + 2^n$, then notice that for every $n \in \mathbb{N}$, $f(n+1) = 2^0 + 2^1 + 2^2 + \cdots + 2^n + 2^{n+1} = f(n) + 2^{n+1}$. Thus, we could define f recursively as follows:

$$f(0) = 2^0 = 1;$$
$$\text{for every } n \in \mathbb{N},\ f(n+1) = f(n) + 2^{n+1}.$$

As a check that this definition is right, let's try it out in the case $n = 3$:

$$\begin{aligned} f(3) &= f(2) + 2^3 \\ &= f(1) + 2^2 + 2^3 \\ &= f(0) + 2^1 + 2^2 + 2^3 \\ &= 2^0 + 2^1 + 2^2 + 2^3 \\ &= 15. \end{aligned}$$

Sums such as the one in the last example come up often enough that there is a special notation for them. If $a_0, a_1, \ldots, a_n$ is a list of numbers, then the sum of these numbers is written $\sum_{i=0}^{n} a_i$. This is read "the sum as i goes from 0 to n of a_i." For example, we can use this notation to write the sum in the last example:

$$\sum_{i=0}^{n} 2^i = 2^0 + 2^1 + 2^2 + \cdots + 2^n.$$

More generally, if $n \geq m$, then

$$\sum_{i=m}^{n} a_i = a_m + a_{m+1} + a_{m+2} + \cdots + a_n.$$

For example,

$$\begin{aligned} \sum_{i=3}^{6} i^2 &= 3^2 + 4^2 + 5^2 + 6^2 \\ &= 9 + 16 + 25 + 36 = 86. \end{aligned}$$

The letted i in these formulas is a bound variable and therefore can be replaced by a new variable without changing the meaning of the formula.

Now let's try giving a recursive definition for this notation. We let m be an arbitrary integer, and then proceed by recursion on n. Just as the base case for an induction proof need not be $n = 0$, the base for a recursive definition can also be a number other than 0. In this case we are only interested in $n \geq m$, so we take $n = m$ as the base for our recursion:

$$\sum_{i=m}^{m} a_i = a_m;$$

$$\text{for every } n \geq m, \sum_{i=m}^{n+1} a_i = \sum_{i=m}^{n} a_i + a_{n+1}.$$

Trying this definition out on the previous example, we get

$$\begin{aligned}\sum_{i=3}^{6} i^2 &= \sum_{i=3}^{5} i^2 + 6^2 \\ &= \sum_{i=3}^{4} i^2 + 5^2 + 6^2 \\ &= \sum_{i=3}^{3} i^2 + 4^2 + 5^2 + 6^2 \\ &= 3^2 + 4^2 + 5^2 + 6^2,\end{aligned}$$

just as we wanted.

Clearly induction and recursion are closely related, so it shouldn't be surprising that if a concept has been defined by recursion, then proofs involving this concept are often best done by induction. For example, in Section 6.1 we saw some proofs by induction that involved summations and exponentiation, and now we have seen that summations and exponentiation can be defined recursively. Because the factorial function can also be defined recursively, proofs involving factorials also often use induction.

Example 6.3.1. Prove that for every $n \geq 4$, $n! > 2^n$.

Scratch work

Because the problem involves factorial and exponentiation, both of which are defined recursively, induction seems like a good method to use. The base case will be $n = 4$, and it is just a matter of simple arithmetic to check that the inequality is true in this case. For the induction step, our inductive hypothesis will be $n! > 2^n$, and we must prove that $(n + 1)! > 2^{n+1}$. Of course, the way to relate the inductive hypothesis to the goal is to use the recursive definitions of factorial and exponentiation, which tell us that $(n + 1)! = (n + 1) \cdot n!$ and $2^{n+1} = 2^n \cdot 2$. Once these equations are plugged in, the rest is fairly straightforward.

Solution

Theorem. *For every $n \geq 4$, $n! > 2^n$.*

Proof. By mathematical induction.

Base case: When $n = 4$ we have $n! = 24 > 16 = 2^n$.
Induction step: Let $n \geq 4$ be arbitrary and suppose that $n! > 2^n$. Then

$$\begin{aligned}(n+1)! &= (n+1) \cdot n! \\ &> (n+1) \cdot 2^n \qquad \text{(by inductive hypothesis)} \\ &> 2 \cdot 2^n = 2^{n+1}.\end{aligned}$$

□

Example 6.3.2. Prove that for every real number a and all natural numbers m and n, $a^{m+n} = a^m \cdot a^n$.

Scratch work

There are three universal quantifiers here, and we'll treat the first two differently from the third. We let a and m be arbitrary and then use mathematical induction to prove that $\forall n \in \mathbb{N}(a^{m+n} = a^m \cdot a^n)$. The key algebraic fact in the induction step will be the formula $a^{n+1} = a^n \cdot a$ from the recursive definition of exponentiation.

Solution

Theorem. *For every real number a and all natural numbers m and n,* $a^{m+n} = a^m \cdot a^n$.

Proof. Let a be an arbitrary real number and m an arbitrary natural number. We now proceed by induction on n.

Base case: When $n = 0$, we have $a^{m+n} = a^{m+0} = a^m = a^m \cdot 1 = a^m \cdot a^0 = a^m \cdot a^n$.

Induction step. Suppose $a^{m+n} = a^m \cdot a^n$. Then

$$\begin{aligned}a^{m+(n+1)} &= a^{(m+n)+1} \\ &= a^{m+n} \cdot a \qquad \text{(by definition of exponentiation)} \\ &= a^m \cdot a^n \cdot a \qquad \text{(by inductive hypothesis)} \\ &= a^m \cdot a^{n+1} \qquad \text{(by definition of exponentiation).}\end{aligned}$$

□

Example 6.3.3. A sequence of numbers $a_0, a_1, a_2, \ldots$ is defined recursively as follows:

$$a_0 = 0;$$
$$\text{for every } n \in \mathbb{N},\ a_{n+1} = 2a_n + 1.$$

Find a formula for a_n and prove that your formula is correct.

Scratch work

It's probably a good idea to start out by computing the first few terms in the sequence. We already know $a_0 = 0$, so plugging in $n = 0$ in the second equation we get $a_1 = 2a_0 + 1 = 0 + 1 = 1$. Thus, plugging in $n = 1$, we get $a_2 = 2a_1 + 1 = 2 + 1 = 3$. Continuing in this way we get the following table of values.

n	a_n
0	0
1	1
2	3
3	7
4	15
5	31
6	63
$\vdots$	$\vdots$

Aha! The numbers we're getting are one less than the powers of 2. It looks like the formula is probably $a_n = 2^n - 1$, but we can't be sure this is right unless we prove it. Fortunately, it is fairly easy to prove the formula by induction.

Solution

Theorem. *If the sequence $a_0, a_1, a_2, \ldots$ is defined by the recursive definition given earlier, then for every natural number n, $a_n = 2^n - 1$.*

Proof. By induction.

Base case: $a_0 = 0 = 2^0 - 1$.

Induction step: Suppose $a_n = 2^n - 1$. Then

$$\begin{aligned} a_{n+1} &= 2a_n + 1 && \text{(definition of } a_{n+1}\text{)} \\ &= 2(2^n - 1) + 1 && \text{(inductive hypothesis)} \\ &= 2^{n+1} - 2 + 1 = 2^{n+1} - 1. \end{aligned}$$

□

We end this section with a rather unusual example. We'll prove that for every real number $x > -1$ and every natural number n, $(1 + x)^n > nx$. A natural way to proceed would be to let $x > -1$ be arbitrary, and then use induction on n. In the induction step we assume that $(1 + x)^n > nx$, and then try to prove that $(1 + x)^{n+1} > (n + 1)x$. Because we've assumed $x > -1$, we have $1 + x > 0$, so we can multiply both sides of the inductive hypothesis $(1 + x)^n > nx$ by $1 + x$ to get

$$\begin{aligned} (1 + x)^{n+1} &= (1 + x)(1 + x)^n \\ &> (1 + x)nx \\ &= nx + nx^2. \end{aligned}$$

But the conclusion we need for the induction step is $(1+x)^{n+1} > (n+1)x$, and it's not clear how to get this conclusion from the inequality we've derived.

Our solution to this difficulty will be to replace our original problem with a problem that appears to be harder but is actually easier. Instead of proving the inequality $(1+x)^n > nx$ directly, we'll prove $(1+x)^n \geq 1+nx$, and then observe that since $1+nx > nx$, it follows immediately that $(1+x)^n > nx$. You might think that if we had difficulty proving $(1+x)^n > nx$, we'll surely have more difficulty proving the stronger statement $(1+x)^n \geq 1+nx$. But it turns out that the approach we tried unsuccessfully on the original problem works perfectly on the new problem!

Theorem 6.3.4. *For every $x > -1$ and every natural number n, $(1+x)^n > nx$.*

Proof. Let $x > -1$ be arbitrary. We will prove by induction that for every natural number n, $(1+x)^n \geq 1+nx$, from which it clearly follows that $(1+x)^n > nx$.

Base case: If $n = 0$, then $(1+x)^n = (1+x)^0 = 1 = 1+0 = 1+nx$.

Induction step: Suppose $(1+x)^n \geq 1+nx$. Then

$$\begin{aligned}
(1+x)^{n+1} &= (1+x)(1+x)^n \\
&\geq (1+x)(1+nx) && \text{(by inductive hypothesis)} \\
&= 1+x+nx+nx^2 \\
&\geq 1+(n+1)x && \text{(since } nx^2 \geq 0\text{).}
\end{aligned}$$

□

Exercises

*1. Find a formula for $\sum_{i=1}^{n} \frac{1}{i(i+1)}$ and prove that your formula is correct.

2. Prove that for all $n \geq 1$,

$$\sum_{i=1}^{n} \frac{1}{i(i+1)(i+2)} = \frac{n^2+3n}{4(n+1)(n+2)}.$$

3. Prove that for all $n \geq 2$,

$$\sum_{i=2}^{n} \frac{1}{(i-1)(i+1)} = \frac{3n^2-n-2}{4n(n+1)}.$$

4. Prove that for all $n \in \mathbb{N}$,

$$\sum_{i=0}^{n} (2i+1)^2 = \frac{(n+1)(2n+1)(2n+3)}{3}.$$

5. Suppose r is a real number and $r \neq 1$. Prove that for all $n \in \mathbb{N}$,

$$\sum_{i=0}^{n} r^i = \frac{r^{n+1} - 1}{r - 1}.$$

(Note that this exercise generalizes Example 6.1.1 and exercise 7 of Section 6.1.)

*6. Prove that for all $n \geq 1$,

$$\sum_{i=1}^{n} \frac{1}{i^2} \leq 2 - \frac{1}{n}.$$

7. (a) Suppose $a_0, a_1, a_2, \ldots, a_n$ and $b_0, b_1, b_2, \ldots, b_n$ are two sequences of real numbers. Prove that

$$\sum_{i=0}^{n}(a_i + b_i) = \sum_{i=0}^{n} a_i + \sum_{i=0}^{n} b_i.$$

(b) Suppose c is a real number and $a_0, a_1, \ldots, a_n$ is a sequence of real numbers. Prove that

$$c \cdot \sum_{i=0}^{n} a_i = \sum_{i=0}^{n}(c \cdot a_i).$$

*8. The *harmonic numbers* are the numbers H_n for $n \geq 1$ defined by the formula

$$H_n = \sum_{i=1}^{n} \frac{1}{i}.$$

(a) Prove that for all natural numbers n and m, if $n \geq m$ then $H_n - H_m \geq \frac{n-m}{n}$. (Hint: Let m be an arbitrary natural number and then proceed by induction on n, with $n = m$ as the base case of the induction.)
(b) Prove that for all $n \geq 0$, $H_{2^n} \geq 1 + n/2$.
(c) (For those who have studied calculus) Show that $\lim_{n\to\infty} H_n = \infty$, so $\sum_{i=1}^{\infty} \frac{1}{i}$ diverges.

9. Let H_n be defined as in exercise 8. Prove that for all $n \geq 2$,

$$\sum_{k=1}^{n-1} H_k = nH_n - n.$$

10. Find a formula for $\sum_{i=1}^{n}(i \cdot (i!))$ and prove that your formula is correct.
11. Find a formula for $\sum_{i=0}^{n} \frac{i}{(i+1)!}$ and prove that your formula is correct.

*12. (a) Prove that for all $n \in \mathbb{N}$, $2^n > n$.
(b) Prove that for all $n \geq 9$, $n! \geq (2^n)^2$.
(c) Prove that for all $n \in \mathbb{N}$, $n! \leq 2^{(n^2)}$.

13. Suppose k is a positive integer.
    (a) Prove that for all $n \in \mathbb{N}$, $(k^2 + n)! \geq k^{2n}$.
    (b) Prove that for all $n \geq 2k^2$, $n! \geq k^n$. (Hint: Use induction, and for the base case use part (a). Note that in the language of exercise 16 of Section 5.1, this shows that if $f(n) = k^n$ and $g(n) = n!$, then $f \in O(g)$.)
14. Prove that for every real number a and all natural numbers m and n, $(a^m)^n = a^{mn}$.

*15. A sequence $a_0, a_1, a_2, \ldots$ is defined recursively as follows:

$$a_0 = 0;$$
$$\text{for every } n \in \mathbb{N},\ a_{n+1} = 2a_n + n.$$

Prove that for all $n \in \mathbb{N}$, $a_n = 2^n - n - 1$.

16. A sequence $a_0, a_1, a_2, \ldots$ is defined recursively as follows:

$$a_0 = 2;$$
$$\text{for every } n \in \mathbb{N},\ a_{n+1} = (a_n)^2.$$

Find a formula for a_n and prove that your formula is correct.

17. A sequence $a_1, a_2, a_3, \ldots$ is defined recursively as follows:

$$a_1 = 1;$$
$$\text{for every } n \geq 1,\ a_{n+1} = \frac{a_n}{a_n + 1}.$$

Find a formula for a_n and prove that your formula is correct.

*18. For $n \geq k \geq 0$, the quantity $\binom{n}{k}$ is defined as follows:

$$\binom{n}{k} = \frac{n!}{k! \cdot (n-k)!}.$$

(a) Prove that for all $n \in \mathbb{N}$, $\binom{n}{0} = \binom{n}{n} = 1$.
(b) Prove that for all $n \geq k > 0$, $\binom{n+1}{k} = \binom{n}{k} + \binom{n}{k-1}$.
(c) If A is a set and $k \in \mathbb{N}$, let $\mathscr{P}_k(A)$ be the set of all subsets of A that have k elements. Prove that if A has n elements and $n \geq k \geq 0$, then $\mathscr{P}_k(A)$ has $\binom{n}{k}$ elements. (Hint: Prove by induction that $\forall n \in \mathbb{N} \forall A[A$ is a set with n elements $\rightarrow \forall k(n \geq k \geq 0 \rightarrow \mathscr{P}_k(A)$ has $\binom{n}{k}$ elements)]. Imitate exercises 10 and 11 of Section 6.2. In fact, this exercise generalizes exercise 11 of Section 6.2. This exercise shows that $\binom{n}{k}$ is the number of ways of choosing k elements out of a set of size n, so it is sometimes called *n choose k*.)
(d) Prove that for all real numbers x and y and every natural number n,

$$(x+y)^n = \sum_{k=0}^{n} \binom{n}{k} x^{n-k} y^k.$$

(This is called the *binomial theorem*, so the numbers $\binom{n}{k}$ are sometimes called *binomial coefficients*.)

Note: Parts (a) and (b) show that we can compute the numbers $\binom{n}{k}$ conveniently by using a triangular array as in Figure 1. This array is called *Pascal's triangle*, after the French mathematician Blaise Pascal (1623–1662). Each row of the triangle corresponds to a particular value of n, and it lists the values of $\binom{n}{k}$ for all k from 0 to n. Part (a) shows that the first and last number in every row is 1. Part (b) shows that every other number is the sum of the two numbers above it. For example, the lines in Figure 1 illustrate that $\binom{3}{2} = 3$ is the sum of $\binom{2}{1} = 2$ and $\binom{2}{2} = 1$.

$$
\begin{array}{lc}
n = 0: & 1 \\
n = 1: & 1 \quad 1 \\
n = 2: & 1 \quad 2 \quad 1 \\
n = 3: & 1 \quad 3 \quad 3 \quad 1 \\
n = 4: & 1 \quad 4 \quad 6 \quad 4 \quad 1 \\
 & \vdots
\end{array}
$$

Figure 1: Pascal's triangle

19. For the meaning of the notation used in this exercise, see exercise 18.
    (a) Prove that for all $n \in \mathbb{N}$, $\sum_{k=0}^{n} \binom{n}{k} = 2^n$. (Hint: You can do this by induction using parts (a) and (b) of exercise 18, or you can combine part (c) of exercise 18 with exercise 10 of Section 6.2, or you can plug something in for x and y in part (d) of exercise 18.)
    (b) Prove that for all $n \geq 1$, $\sum_{k=0}^{n} (-1)^k \binom{n}{k} = 0$.

*20. A sequence $a_0, a_1, a_2, \ldots$ is defined recursively as follows:

$$a_0 = 0;$$

$$\text{for every } n \in \mathbb{N},\ a_{n+1} = (a_n)^2 + \frac{1}{4}.$$

Prove that for all $n \geq 1$, $0 < a_n < 1$.

21. Explain the paradox in the proof of Theorem 6.3.4, in which we made the proof easier by changing the goal to a statement that looked like it would be harder to prove.

6.4. Strong Induction

In the induction step of a proof by mathematical induction, we prove that a natural number has some property based on the assumption that the previous number has the same property. In some cases this assumption isn't strong enough to make the proof work, and we need to assume that *all* smaller natural numbers have the property. This is the idea behind a variant of mathematical induction sometimes called *strong induction*:

To prove a goal of the form $\forall n \in \mathbb{N} P(n)$:

Prove that $\forall n[(\forall k < n P(k)) \rightarrow P(n)]$, where both n and k range over the natural numbers in this statement. Of course, the most direct way to prove this is to let n be an arbitrary natural number, assume that $\forall k < n P(k)$, and then prove $P(n)$.

Note that no base case is necessary in a proof by strong induction. All that is needed is a modified form of the induction step in which we prove that if every natural number smaller than n has the property P, then n has the property P. In a proof by strong induction, we refer to the assumption that every natural number smaller than n has the property P as the *inductive hypothesis*.

To see why strong induction works, it might help if we first review briefly why ordinary induction works. Recall that a proof by ordinary induction enables us to go through all the natural numbers in order and see that each of them has some property P. The base case gets the process started, and the induction step shows that the process can always be continued from one number to the next. But note that in this process, by the time we check that some natural number n has the property P, we've already checked that *all smaller numbers* have the property. In other words, we already know that $\forall k < n P(k)$. The idea behind strong induction is that we should be allowed to use this information in our proof of $P(n)$.

Let's work out the details of this idea more carefully. Suppose that we've followed the strong induction proof strategy and proven the statement $\forall n[(\forall k < n P(k)) \rightarrow P(n)]$. Then, plugging in 0 for n, we can conclude that $(\forall k < 0 P(k)) \rightarrow P(0)$. But because there are no natural numbers smaller than 0, the statement $\forall k < 0 P(k)$ is vacuously true. Therefore, by modus ponens, $P(0)$ is true. (This explains why the base case doesn't have to be checked separately in a proof by strong induction; the base case $P(0)$ actually follows from the modified form of the induction step used in strong induction.) Similarly, plugging in 1 for n we can conclude that $(\forall k < 1 P(k)) \rightarrow P(1)$. The only natural number smaller than 1 is 0, and we've just shown that $P(0)$ is true, so the statement $\forall k < 1 P(k)$ is true. Therefore, by modus ponens, $P(1)$ is also true. Now plug in 2 for n to get the statement $(\forall k < 2P(k)) \rightarrow P(2)$. Since $P(0)$ and $P(1)$ are both true, the statement $\forall k < 2P(k)$ is true, and therefore by modus ponens, $P(2)$ is true. Continuing in this way we can show that $P(n)$ is true for every natural number n, as required. For an alternative justification of the method of strong induction, see exercise 1.

As our first example of the method of strong induction, we prove an important fact of number theory known as the *division algorithm.*

Theorem 6.4.1. (Division algorithm) *For all natural numbers n and m, if $m > 0$, then there are natural numbers q and r such that $n = mq + r$ and $r < m$.* (The numbers q and r are called the *quotient* and *remainder* when n is divided by m.)

Scratch work

We let m be an arbitrary positive integer and then use strong induction to prove that $\forall n \exists q \exists r(n = mq + r \wedge r < m)$. According to the description of strong induction, this means that we should let n be an arbitrary natural number, assume that $\forall k < n \exists q \exists r(k = mq + r \wedge r < m)$, and prove that $\exists q \exists r(n = mq + r \wedge r < m)$.

Our goal is an existential statement, so we should try to come up with values of q and r with the required properties. If $n < m$ then this is easy because we can just let $q = 0$ and $r = n$. But if $n \geq m$, then this won't work, since we must have $r < m$, so we must do something different in this case. As usual in induction proofs, we look to the inductive hypothesis. The inductive hypothesis starts with $\forall k < n$, so to apply it we should plug in some natural number smaller than n for k, but what should we plug in? The reference to division in the statement of the theorem provides a hint. If we think of division as repeated subtraction, then dividing n by m involves subtracting m from n repeatedly. The first step in this process would be to compute $n - m$, which is a natural number smaller than n. Perhaps we should plug in $n - m$ for k. It's not entirely clear where this will lead, but it's worth a try. In fact, as you'll see in the proof, once we take this step the desired conclusion follows almost immediately.

Notice that we are using the fact that a quotient and remainder exist for some natural number smaller than n to prove that they exist for n, but this smaller number is not $n - 1$, it's $n - m$. This is why we're using strong induction rather than ordinary induction for this proof.

Proof. We let m be an arbitrary positive integer and then proceed by strong induction on n.

Suppose n is a natural number, and for every $k < n$ there are natural numbers q and r such that $k = mq + r$ and $r < m$.

Case 1. $n < m$. Let $q = 0$ and $r = n$. Then clearly $n = mq + r$ and $r < m$.

Case 2. $n \geq m$. Let $k = n - m < n$ and note that since $n \geq m$, k is a natural number. By inductive hypothesis we can choose q' and r' such that $k = mq' + r'$ and $r' < m$. Then $n - m = mq' + r'$, so $n = mq' + r' + m = m(q' + 1) + r'$. Thus, if we let $q = q' + 1$ and $r = r'$, then we have $n = mq + r$ and $r < m$, as required. □

The division algorithm can also be extended to negative integers n, and it can be shown that for every m and n the quotient and remainder q and r are unique. For more on this, see exercise 13.

Our next example is another important theorem of number theory. We used this theorem in our proof in the introduction that there are infinitely many primes. For more on this theorem, see exercise 18.

Theorem 6.4.2. *Every integer $n > 1$ is either prime or a product of primes.*

Scratch work

We write the goal in the form $\forall n \in \mathbb{N}[n > 1 \rightarrow (n \text{ is prime} \vee n \text{ is a product of primes})]$ and then use strong induction. Thus, our inductive hypothesis is $\forall k < n[k > 1 \rightarrow (k \text{ is prime} \vee k \text{ is a product of primes})]$, and we must prove that $n > 1 \rightarrow (n \text{ is prime} \vee n \text{ is a product of primes})$. Of course, we start by assuming $n > 1$, and according to our strategies for proving disjunctions, a good way to complete the proof would be to assume that n is not prime and prove that it must be a product of primes. Because the assumption that n is not prime means $\exists a \exists b(n = ab \wedge a < n \wedge b < n)$, we immediately use existential instantiation to introduce the new variables a and b into the proof. Applying the inductive hypothesis to a and b now leads to the desired conclusion.

Proof. We use strong induction. Suppose $n > 1$, and suppose that for every integer k, if $1 < k < n$ then k is either prime or a product of primes. Of course, if n is prime then there is nothing to prove, so suppose n is not prime. Then we can choose natural numbers a and b such that $n = ab$, $a < n$, and $b < n$. Note that since $a < n = ab$, it follows that $b > 1$, and similarly we must have $a > 1$. Thus, by inductive hypothesis, each of a and b is either prime or a product of primes. But then since $n = ab$, n is a product of primes. □

The method of recursion studied in the last section also has a strong form. As an example of this, consider the following definition of a sequence of numbers, called the *Fibonacci* numbers after the Italian mathematician Fibonacci (ca 1174–1250) who first defined them.

$$F_0 = 0;$$

$$F_1 = 1;$$

$$\text{for every } n \geq 2,\ F_n = F_{n-2} + F_{n-1}.$$

For example, plugging in $n = 2$ in the last equation we find that $F_2 = F_0 + F_1 = 0 + 1 = 1$. Similarly, $F_3 = F_1 + F_2 = 1 + 1 = 2$, $F_4 = F_2 + F_3 = 1 + 2 = 3$, and so on. Note that, starting with F_2, each Fibonacci number is computed using, not just the previous number in the sequence, but also the one before that. This is the sense in which the recursion is strong. It shouldn't be surprising, therefore, that proofs involving the Fibonacci numbers often require strong induction rather than ordinary induction.

To illustrate this we'll prove the following remarkable formula for the Fibonacci numbers:

$$F_n = \frac{\left(\frac{1+\sqrt{5}}{2}\right)^n - \left(\frac{1-\sqrt{5}}{2}\right)^n}{\sqrt{5}}.$$

It is hard at first to believe that this formula is right. After all, the Fibonacci numbers are integers, and it is not at all clear that this formula will give an integer value. And what do the Fibonacci numbers have to do with $\sqrt{5}$? Nevertheless, a proof by strong induction shows that the formula is correct. (To see how this formula could be derived, see exercise 8.)

Theorem 6.4.3. *If F_n is the nth Fibonacci number, then*

$$F_n = \frac{\left(\frac{1+\sqrt{5}}{2}\right)^n - \left(\frac{1-\sqrt{5}}{2}\right)^n}{\sqrt{5}}.$$

Scratch work

Because F_0 and F_1 are defined separately from F_n for $n \geq 2$, we check the formula for these cases separately. For $n \geq 2$, the definition of F_n suggests that we should use the assumption that the formula is correct for F_{n-2} and F_{n-1} to prove that it is correct for F_n. Because we need to know that the formula works for *two* previous cases, we must use strong induction rather than ordinary induction. The rest of the proof is straightforward, although the algebra gets a little messy.

Proof. We use strong induction. Let n be an arbitrary natural number, and suppose that for all $k < n$,

$$F_k = \frac{\left(\frac{1+\sqrt{5}}{2}\right)^k - \left(\frac{1-\sqrt{5}}{2}\right)^k}{\sqrt{5}}.$$

Case 1. $n = 0$. Then

$$\frac{\left(\frac{1+\sqrt{5}}{2}\right)^n - \left(\frac{1-\sqrt{5}}{2}\right)^n}{\sqrt{5}} = \frac{\left(\frac{1+\sqrt{5}}{2}\right)^0 - \left(\frac{1-\sqrt{5}}{2}\right)^0}{\sqrt{5}}$$

$$= \frac{1-1}{\sqrt{5}} = 0 = F_0.$$

Case 2. $n = 1$. Then

$$\frac{\left(\frac{1+\sqrt{5}}{2}\right)^n - \left(\frac{1-\sqrt{5}}{2}\right)^n}{\sqrt{5}} = \frac{\left(\frac{1+\sqrt{5}}{2}\right)^1 - \left(\frac{1-\sqrt{5}}{2}\right)^1}{\sqrt{5}}$$
$$= \frac{\sqrt{5}}{\sqrt{5}} = 1 = F_1.$$

Case 3. $n \geq 2$. Then applying the inductive hypothesis to $n-2$ and $n-1$, we get

$$\begin{aligned}
F_n &= F_{n-2} + F_{n-1} \\
&= \frac{\left(\frac{1+\sqrt{5}}{2}\right)^{n-2} - \left(\frac{1-\sqrt{5}}{2}\right)^{n-2}}{\sqrt{5}} + \frac{\left(\frac{1+\sqrt{5}}{2}\right)^{n-1} - \left(\frac{1-\sqrt{5}}{2}\right)^{n-1}}{\sqrt{5}} \\
&= \frac{\left[\left(\frac{1+\sqrt{5}}{2}\right)^{n-2} + \left(\frac{1+\sqrt{5}}{2}\right)^{n-1}\right] - \left[\left(\frac{1-\sqrt{5}}{2}\right)^{n-2} + \left(\frac{1-\sqrt{5}}{2}\right)^{n-1}\right]}{\sqrt{5}} \\
&= \frac{\left(\frac{1+\sqrt{5}}{2}\right)^{n-2}\left[1 + \frac{1+\sqrt{5}}{2}\right] - \left(\frac{1-\sqrt{5}}{2}\right)^{n-2}\left[1 + \frac{1-\sqrt{5}}{2}\right]}{\sqrt{5}}.
\end{aligned}$$

Now note that

$$\left(\frac{1+\sqrt{5}}{2}\right)^2 = \frac{1 + 2\sqrt{5} + 5}{4} = \frac{6 + 2\sqrt{5}}{4} = \frac{3 + \sqrt{5}}{2} = 1 + \frac{1+\sqrt{5}}{2},$$

and similarly

$$\left(\frac{1-\sqrt{5}}{2}\right)^2 = 1 + \frac{1-\sqrt{5}}{2}.$$

Substituting into the formula for F_n, we get

$$\begin{aligned}
F_n &= \frac{\left(\frac{1+\sqrt{5}}{2}\right)^{n-2}\left(\frac{1+\sqrt{5}}{2}\right)^2 - \left(\frac{1-\sqrt{5}}{2}\right)^{n-2}\left(\frac{1-\sqrt{5}}{2}\right)^2}{\sqrt{5}} \\
&= \frac{\left(\frac{1+\sqrt{5}}{2}\right)^n - \left(\frac{1-\sqrt{5}}{2}\right)^n}{\sqrt{5}}.
\end{aligned}$$

□

Notice that in the proof of Theorem 6.4.3 we had to treat the cases $n = 0$ and $n = 1$ separately. The role that these cases play in the proof is similar to the role played by the base case in a proof by ordinary mathematical induction. Although we have said that proofs by strong induction don't need base cases, it is not uncommon to find some initial cases treated separately in such proofs.

An important property of the natural numbers that is related to mathematical induction is the fact that every nonempty set of natural numbers has a smallest element. This is sometimes called the *well-ordering principle*, and we can prove it using strong induction.

Theorem 6.4.4. (Well-ordering principle) *Every nonempty set of natural numbers has a smallest element.*

Scratch work

Our goal is $\forall S \subseteq \mathbb{N}(S \neq \varnothing \rightarrow S$ has a smallest element). After letting S be an arbitrary subset of $\mathbb{N}$, we'll prove the contrapositive of the conditional statement. In other words, we will assume that S has no smallest element and prove that $S = \varnothing$. The way induction comes into it is that, for a set $S \subseteq \mathbb{N}$, to say that $S = \varnothing$ is the same as saying that $\forall n \in \mathbb{N}(n \notin S)$. We'll prove this last statement by strong induction.

Proof. Suppose $S \subseteq \mathbb{N}$, and S does not have a smallest element. We will prove that $\forall n \in \mathbb{N}(n \notin S)$, so $S = \varnothing$. Thus, if $S \neq \varnothing$ then S must have a smallest element.

To prove that $\forall n \in \mathbb{N}(n \notin S)$, we use strong induction. Suppose that $n \in \mathbb{N}$ and $\forall k < n(k \notin S)$. Clearly if $n \in S$ then n would be the smallest element of S, and this would contradict the assumption that S has no smallest element. Therefore $n \notin S$. □

Sometimes, proofs that could be done by induction are written instead as applications of the well-ordering principle. As an example of the use of the well-ordering principle in a proof, we present a proof that $\sqrt{2}$ is irrational. See exercise 2 for an alternative approach to this proof using strong induction. See exercise 16 for another application of the well-ordering principle.

Theorem 6.4.5. $\sqrt{2}$ *is irrational.*

Scratch work

Because *irrational* means "not rational," our goal is a negative statement, so proof by contradiction is a logical method to use. Thus, we assume $\sqrt{2}$ is rational and try to reach a contradiction. The assumption that $\sqrt{2}$ is rational means that there exist integers p and q such that $p/q = \sqrt{2}$, and since $\sqrt{2}$ is positive, we may as well restrict our attention to positive p and q. Because this is an existential statement, our next step should probably be to choose positive integers p and q such that $p/q = \sqrt{2}$. As you will see in the proof, simple algebraic manipulations with the equation $p/q = \sqrt{2}$ do not lead to any obvious contradictions, but they do lead to the conclusion that p and q must both be even. Thus, in the fraction p/q we can cancel a 2 from both numerator and denominator, getting a new fraction with smaller numerator and denominator that is equal to $\sqrt{2}$.

How can we derive a contradiction from this conclusion? The key idea is to note that our reasoning would apply to *any* fraction that is equal to $\sqrt{2}$. Thus, in any such fraction we can cancel a factor of 2 from numerator and denominator, and therefore there can be no smallest possible numerator or denominator for such a fraction. But this would violate the well-ordering principle! Thus, we have our contradiction.

This idea is spelled out more carefully in the following proof, in which we've applied the well-ordering principle to the set of all possible denominators of fractions equal to $\sqrt{2}$. We have chosen to put this application of the well-ordering principle at the beginning of the proof, because this seems to give the shortest and most direct proof. Readers of the proof might be puzzled at first about why we're using the well-ordering principle (unless they've read this scratch work!), but after the algebraic manipulations with the equation $p/q = \sqrt{2}$ are completed, the contradiction appears almost immediately. This is a good example of how a clever, carefully planned step early in a proof can lead to a wonderful punch line at the end of the proof.

Proof. Suppose that $\sqrt{2}$ is rational. This means that $\exists q \in \mathbb{Z}^+ \exists p \in \mathbb{Z}^+(p/q = \sqrt{2})$, so the set $S = \{q \in \mathbb{Z}^+ \mid \exists p \in \mathbb{Z}^+(p/q = \sqrt{2})\}$ is nonempty. By the well-ordering principle we can let q be the smallest element of S. Since $q \in S$, we can choose some $p \in \mathbb{Z}^+$ such that $p/q = \sqrt{2}$. Therefore $p^2/q^2 = 2$, so $p^2 = 2q^2$ and therefore p^2 is even. We now apply the theorem from Example 3.4.2, which says that for any integer x, x is even iff x^2 is even. Since p^2 is even, p must be even, so we can choose some $\bar{p} \in \mathbb{Z}^+$ such that $p = 2\bar{p}$. Therefore $p^2 = 4\bar{p}^2$, and substituting this into the equation $p^2 = 2q^2$ we get $4\bar{p}^2 = 2q^2$, so $2\bar{p}^2 = q^2$ and therefore q^2 is even. Appealing to Example 3.4.2 again, this means q must be even, so we can choose some $\bar{q} \in \mathbb{Z}^+$ such that $q = 2\bar{q}$. But then $\sqrt{2} = p/q = (2\bar{p})/(2\bar{q}) = \bar{p}/\bar{q}$, so $\bar{q} \in S$. Clearly $\bar{q} < q$, so this contradicts the fact that q was chosen to be the *smallest* element of S. Therefore $\sqrt{2}$ is irrational. □

Exercises

*1. This exercise gives an alternative way to justify the method of strong induction. All variables in this exercise range over $\mathbb{N}$. Suppose

$P(n)$ is a statement about a natural number n, and suppose that $\forall n[(\forall k < n P(k)) \to P(n)]$. Let $Q(n)$ be the statement $\forall k < n P(k)$.

(a) Prove $\forall n Q(n) \leftrightarrow \forall n P(n)$ without using induction.

(b) Prove $\forall n Q(n)$ by *ordinary* induction. Thus, by part (a), $\forall n P(n)$ is true.

2. This exercise gives an alternative way of writing the proof that $\sqrt{2}$ is irrational. Use strong induction to prove that $\forall q \in \mathbb{N}[q > 0 \to \neg\exists p \in \mathbb{N}(p/q = \sqrt{2})]$.

*3. (a) Prove that $\sqrt{6}$ is irrational.

(b) Prove that $\sqrt{2} + \sqrt{3}$ is irrational.

4. The Martian monetary system uses colored beads instead of coins. A blue bead is worth 3 Martian credits, and a red bead is worth 7 Martian credits. Thus, three blue beads are worth 9 credits, and a blue and red bead together are worth 10 credits, but no combination of blue and red beads is worth 11 credits. Prove that for all $n \geq 12$, there is some combination of blue and red beads that is worth n credits.

5. Suppose that x is a real number, $x \neq 0$, and $x + 1/x$ is an integer. Prove that for all $n \geq 1$, $x^n + 1/x^n$ is an integer.

*6. Let F_n be the nth Fibonacci number. All variables in this exercise range over $\mathbb{N}$

(a) Prove that for all n, $\sum_{i=0}^{n} F_i = F_{n+2} - 1$.

(b) Prove that for all n, $\sum_{i=0}^{n} (F_i)^2 = F_n F_{n+1}$.

(c) Prove that for all n, $\sum_{i=0}^{n} F_{2i+1} = F_{2n+2}$.

(d) Find a formula for $\sum_{i=0}^{n} F_{2i}$ and prove that your formula is correct.

7. Let F_n be the nth Fibonacci number. All variables in this exercise range over $\mathbb{N}$.

(a) Prove that for all $m \geq 1$ and all n, $F_{m+n} = F_{m-1}F_n + F_m F_{n+1}$.

(b) Prove that for all $m \geq 1$ and all $n \geq 1$, $F_{m+n} = F_{m+1}F_{n+1} - F_{m-1}F_{n-1}$.

(c) Prove that for all n, $(F_n)^2 + (F_{n+1})^2 = F_{2n+1}$ and $(F_{n+2})^2 - (F_n)^2 = F_{2n+2}$.

(d) Prove that for all m and n, if $m \mid n$ then $F_m \mid F_n$.

(e) See exercise 18 of Section 6.3 for the meaning of the notation used in this exercise. Prove that for all $n \geq 1$,

$$F_{2n-1} = \binom{2n-2}{0} + \binom{2n-3}{1} + \binom{2n-4}{2} + \cdots + \binom{n-1}{n-1}$$
$$= \sum_{i=0}^{n-1} \binom{2n-i-2}{i}$$

and

$$F_{2n} = \binom{2n-1}{0} + \binom{2n-2}{1} + \binom{2n-3}{2} + \cdots + \binom{n}{n-1}$$
$$= \sum_{i=0}^{n-1} \binom{2n-i-1}{i}.$$

*8. A sequence of numbers $a_0, a_1, a_2, \ldots$ is called a *generalized Fibonacci sequence*, or a *Gibonacci sequence* for short, if for every $n \geq 2$, $a_n = a_{n-2} + a_{n-1}$. Thus, a Gibonacci sequence satisfies the same *recurrence relation* as the Fibonacci numbers, but it may start out differently.

(a) Suppose c is a real number and $\forall n \in \mathbb{N}(a_n = c^n)$. Prove that a_0, $a_1, a_2, \ldots$ is a Gibonacci sequence iff either $c = (1+\sqrt{5})/2$ or $c = (1-\sqrt{5})/2$.

(b) Suppose s and t are real numbers, and for all $n \in \mathbb{N}$,

$$a_n = s\left(\frac{1+\sqrt{5}}{2}\right)^n + t\left(\frac{1-\sqrt{5}}{2}\right)^n.$$

Prove that $a_0, a_1, a_2, \ldots$ is a Gibonacci sequence.

(c) Suppose $a_0, a_1, a_2, \ldots$ is a Gibonacci sequence. Prove that there are real numbers s and t such that for all $n \in \mathbb{N}$,

$$a_n = s\left(\frac{1+\sqrt{5}}{2}\right)^n + t\left(\frac{1-\sqrt{5}}{2}\right)^n.$$

(Hint: First show that there are real numbers s and t such that the formula above is correct for a_0 and a_1. Then show that with this choice of s and t, the formula is correct for all n.)

9. The *Lucas numbers* (named for the French mathematician Edouard Lucas (1842–1891)) are the numbers $L_0, L_1, L_2, \ldots$ defined as follows:

$L_0 = 2;$

$L_1 = 1;$

for every $n \geq 2$, $L_n = L_{n-2} + L_{n-1}$.

Find a formula for L_n and prove that your formula is correct. (Hint: Apply exercise 8.)

*10. A sequence $a_0, a_1, a_2, \ldots$ is defined recursively as follows:

$a_0 = -1;$

$a_1 = 0;$

for every $n \geq 2$, $a_n = 5a_{n-1} - 6a_{n-2}$.

Find a formula for a_n and prove that your formula is correct. (Hint: Imitate exercise 8.)

11. A sequence $a_0, a_1, a_2, \ldots$ is defined recursively as follows:

$$a_0 = 0;$$
$$a_1 = 1;$$
$$a_2 = 1;$$
$$\text{for every } n \geq 3,\ a_n = \frac{1}{2}a_{n-3} + \frac{3}{2}a_{n-2} + \frac{1}{2}a_{n-1}.$$

Prove that for all $n \in \mathbb{N}$, $a_n = F_n$, the nth Fibonacci number.

12. For each positive integer n, let $A_n = \{1, 2, \ldots, n\}$, and let $P_n = \{X \in \mathscr{P}(A_n) \mid X \text{ does not contain two consecutive integers}\}$. For example, $P_3 = \{\varnothing, \{1\}, \{2\}, \{3\}, \{1, 3\}\}$; P_3 does not contain the sets $\{1, 2\}$, $\{2, 3\}$, and $\{1, 2, 3\}$ because each contains at least one pair of consecutive integers. Prove that for every n, the number of elements in P_n is F_{n+2}, the $(n+2)^{\text{th}}$ Fibonacci number. (For example, the number of elements in P_3 is $5 = F_5$. Hint: Which elements of P_n contain n? Which don't? The answers to both questions are related to the elements of P_m, for certain $m < n$.)

13. Suppose n and m are integers and $m > 0$.
    (a) Prove that there are integers q and r such that $n = mq + r$ and $0 \leq r < m$. (Hint: If $n \geq 0$, then this follows from Theorem 6.4.1. If $n < 0$, then start by applying Theorem 6.4.1 to $-n$ and m.)
    (b) Prove that the integers q and r in part (a) are unique. In other words, show that if q' and r' are integers such that $n = mq' + r'$ and $0 \leq r' < m$, then $q = q'$ and $r = r'$.
    (c) Prove that, as claimed in Section 3.4, every integer is either even or odd but not both.

*14. Suppose k is a positive integer. Prove that there is some positive integer a such that for all $n > a$, $2^n \geq n^k$. (In the language of exercise 16 of Section 5.1, this means that if $f(n) = n^k$ and $g(n) = 2^n$ then $f \in O(g)$. Hint: By the division algorithm, for any natural number n there are natural numbers q and r such that $n = kq + r$ and $0 \leq r < k$. Therefore $2^n \geq 2^{kq} = (2^q)^k$. To choose a, figure out how large q has to be to guarantee that $2^q \geq n$. You may find Example 6.1.3 useful.)

15. (a) Suppose k is a positive integer, $a_1, a_2, \ldots, a_k$ are real numbers, and $f_1, f_2, \ldots, f_k$, and g are all functions from $\mathbb{Z}^+$ to $\mathbb{R}$. Also, suppose that $f_1, f_2, \ldots, f_k$ are all elements of $O(g)$. (See exercise 16 of Section 5.1 for the meaning of the notation used here.) Define $f : \mathbb{Z}^+ \to \mathbb{R}$ by the formula $f(n) = a_1 f_1(n) + a_2 f_2(n) + \cdots + a_k f_k(n)$. Prove that $f \in O(g)$. (Hint: Use induction on k, and exercise 16(c) of Section 5.1.)

(b) Let $g : \mathbb{Z}^+ \to \mathbb{R}$ be defined by the formula $g(n) = 2^n$. Suppose a_0, a_1, a_2, ..., a_k are real numbers, and define $f : \mathbb{Z}^+ \to \mathbb{R}$ by the formula $f(n) = a_0 + a_1 n + a_2 n^2 + \cdots + a_k n^k$. (Such a function is called a *polynomial*.) Prove that $f \in O(g)$. (Hint: Use exercise 14 and part (a).)

16. Suppose a and b are positive integers. Let $S = \{x \in \mathbb{Z}^+ \mid \exists s \in \mathbb{Z} \exists t \in \mathbb{Z}(x = as + bt)\}$. Note that $S \neq \varnothing$ since, for example, $a = a \cdot 1 + b \cdot 0$, and therefore $a \in S$. Thus, by the well-ordering principle, we can let d be the smallest element of S.
    (a) Prove that $d \mid a$ and $d \mid b$. (Hint: Use the division algorithm to choose integers q and r such that $a = dq + r$ and $0 \leq r < d$. Now show that $r = 0$).
    (b) Prove that if c is any integer such that $c \mid a$ and $c \mid b$, then $c \mid d$. (Note that it follows that $c \leq d$, so d is the *greatest common divisor* of a and b).
17. (a) Suppose a, b, and p are positive integers and p is prime. Prove that if $p \mid ab$ then either $p \mid a$ or $p \mid b$. (Hint: Let d be the greatest common divisor of a and p. By exercise 16, $d = as + pt$ for some integers s and t. Since p is prime, there are not many possibilities for the value of d. What are they?)
    (b) Suppose $a_1, a_2, \ldots, a_n$ is a sequence of positive integers and p is a prime number. Prove that if $p \mid (a_1 a_2 \ldots a_n)$, then $p \mid a_i$ for some i, $1 \leq i \leq n$. (Hint: Use part (a) and induction.)

*18. Suppose $p_1, p_2, \ldots, p_j$ and $q_1, q_2, \ldots, q_k$ are two sequences of prime numbers and $p_1 p_2 \ldots p_j = q_1 q_2 \ldots q_k$. Suppose also that both sequences are nondecreasing; that is, $p_1 \leq p_2 \leq \ldots \leq p_j$ and $q_1 \leq q_2 \leq \ldots \leq q_k$. Prove that the two sequences must be the same. In other words, $j = k$ and $p_i = q_i$ for all i, $1 \leq i \leq j$. (Hint: Apply exercise 17 and use induction on either j or k. Note that this shows that the factorization of an integer $n > 1$ into primes in Theorem 6.4.2 is unique.)

19. A sequence $a_0, a_1, a_2, \ldots$ is defined recursively as follows:

$$a_0 = 1;$$

$$\text{for every } n \in \mathbb{N},\ a_{n+1} = 1 + \sum_{i=0}^{n} a_i.$$

Find a formula for a_n and prove that your formula is correct.

*20. A sequence $a_0, a_1, a_2, \ldots$ is defined recursively as follows:

$$a_0 = 1;$$

$$\text{for every } n \in \mathbb{N},\ a_{n+1} = 1 + \frac{1}{a_n}.$$

Find a formula for a_n and prove that your formula is correct. (Hint: These numbers are related to the Fibonacci numbers.)

6.5. Closures Again

In Chapter 4 we promised to give an alternative treatment of transitive closures of relations using mathematical induction. In this section we fulfill this promise.

Recall that if R is a relation on a set A, then the transitive closure of R is the smallest relation S on A such that $R \subseteq S$ and S is transitive. In this section we'll find this relation S by starting with R and then adding only those ordered pairs that *must* be added if we want to end up with a transitive relation. We begin with a sketchy description of how we'll do this, motivated by the examples in Section 4.5. Then we'll use recursion and induction to make this sketchy idea precise and prove that it works.

The examples in Section 4.5 suggested that if $a_0, a_1, \ldots, a_n$ is a list of elements of A such that $(a_0, a_1) \in R, (a_1, a_2) \in R, \ldots, (a_{n-1}, a_n) \in R$, then to create a transitive relation S extending R we must have $(a_0, a_n) \in S$. Let's rephrase this idea in terms of composition of relations. Because $(a_0, a_1) \in R$ and $(a_1, a_2) \in R$, by the definition of composition, $(a_0, a_2) \in R \circ R$. Similarly, from $(a_0, a_2) \in R \circ R$ and $(a_2, a_3) \in R$ it follows that $(a_0, a_3) \in R \circ (R \circ R)$. It is natural to call this last relation R^3. Note that by Theorem 4.2.5, composition of relations is associative, so there is no ambiguity if we leave the parentheses out of the definition of R^3 and write $R^3 = R \circ R \circ R$. Thus, we have $(a_0, a_3) \in R^3$, and because $(a_3, a_4) \in R$, it follows that $(a_0, a_4) \in R \circ R^3 = R \circ R \circ R \circ R = R^4$. Continuing in this way we'll eventually reach the conclusion that $(a_0, a_n) \in R^n = R \circ R \circ \cdots \circ R$, where there are n R's in the last composition. We'll show that the ordered pairs that must be added to R to create a transitive relation are the elements of R^n for every positive integer n.

The use of an ellipsis in the last paragraph suggests that it might be best to define R^n by recursion. Here's the precise definition:

$$R^1 = R;$$
$$\text{for every } n \geq 1,\ R^{n+1} = R^n \circ R.$$

Before using this definition to construct the transitive closure of R, we prove a lemma about it. Of course, the proof will be done by induction!

Lemma 6.5.1. *For all positive integers m and n, $R^{m+n} = R^m \circ R^n$.*

Proof. We let m be an arbitrary positive integer and then proceed by induction on n.

Base case: When $n = 1$ we have

$$\begin{aligned} R^{m+1} &= R^m \circ R \quad &&\text{(by definition of } R^{m+1}\text{)} \\ &= R^m \circ R^1 \quad &&\text{(by definition of } R^1\text{).} \end{aligned}$$

Induction step: Let n be an arbitrary positive integer and suppose $R^{m+n} = R^m \circ R^n$. Then

$$\begin{aligned} R^{m+(n+1)} &= R^{(m+n)+1} \\ &= R^{m+n} \circ R \quad &&\text{(by definition of } R^{(m+n)+1}\text{)} \\ &= (R^m \circ R^n) \circ R \quad &&\text{(by inductive hypothesis)} \\ &= R^m \circ (R^n \circ R) \quad &&\text{(by associativity of composition)} \\ &= R^m \circ R^{n+1} \quad &&\text{(by definition of } R^{n+1}\text{).} \end{aligned}$$

□

We can now say precisely how to form the transitive closure of R.

Theorem 6.5.2. *The transitive closure of R is $\bigcup_{n\in\mathbb{Z}^+} R^n$.*

Proof. Let $S = \bigcup_{n\in\mathbb{Z}^+} R^n$. Clearly $R = R^1 \subseteq S$. To see that S is transitive, suppose $(x, y) \in S$ and $(y, z) \in S$. Then by the definition of S, we can choose positive integers n and m such that $(x, y) \in R^n$ and $(y, z) \in R^m$. But then by Lemma 6.5.1, $(x, z) \in R^m \circ R^n = R^{m+n}$, so $(x, z) \in \bigcup_{n\in\mathbb{Z}^+} R^n = S$. Thus S is transitive.

Finally, suppose $R \subseteq T \subseteq A \times A$ and T is transitive. We must show that $S \subseteq T$, and clearly by the definition of S it suffices to show that $\forall n \in \mathbb{Z}^+(R^n \subseteq T)$. We prove this by induction on n.

We have assumed $R \subseteq T$, so when $n = 1$ we have $R^n = R^1 = R \subseteq T$. For the induction step, suppose n is a positive integer and $R^n \subseteq T$. Now suppose $(x, y) \in R^{n+1}$. Then by definition of R^{n+1} we can choose some $z \in A$ such that $(x, z) \in R$ and $(z, y) \in R^n$. By assumption $R \subseteq T$, and by inductive hypothesis $R^n \subseteq T$. Therefore, $(x, z) \in T$ and $(z, y) \in T$, so since T is transitive, $(x, y) \in T$. Since (x, y) was an arbitrary element of R^{n+1}, this shows that $R^{n+1} \subseteq T$. □

Commentary. Because the proof must refer to the set $\bigcup_{n\in\mathbb{Z}^+} R^n$ often, it is convenient to give this set a name right at the beginning of the proof. According to the definition of transitive closure we must prove three things: $R \subseteq S$, S is transitive, and for all T, if $R \subseteq T \subseteq A \times A$ and T is transitive, then $S \subseteq T$. Of course, we prove them one at a time.

The proof of the first of these goals is not spelled out. As usual, if you don't see why it is true you should work out the details of the proof yourself. The second goal is to prove that S is transitive, and the proof is based on the definition of transitive. We let x, y, and z be arbitrary, assume $(x, y) \in S$ and $(y, z) \in S$, and prove that $(x, z) \in S$. According to the definition of S, the statement $(x, y) \in S$ means $\exists n \in \mathbb{Z}^+((x, y) \in R^n)$, so we immediately introduce the variable n to stand for a positive integer such that $(x, y) \in R^n$. The assumption that $(y, z) \in S$ is handled similarly. The goal $(x, z) \in S$ is also an existential statement, so to prove it we must find a positive integer k such that $(x, z) \in R^k$. We use Lemma 6.5.1 to show that $k = m + n$ works.

Finally, for the third goal we use the natural strategy of letting T be arbitrary, assuming that $R \subseteq T \subseteq A \times A$ and T is transitive, and then proving that $S \subseteq T$. Once again, if you don't see why the conclusion $S \subseteq T$ follows from $\forall n \in \mathbb{Z}^+(R^n \subseteq T)$, as claimed in the proof, you should work out the details of the proof yourself. This last goal is proven by induction, as you might expect based on the recursive nature of the definition of R^n. For the induction step, we let n be an arbitrary positive integer, assume that $R^n \subseteq T$, and prove that $R^{n+1} \subseteq T$. To prove that $R^{n+1} \subseteq T$ we take an arbitrary element of R^{n+1} and prove that it must be an element of T. Writing out the recursive definition of R^{n+1} gives us a way to use the inductive hypothesis, which, as usual, is the key to completing the induction step.

We end this chapter by returning once again to one of the proofs in the introduction. Recall that in our first proof in the introduction we used the formula

$$(2^b - 1) \cdot (1 + 2^b + 2^{2b} + \cdots + 2^{(a-1)b}) = 2^{ab} - 1.$$

We discussed this proof again in Section 3.7 and promised to give a more careful proof of this formula after we had discussed mathematical induction. We are ready now to give this more careful proof. Of course, we can also state the formula more precisely now, using summation notation.

Theorem 6.5.3. *For all positive integers a and b,*

$$(2^b - 1) \cdot \sum_{k=0}^{a-1} 2^{kb} = 2^{ab} - 1.$$

Proof. We let b be an arbitrary positive integer and then proceed by induction on a.

Base case: When $a = 1$ we have

$$\begin{aligned}
(2^b - 1) \cdot \sum_{k=0}^{a-1} 2^{kb} &= (2^b - 1) \cdot \sum_{k=0}^{0} 2^{kb} \\
&= (2^b - 1) \cdot 1 \\
&= 2^{ab} - 1.
\end{aligned}$$

Induction step: Suppose $a \geq 1$ and $(2^b - 1) \cdot \sum_{k=0}^{a-1} 2^{kb} = 2^{ab} - 1$. Then

$$\begin{aligned}(2^b - 1) \cdot \sum_{k=0}^{a} 2^{kb} &= (2^b - 1) \cdot \left(\sum_{k=0}^{a-1} 2^{kb} + 2^{ab}\right) \\ &= (2^b - 1) \cdot \sum_{k=0}^{a-1} 2^{kb} + 2^b \cdot 2^{ab} - 2^{ab} \\ &= 2^{ab} - 1 + 2^{b+ab} - 2^{ab} \\ &= 2^{(a+1)b} - 1.\end{aligned}$$

□

Exercises

*1. Suppose $f : A \to A$. A set $C \subseteq A$ is said to be *closed under* f if $\forall x \in C(f(x) \in C)$. Now suppose $B \subseteq A$. The *closure of* B *under* f is the smallest set C such that $B \subseteq C \subseteq A$ and C is closed under f, if there is such a smallest set. In this problem you will give two different proofs that the closure of B under f exists.
   (a) Let $\mathcal{F} = \{C \mid B \subseteq C \subseteq A$ and C is closed under $f\}$. Prove that $\mathcal{F} \neq \varnothing$, and then prove that $\cap\mathcal{F}$ is the closure of B under f.
   (b) Define sets B_n, for $n \geq 1$, as follows:

   $B_1 = B$;

   for every $n \geq 1$, $B_{n+1} = f(B_n) = \{f(x) \mid x \in B_n\}$.

   Prove that $\cup_{n \in \mathbb{Z}^+} B_n$ is the closure of B under f.

2. Let $f : \mathbb{R} \to \mathbb{R}$ be defined by the formula $f(x) = x + 1$. What is the closure of the set $\{0\}$ under f? (See exercise 1.)
3. Suppose $\mathcal{F}$ is a family of functions from A to A, and $B \subseteq A$. The *closure of* B *under* $\mathcal{F}$ is the smallest set C such that $B \subseteq C \subseteq A$ and for all $f \in \mathcal{F}$, C is closed under f, if there is such a smallest set. Prove that the closure of B under $\mathcal{F}$ exists.
4. Suppose $f : A \times A \to A$. If $(x, y) \in A \times A$, then the result of applying f to (x, y) should be written $f((x, y))$, but it is customary to leave out one set of parentheses and just write $f(x, y)$. A set $C \subset A$ is said to be *closed under* f if $\forall x \in C \forall y \in C(f(x, y) \in C)$. Now suppose $B \subseteq A$. The *closure of* B *under* f is the smallest set C such that $B \subseteq C \subseteq A$ and C is closed under f, if there is such a smallest set. Prove that the closure of B under f exists.

*5. Let $f : \mathbb{Z} \times \mathbb{Z} \to \mathbb{Z}$ be defined by the formula $f(x, y) = xy$. Let P be the set of all prime numbers. What is the closure of P under f? (See exercise 4.)

6. Suppose $\mathcal{F}$ is a family of functions from $A \times A$ to A, and $B \subseteq A$. Combining the ideas in exercises 3 and 4, we say that the *closure of B under $\mathcal{F}$* is the smallest set C such that $B \subseteq C \subseteq A$ and for all $f \in \mathcal{F}$, C is closed under f, if there is such a smallest set.
   (a) Prove that the closure of B under $\mathcal{F}$ exists.
   (b) Let $f : \mathbb{R} \times \mathbb{R} \to \mathbb{R}$ and $g : \mathbb{R} \times \mathbb{R} \to \mathbb{R}$ be defined by the formulas $f(x, y) = x + y$ and $g(x, y) = xy$. Prove that the closure of $\mathbb{Q} \cup \{\sqrt{2}\}$ under $\{f, g\}$ is the set $\{a + b\sqrt{2} \mid a \in \mathbb{Q}, b \in \mathbb{Q}\}$. (This set is called $\mathbb{Q}$ *with* $\sqrt{2}$ *adjoined*, and is denoted $\mathbb{Q}(\sqrt{2})$.)
   (c) With f and g defined as in part (b), what is the closure of $\mathbb{Q} \cup \{\sqrt[3]{2}\}$ under $\{f, g\}$?
7. Suppose R and S are relations on A and $R \subseteq S$. Prove that for every positive integer n, $R^n \subseteq S^n$.

*8. Suppose R and S are relations on A and n is a positive integer.
   (a) What is the relationship between $R^n \cap S^n$ and $(R \cap S)^n$? Justify your conclusions with proofs or counterexamples.
   (b) What is the relationship between $R^n \cup S^n$ and $(R \cup S)^n$? Justify your conclusions with proofs or counterexamples.

9. Suppose R is a relation on A and S is the transitive closure of R. If $(a, b) \in S$, then by Theorem 6.5.2 there is some positive integer n such that $(a, b) \in R^n$, and therefore by the well-ordering principle (Theorem 6.4.4), there must be a smallest such n. We define the *distance* from a to b to be the smallest positive integer n such that $(a, b) \in R^n$, and we write $d(a, b)$ to denote this distance.
   (a) Suppose that $(a, b) \in S$ and $(b, c) \in S$ (and therefore $(a, c) \in S$, since S is transitive). Prove that $d(a, c) \leq d(a, b) + d(b, c)$.
   (b) Suppose $(a, c) \in S$ and $0 < m < d(a, c)$. Prove that there is some $b \in A$ such that $d(a, b) = m$ and $d(b, c) = d(a, c) - m$.

*10. Suppose R is a relation on A and S is the transitive closure of R. For each positive integer n, let $J_n = \{0, 1, 2, \ldots, n\}$. If $a \in A$ and $b \in A$, we will say that a function $f : J_n \to A$ is an *R-path from a to b* of *length n* if $f(0) = a$, $f(n) = b$, and for all $i < n$, $(f(i), f(i + 1)) \in R$.
   (a) Prove that for all $n \in \mathbb{Z}^+$, $R^n = \{(a, b) \in A \times A \mid$ there is an R-path from a to b of length $n\}$.
   (b) Prove that $S = \{(a, b) \in A \times A \mid$ there is an R-path from a to $b\}$.

11. Suppose R is a relation on A and S is the transitive closure of R. If f is an R-path, then we say that the path is *simple* if f is one-to-one. (See exercise 10 for the definition of *R-path*.)
   (a) Prove that for all $n \in \mathbb{Z}^+$, $R^n \setminus i_A \subseteq \{(a, b) \in A \times A \mid$ there is a simple R-path from a to b of length at most $n\}$.

(b) Prove that $S \setminus i_A = \{(a, b) \in A \times A \mid$ there is a simple R-path from a to $b\}$.

12. Suppose R is a relation on A. In this problem we find a relationship between distance, as defined in exercise 9, and the lengths of R-paths, which were discussed in exercises 10 and 11.
    (a) Suppose $d(a, b) = n$ and $a \neq b$. Prove that there is a simple R-path from a to b of length n.
    (b) Suppose $d(a, a) = n$. Prove that there is an R-path f from a to a of length n such that $\forall i < n \forall j < n(f(i) = f(j) \to i = j)$. (In other words, f is simple, except for the fact that $f(0) = f(n) = a$.)
13. Suppose R is a relation on A, S is the transitive closure of R, and A has m elements. Prove that

$$S = R \cup R^2 \cup \ldots \cup R^m = \cup\{R^n \mid 1 \leq n \leq m\}.$$

(Hint: Use exercise 12. What is the maximum possible length of a simple R-path?)

*14. There is another proof in the introduction that could be written more rigorously using induction. Recall that in the proof of Theorem 4 in the introduction we used the fact that if n is a positive integer, $x = (n + 1)! + 2$, and $0 \leq i \leq n - 1$, then $(i + 2) \mid (x + i)$. Use induction to prove this. (We used this fact to show that $x + i$ is not prime.)

7

Infinite Sets

7.1. Equinumerous Sets

In this chapter, we'll discuss a method of comparing the sizes of infinite sets. Surprisingly, we'll find that, in a sense, infinity comes in different sizes! By now, you should be fairly proficient at reading and writing proofs, so we'll give less discussion of the strategy behind proofs and leave more proofs as exercises.

For finite sets, we determine the size of a set by counting. What does it mean to count the number of elements in a set? When you count the elements in a set A, you point to the elements of A in turn while saying the words *one, two*, and so forth. We could think of this process as defining a function f from the set $\{1, 2, \ldots, n\}$ to A, for some natural number n. For each $i \in \{1, 2, \ldots, n\}$, we let $f(i)$ be the element of A you're pointing to when you say "i." Because every element of A gets pointed to exactly once, the function f is one-to-one and onto. Thus, counting the elements of A is simply a method of establishing a one-to-one correspondence between the sets $\{1, 2, \ldots, n\}$ and A, for some natural number n. One-to-one correspondence is the key idea behind measuring the sizes of sets, and sets of the form $\{1, 2, \ldots, n\}$ are the standards against which we measure the sizes of finite sets. This suggests the following definition.

Definition 7.1.1. Suppose A and B are sets. We'll say that A is *equinumerous* with B if there is a function $f : A \rightarrow B$ that is one-to-one and onto. We'll write $A \sim B$ to indicate that A is equinumerous with B. For each natural number n, let $I_n = \{i \in \mathbb{Z}^+ \mid i \leq n\}$. A set A is called *finite* if there is a natural number n such that $I_n \sim A$. Otherwise, A is *infinite*.

You are asked in exercise 6 to show that if A is finite, then there is *exactly one* n such that $I_n \sim A$. Thus, it makes sense to define the *number of elements* of a finite set A to be the unique n such that $I_n \sim A$. This number is also sometimes called the *cardinality* of A, and it is denoted $|A|$. Note that according to this definition, $\varnothing$ is finite and $|\varnothing| = 0$.

The definition of equinumerous can also be applied to infinite sets, with results that are sometimes surprising. For example, you might think that $\mathbb{Z}^+$ could not be equinumerous with $\mathbb{Z}$ because $\mathbb{Z}$ includes all the positive integers, plus all the negative integers and zero as well. But consider the function $f : \mathbb{Z}^+ \to \mathbb{Z}$ defined as follows:

$$f(n) = \begin{cases} \dfrac{n}{2} & \text{if } n \text{ is even} \\[2ex] \dfrac{1-n}{2} & \text{if } n \text{ is odd.} \end{cases}$$

This notation means that for every positive integer n, if n is even then $f(n) = n/2$ and if n is odd then $f(n) = (1 - n)/2$. The table of values for f in Figure 1 reveals a pattern that suggests that f might be one-to-one and onto.

n	$f(n)$
1	0
2	1
3	−1
4	2
5	−2
6	3
7	−3
⋮	⋮

Figure 1

To check this more carefully, first note that for every positive integer n, if n is even then $f(n) = n/2 > 0$, and if n is odd then $f(n) = (1 - n)/2 \leq 0$. Now suppose n_1 and n_2 are positive integers and $f(n_1) = f(n_2)$. If $f(n_1) = f(n_2) > 0$ then n_1 and n_2 must both be even, so the equation $f(n_1) = f(n_2)$ means $n_1/2 = n_2/2$, and therefore $n_1 = n_2$. Similarly, if $f(n_1) = f(n_2) \leq 0$ then n_1 and n_2 are both odd, so we get $(1 - n_1)/2 = (1 - n_2)/2$, and once again it follows that $n_1 = n_2$. Thus, f is one-to-one.

To see that f is onto, let m be an arbitrary integer. If $m > 0$ then let $n = 2m$, an even positive integer, and if $m \leq 0$ then let $n = 1 - 2m$, an odd positive integer. In both cases it is easy to verify that $f(n) = m$. Thus, f is onto as well as one-to-one, so according to Definition 7.1.1, $\mathbb{Z}^+ \sim \mathbb{Z}$.

Note that the function f had to be chosen very carefully. There are many other functions from $\mathbb{Z}^+$ to $\mathbb{Z}$ that are one-to-one but not onto, onto but not one-to-one, or neither one-to-one nor onto, but this does not contradict our claim that $\mathbb{Z}^+ \sim \mathbb{Z}$. According to Definition 7.1.1, to show that $\mathbb{Z}^+ \sim \mathbb{Z}$ we need only show that there is at least one function from $\mathbb{Z}^+$ to $\mathbb{Z}$ that is both one-to-one and onto, and of course to prove this it suffices to give an example of such a function.

Perhaps an even more surprising example is that $\mathbb{Z}^+ \times \mathbb{Z}^+ \sim \mathbb{Z}^+$. To show this we must come up with a one-to-one, onto function $f : \mathbb{Z}^+ \times \mathbb{Z}^+ \to \mathbb{Z}^+$. An element of the domain of this function would be an ordered pair (i, j), where i and j are positive integers. The result of applying f to this pair should be written $f((i, j))$, but it is customary to leave out one pair of parentheses and just write $f(i, j)$. Exercise 12 asks you to show that the following formula defines a function from $\mathbb{Z}^+ \times \mathbb{Z}^+$ to $\mathbb{Z}^+$ that is one-to-one and onto:

$$f(i, j) = \frac{(i + j - 2)(i + j - 1)}{2} + i.$$

Once again, the table of values in Figure 2 may help you understand this example.

			j		
$f(i, j)$	1	2	3	4	5
1	1	2	4	7	11
2	3	5	8	12	
i 3	6	9	13		
4	10	14			...
5	15				

Figure 2

Theorem 7.1.2. *Suppose $A \sim B$ and $C \sim D$. Then:*

1. *$A \times C \sim B \times D$.*
2. *If A and C are disjoint and B and D are disjoint, then $A \cup C \sim B \cup D$.*

Proof. Since $A \sim B$ and $C \sim D$, we can choose functions $f : A \to B$ and $g : C \to D$ that are one-to-one and onto.

1. Define $h : A \times C \to B \times D$ by the formula

$$h(a, c) = (f(a), g(c)).$$

To see that h is one-to-one, suppose $h(a_1, c_1) = h(a_2, c_2)$. This means that $(f(a_1), g(c_1)) = (f(a_2), g(c_2))$, so $f(a_1) = f(a_2)$ and $g(c_1) = g(c_2)$.

Since f and g are both one-to-one, it follows that $a_1 = a_2$ and $c_1 = c_2$, so $(a_1, c_1) = (a_2, c_2)$.

To see that h is onto, suppose $(b, d) \in B \times D$. Then since f and g are both onto, we can choose $a \in A$ and $c \in C$ such that $f(a) = b$ and $g(c) = d$. Therefore $h(a, c) = (f(a), g(c)) = (b, d)$, as required. Thus h is one-to-one and onto, so $A \times C \sim B \times D$.

2. Suppose A and C are disjoint and B and D are disjoint. You are asked in exercise 13 to show that $f \cup g$ is a one-to-one, onto function from $A \cup C$ to $B \cup D$, so $A \cup C \sim B \cup D$. □

It is not hard to show that $\sim$ is reflexive, symmetric, and transitive, so it is an equivalence relation. In other words, we have the following theorem:

Theorem 7.1.3. *For any sets A, B, and C:*

1. $A \sim A$.
2. *If* $A \sim B$ *then* $B \sim A$.
3. *If* $A \sim B$ *and* $B \sim C$ *then* $A \sim C$.

Proof.

1. The identity function i_A is a one-to-one, onto function from A to A.
2. Suppose $A \sim B$. Then we can choose some function $f : A \to B$ that is one-to-one and onto. By Theorem 5.3.4, f^{-1} is a function from B to A. But now note that $(f^{-1})^{-1} = f$, which is a function from A to B, so by Theorem 5.3.4. again, f^{-1} is also one-to-one and onto. Therefore $B \sim A$.
3. Suppose $A \sim B$ and $B \sim C$. Then we can choose one-to-one, onto functions $f : A \to B$ and $g : B \to C$. By Theorem 5.2.5, $g \circ f : A \to C$ is one-to-one and onto, so $A \sim C$. □

Theorems 7.1.2 and 7.1.3 are often helpful in showing that sets are equinumerous. For example, we showed earlier that $\mathbb{Z}^+ \times \mathbb{Z}^+ \sim \mathbb{Z}^+$ and $\mathbb{Z}^+ \sim \mathbb{Z}$, so by part 3 of Theorem 7.1.3 it follows that $\mathbb{Z}^+ \times \mathbb{Z}^+ \sim \mathbb{Z}$. Part 2 tells us that we need not distinguish between the statements "A is equinumerous with B" and "B is equinumerous with A", because they are equivalent. For example, we already know that $\mathbb{Z}^+ \times \mathbb{Z}^+ \sim \mathbb{Z}^+$, so we can also write $\mathbb{Z}^+ \sim \mathbb{Z}^+ \times \mathbb{Z}^+$. By part 1 of Theorem 7.1.2, $\mathbb{Z}^+ \times \mathbb{Z}^+ \sim \mathbb{Z} \times \mathbb{Z}$, so we also have $\mathbb{Z}^+ \sim \mathbb{Z} \times \mathbb{Z}$.

We have now found three sets, $\mathbb{Z}$, $\mathbb{Z}^+ \times \mathbb{Z}^+$, and $\mathbb{Z} \times \mathbb{Z}$, that are equinumerous with $\mathbb{Z}^+$. Such sets are especially important and have a special name.

Definition 7.1.4. A set A is called *denumerable* if $\mathbb{Z}^+ \sim A$. It is called *countable* if it is either finite or denumerable. Otherwise, it is *uncountable*.

You might think of the countable sets as those sets whose elements can be *counted* by pointing to all of them, one by one, while naming positive integers in order. If the counting process ends at some point, then the set is finite; and if it never ends, then the set is denumerable. The following theorem gives two more ways of thinking about countable sets.

Theorem 7.1.5. *Suppose A is a set. The following statements are equivalent:*

1. *A is countable.*
2. *Either $A = \varnothing$ or there is a function $f : \mathbb{Z}^+ \to A$ that is onto.*
3. *There is a function $f : A \to \mathbb{Z}^+$ that is one-to-one.*

Proof. $1 \to 2$. Suppose A is countable. If A is denumerable, then there is a function $f : \mathbb{Z}^+ \to A$ that is one-to-one and onto, so clearly statement 2 is true. Now suppose A is finite. If $A = \varnothing$ then there is nothing more to prove, so suppose $A \neq \varnothing$. Then we can choose some element $a_0 \in A$. Let $g : I_n \to A$ be a one-to-one, onto function, where n is the number of elements of A. Now define $f : \mathbb{Z}^+ \to A$ as follows:

$$f(i) = \begin{cases} g(i) & \text{if } i \leq n \\ a_0 & \text{if } i > n. \end{cases}$$

It is easy to check now that f is onto, as required.

$2 \to 3$. Suppose that either $A = \varnothing$ or there is an onto function from $\mathbb{Z}^+$ to A. We consider these two possibilities in turn. If $A = \varnothing$, then the empty set is a one-to-one function from A to $\mathbb{Z}^+$. Now suppose $g : \mathbb{Z}^+ \to A$, and g is onto. Then for each $a \in A$, the set $\{n \in \mathbb{Z}^+ \mid g(n) = a\}$ is not empty, so by the well-ordering principle it must have a smallest element. Thus, we can define a function $f : A \to \mathbb{Z}^+$ by the formula

$$f(a) = \text{the smallest } n \in \mathbb{Z}^+ \text{ such that } g(n) = a.$$

Note that for each $a \in A$, $g(f(a)) = a$, so $g \circ f = i_A$. But then by Theorem 5.3.3, it follows that f is one-to-one, as required.

$3 \to 1$. Suppose $g : A \to \mathbb{Z}^+$ and g is one-to-one. Let $B = \text{Ran}(g) \subseteq \mathbb{Z}^+$. If we think of g as a function from A to B, then it is one-to-one and onto, so $A \sim B$. Thus, it suffices to show that B is countable, since by Theorem 7.1.3 it follows from this that A is also countable.

Suppose B is not finite. We must show that B is denumerable, which we can do by defining a one-to-one, onto function $f : \mathbb{Z}^+ \to B$. The idea behind the definition is simply to let $f(n)$ be the nth element of B, for each $n \in \mathbb{Z}^+$. (Recall that $B \subseteq \mathbb{Z}^+$, so we can use the ordering of the positive integers to make sense of the idea of the nth element of B.) For a more careful definition of f and the proof that f is one-to-one and onto, see exercise 14. □

If A is countable and $A \neq \varnothing$, then by Theorem 7.1.5 there is a function $f : \mathbb{Z}^+ \to A$ that is onto. If, for every $n \in \mathbb{Z}^+$, we let $a_n = f(n)$, then the fact that f is onto means that every element of A appears at least once in the list $a_1, a_2, a_3, \ldots$. In other words, $A = \{a_1, a_2, a_3, \ldots\}$. Countability of a set A is often used in this way to enable us to write the elements of A in a list, indexed by the positive integers. In fact, you might want to think of countability for nonempty sets as meaning *listability*. Of course, if A is denumerable, then the function f can be taken to be one-to-one, which means that each element of A will appear only once in the list $a_1, a_2, a_3, \ldots$. For an example of an application of countability in which the elements of a countable set are written in a list, see exercise 17.

Theorem 7.1.5 is also sometimes useful for proving that a set is denumerable, as the proof of our next theorem shows.

Theorem 7.1.6. *$\mathbb{Q}$ is denumerable.*

Proof. Let $f : \mathbb{Z} \times \mathbb{Z}^+ \to \mathbb{Q}$ be defined as follows:

$$f(p, q) = p/q.$$

Clearly f is onto, since by definition all rational numbers can be written as fractions, but note that f is not one-to-one. For example, $f(1, 2) = f(2, 4) = 1/2$. Since $\mathbb{Z}^+ \sim \mathbb{Z}$, by Theorem 7.1.2 we have $\mathbb{Z}^+ \times \mathbb{Z}^+ \sim \mathbb{Z} \times \mathbb{Z}^+$, and since we already know that $\mathbb{Z}^+ \times \mathbb{Z}^+$ is denumerable, it follows that $\mathbb{Z} \times \mathbb{Z}^+$ is also denumerable. Thus, we can choose a one-to-one, onto function $g : \mathbb{Z}^+ \to \mathbb{Z} \times \mathbb{Z}^+$. By Theorem 5.2.5, $f \circ g : \mathbb{Z}^+ \to \mathbb{Q}$ is onto, so by Theorem 7.1.5, $\mathbb{Q}$ is countable. Clearly $\mathbb{Q}$ is not finite, so it must be denumerable. □

Although our focus in this chapter is on infinite sets, the methods in this section can be used to prove theorems that are useful for computing the cardinalities of finite sets. We end this section with one example of such a theorem, and give several other examples in the exercises (see exercises 18–28).

Theorem 7.1.7. *Suppose A and B are disjoint finite sets. Then $A \cup B$ is finite, and $|A \cup B| = |A| + |B|$.*

Proof. Let $n = |A|$ and $m = |B|$. Then $A \sim I_n$ and $B \sim I_m$. Notice that if $x \in I_m$ then $1 \leq x \leq m$, and therefore $n + 1 \leq x + n \leq n + m$, so $x + n \in I_{n+m} \setminus I_n$. Thus we can define a function $f : I_m \to I_{n+m} \setminus I_n$ by the formula $f(x) = x + n$. It is easy to check that f is one-to-one and onto, so $I_m \sim I_{n+m} \setminus I_n$. Since $B \sim I_m$, it follows that $B \sim I_{n+m} \setminus I_n$. Applying part 2 of Theorem 7.1.2, we can conclude that $A \cup B \sim I_n \cup (I_{n+m} \setminus I_n) = I_{n+m}$. Therefore $A \cup B$ is finite, and $|A \cup B| = n + m = |A| + |B|$. □

Exercises

*1. Show that the following sets are denumerable.
   (a) $\mathbb{N}$.
   (b) The set of all even integers.

2. Show that the following sets are denumerable:
   (a) $\mathbb{Q} \times \mathbb{Q}$.
   (b) $\mathbb{Q}(\sqrt{2})$. (See exercise 6(b) of Section 6.5 for the meaning of the notation used here.)

3. In this problem we'll use the following notation for intervals of real numbers. If a and b are real numbers and $a < b$, then

$$[a, b] = \{x \in \mathbb{R} \mid a \leq x \leq b\}$$
$$(a, b) = \{x \in \mathbb{R} \mid a < x < b\}$$
$$(a, b] = \{x \in \mathbb{R} \mid a < x \leq b\}$$
$$[a, b) = \{x \in \mathbb{R} \mid a \leq x < b\}.$$

   (a) Show that $[0, 1] \sim [0, 2]$.
   (b) Show that $(-\pi/2, \pi/2) \sim \mathbb{R}$. (Hint: Use a trigonometric function.)
   (c) Show that $(0, 1) \sim \mathbb{R}$.
   (d) Show that $(0, 1] \sim (0, 1)$.

*4. Justify your answer to each question with either a proof or a counterexample.
   (a) Suppose $A \sim B$ and $A \times C \sim B \times D$. Must it be the case that $C \sim D$?
   (b) Suppose $A \sim B$, A and C are disjoint, B and D are disjoint, and $A \cup C \sim B \cup D$. Must it be the case that $C \sim D$?

5. Prove that if $A \sim B$ then $\mathscr{P}(A) \sim \mathscr{P}(B)$.

*6. (a) Prove that for all natural numbers n and m, if $I_n \sim I_m$ then $n = m$. (Hint: Use induction on n.)
   (b) Prove that if A is finite, then there is exactly one natural number n such that $I_n \sim A$.

7. Suppose A and B are sets and A is finite. Prove that $A \sim B$ iff B is also finite and $|A| = |B|$.

*8. (a) Prove that if $n \in \mathbb{N}$ and $A \subseteq I_n$, then A is finite and $|A| \leq n$. Furthermore, if $A \neq I_n$, then $|A| < n$.
   (b) Prove that if A is finite and $B \subseteq A$, then B is also finite, and $|B| \leq |A|$. Furthermore, if $B \neq A$, then $|B| < |A|$.

9. Suppose $B \subseteq A$, $B \neq A$, and $B \sim A$. Prove that A is infinite.

10. Prove that if $n \in \mathbb{N}$, $f : I_n \to B$, and f is onto, then B is finite and $|B| \leq n$.

11. Suppose A and B are finite sets and $f : A \to B$.
   (a) Prove that if $|A| < |B|$ then f is not onto.
   (b) Prove that if $|A| > |B|$ then f is not one-to-one. (This is sometimes called the *Pigeonhole Principle*, because it means that if n pigeons are put into m pigeonholes, where $n > m$, then some pigeonhole must contain more than one pigeon.)
   (c) Prove that if $|A| = |B|$ then f is one-to-one iff f is onto.

*12. Show that the function $f : \mathbb{Z}^+ \times \mathbb{Z}^+ \to \mathbb{Z}^+$ defined by the formula

$$f(i, j) = \frac{(i + j - 2)(i + j - 1)}{2} + i$$

is one-to-one and onto.

13. Complete the proof of part 2 of Theorem 7.1.2 by showing that if $f : A \to B$ and $g : C \to D$ are one-to-one, onto functions, A and C are disjoint, and B and D are disjoint, then $f \cup g$ is a one-to-one, onto function from $A \cup C$ to $B \cup D$.

*14. In this exercise you will complete the proof of $3 \to 1$ of Theorem 7.1.5. Suppose $B \subseteq \mathbb{Z}^+$ and B is infinite. We now define a function $f : \mathbb{Z}^+ \to B$ by recursion as follows:

For all $n \in \mathbb{Z}^+$,

$$f(n) = \text{the smallest element of } B \setminus \{f(m) \mid m \in \mathbb{Z}^+, m < n\}.$$

Of course, the definition is recursive because the specification of $f(n)$ refers to $f(m)$ for all $m < n$.
   (a) Suppose $n \in \mathbb{Z}^+$. The definition of $f(n)$ only makes sense if we can be sure that $B \setminus \{f(m) \mid m \in \mathbb{Z}^+, m < n\} \neq \varnothing$, in which case the well-ordering principle guarantees that it has a smallest element. Prove that $B \setminus \{f(m) \mid m \in \mathbb{Z}^+, m < n\} \neq \varnothing$. (Hint: See exercise 10.)
   (b) Prove that for all $n \in \mathbb{Z}^+$, $f(n) \geq n$.
   (c) Prove that f is one-to-one and onto.

15. Prove that if $B \subseteq A$ and A is countable, then B is countable.

16. Prove that if $B \subseteq A$, A is infinite, and B is finite, then $A \setminus B$ is infinite.

*17. Suppose A is denumerable and R is a partial order on A. Prove that R can be extended to a total order on A. In other words, prove that there is a total order T on A such that $R \subseteq T$. Note that we proved a similar theorem for finite A in Example 6.2.2. (Hint: Since A is denumerable, we can write the elements of A in a list: $A = \{a_1, a_2, a_3, \ldots\}$. Now, using exercise 2 of Section 6.2, recursively define partial orders R_n, for $n \in \mathbb{N}$, so that $R = R_0 \subseteq R_1 \subseteq R_2 \subseteq \ldots$ and $\forall i \in I_n \forall j \in \mathbb{Z}^+((a_i, a_j) \in R_n \vee (a_j, a_i) \in R_n)$. Let $T = \cup_{n \in \mathbb{N}} R_n$.)

18. Suppose A is finite and $B \subseteq A$. By exercise 8, B and $A \setminus B$ are both finite. Prove that $|A \setminus B| = |A| - |B|$. (In particular, if $a \in A$ then $|A \setminus \{a\}| = |A| - 1$. We used this fact in several proofs in Chapter 6; for example, we used it in Examples 6.2.1 and 6.2.2.)

19. Suppose n is a positive integer and for each $i \in I_n$, A_i is a finite set. Also, assume that $\forall i \in I_n \forall j \in I_n(i \neq j \rightarrow A_i \cap A_j = \varnothing)$. Prove that $\cup_{i \in I_n} A_i$ is finite and $|\cup_{i \in I_n} A_i| = \sum_{i=1}^{n} |A_i|$.

*20. (a) Prove that if A and B are finite sets, then $A \times B$ is finite and $|A \times B| = |A| \cdot |B|$. (Hint: Use induction on $|B|$. In other words, prove the following statement by induction: $\forall n \in \mathbb{N} \forall A \forall B$(if A and B are finite and $|B| = n$, then $A \times B$ is finite and $|A \times B| = |A| \cdot n$). You may find Theorem 4.1.3 useful.)

(b) A meal at Alice's Restaurant consists of an entree and a dessert. The entree can be either steak, chicken, pork chops, shrimp, or spaghetti, and dessert can be either ice cream, cake, or pie. How many different meals can you order at Alice's Restaurant?

21. For any sets A and B, the set of all functions from A to B is denoted ${}^{A}B$.

(a) Prove that if $A \sim B$ and $C \sim D$ then ${}^{A}C \sim {}^{B}D$.

(b) Prove that if A, B, and C are sets and $A \cap B = \varnothing$, then ${}^{A \cup B}C \sim {}^{A}C \times {}^{B}C$.

(c) Prove that if A and B are finite sets, then ${}^{A}B$ is finite and $|{}^{A}B| = |B|^{|A|}$. (Hint: Use induction on $|A|$.)

(d) A professor has 20 students in his class, and he has to assign a grade of either A, B, C, D, or F to each student. In how many ways can the grades be assigned?

*22. Suppose $|A| = n$, and let $F = \{f \mid f$ is a one-to-one, onto function from I_n to $A\}$.

(a) Prove that F is finite, and $|F| = n!$. (Hint: Use induction on n.)

(b) Let $L = \{R \mid R$ is a total order on $A\}$. Prove that $F \sim L$, and therefore $|L| = n!$.

(c) Five people are to sit in a row of five seats. In how many ways can they be seated?

23. Suppose A is a finite set and R is an equivalence relation on A. Suppose also that there is some positive integer n such that $\forall x \in A(|[x]_R| = n)$. Prove that A/R is finite and $|A/R| = |A|/n$. (Hint: Use exercise 19.)
24. (a) Suppose that A and B are finite sets. Prove that $A \cup B$ is finite, and $|A \cup B| = |A| + |B| - |A \cap B|$.
    (b) Suppose that A, B, and C are finite sets. Prove that $A \cup B \cup C$ is finite, and $|A \cup B \cup C| = |A| + |B| + |C| - |A \cap B| - |A \cap C| - |B \cap C| + |A \cap B \cap C|$.

*25. In this problem you will prove the *Inclusion–Exclusion Principle*, which generalizes the formulas in exercise 24. Suppose $A_1, A_2, \ldots, A_n$ are finite sets. Let $P = \mathscr{P}(I_n) \setminus \{\varnothing\}$, and for each $S \in P$ let $A_S = \bigcap_{i \in S} A_i$. Prove that $\bigcup_{i \in I_n} A_i$ is finite and $\left|\bigcup_{i \in I_n} A_i\right| = \sum_{S \in P} (-1)^{|S|+1} |A_S|$. (The notation on the right side of this equation denotes the result of running through all sets $S \in P$, computing the number $(-1)^{|S|+1}|A_S|$ for each S, and then adding these numbers. Hint: Use induction on n.)

26. Prove that if A and B are finite sets and $|A| = |B|$, then $|A \triangle B|$ is even.
27. Each customer in a certain bank has a PIN number, which is a sequence of four digits. Show that if the bank has more than 10,000 customers, then some two customers must have the same PIN number. (Hint: See exercise 11.)
28. Alice opened her grade report and exclaimed, "I can't believe Professor Jones flunked me in Probability." "You were in that course?" said Bob. "That's funny, I was in it too, and I don't remember ever seeing you there." "Well," admitted Alice sheepishly, "I guess I did skip class a lot." "Yeah, me too" said Bob. Prove that either Alice or Bob missed at least half of the classes.

7.2. Countable and Uncountable Sets

Often when we perform some set-theoretic operation with countable sets, the result is again a countable set.

Theorem 7.2.1. *Suppose A and B are countable sets. Then:*

1. *$A \times B$ is countable.*
2. *$A \cup B$ is countable.*

Proof. Since A and B are countable, by Theorem 7.1.5 we can choose one-to-one functions $f : A \to \mathbb{Z}^+$ and $g : B \to \mathbb{Z}^+$.

1. Define $h : A \times B \to \mathbb{Z}^+ \times \mathbb{Z}^+$ by the formula

$$h(a, b) = (f(a), g(b)).$$

As in the proof of part 1 Theorem 7.1.2, it is not hard to show that h is one-to-one. Since $\mathbb{Z}^+ \times \mathbb{Z}^+$ is denumerable, we can let $j : \mathbb{Z}^+ \times \mathbb{Z}^+ \to \mathbb{Z}^+$ be a one-to-one, onto function. Then by Theorem 5.2.5, $j \circ h : A \times B \to \mathbb{Z}^+$ is one-to-one, so by Theorem 7.1.5, $A \times B$ is countable.

2. Define $h : A \cup B \to \mathbb{Z}$ as follows:

$$h(x) = \begin{cases} f(x) & \text{if } x \in A \\ -g(x) & \text{if } x \notin A. \end{cases}$$

We claim now that h is one-to-one. To see why, suppose that $h(x_1) = h(x_2)$, for some x_1 and x_2 in $A \cup B$. If $h(x_1) = h(x_2) > 0$, then according to the definition of h, we must have $x_1 \in A$, $x_2 \in A$, and $f(x_1) = h(x_1) = h(x_2) = f(x_2)$. But then since f is one-to-one, $x_1 = x_2$. Similarly, if $h(x_1) = h(x_2) \leq 0$, then we must have $g(x_1) = -h(x_1) = -h(x_2) = g(x_2)$, and then since g is one-to-one, $x_1 = x_2$. Thus, h is one-to-one.

Since $\mathbb{Z}$ is denumerable, we can let $j : \mathbb{Z} \to \mathbb{Z}^+$ be a one-to-one, onto function. As in part 1, we then find that $j \circ h : A \cup B \to \mathbb{Z}^+$ is one-to-one, so $A \cup B$ is countable.

□

As our next theorem shows, part 2 of Theorem 7.2.1 can be extended to unions of more than two sets.

Theorem 7.2.2. *The union of countably many countable sets is countable. In other words, if $\mathcal{F}$ is a family of sets, $\mathcal{F}$ is countable, and also every element of $\mathcal{F}$ is countable, then $\cup\mathcal{F}$ is countable.*

Proof. We will assume first that $\varnothing \notin \mathcal{F}$. At the end of the proof we will discuss the case $\varnothing \in \mathcal{F}$.

If $\mathcal{F} = \varnothing$, then of course $\cup\mathcal{F} = \varnothing$, which is countable. Now suppose $\mathcal{F} \neq \varnothing$. Then, as described after the proof of Theorem 7.1.5, since $\mathcal{F}$ is countable and nonempty we can write the elements of $\mathcal{F}$ in a list, indexed by the positive integers. In other words, we can say that $\mathcal{F} = \{A_1, A_2, A_3, \ldots\}$. Similarly, every element of $\mathcal{F}$ is countable and nonempty (since $\varnothing \notin \mathcal{F}$), so for each positive integer i the elements of A_i can be written in a list. Thus we can write

$$A_1 = \{a_1^1, a_2^1, a_3^1, \ldots\},$$
$$A_2 = \{a_1^2, a_2^2, a_3^2, \ldots\},$$

and, in general,

$$A_i = \{a_1^i, a_2^i, a_3^i, \ldots\}.$$

Note that, by the definition of union, $\cup\mathcal{F} = \{a_j^i \mid i \in \mathbb{Z}^+, j \in \mathbb{Z}^+\}$.

Now define a function $f : \mathbb{Z}^+ \times \mathbb{Z}^+ \to \cup\mathcal{F}$ by the formula

$$f(i, j) = a_j^i \ .$$

Clearly f is onto. Since $\mathbb{Z}^+ \times \mathbb{Z}^+$ is denumerable, we can let $g : \mathbb{Z}^+ \to \mathbb{Z}^+ \times \mathbb{Z}^+$ be a one-to-one, onto function. Then $f \circ g : \mathbb{Z}^+ \to \cup\mathcal{F}$ is onto, so $\cup\mathcal{F}$ is countable.

Finally, suppose $\varnothing \in \mathcal{F}$. Let $\mathcal{F}' = \mathcal{F} \setminus \{\varnothing\}$. Then $\mathcal{F}'$ is also a countable family of countable sets and $\varnothing \notin \mathcal{F}'$, so by the earlier reasoning, $\cup\mathcal{F}'$ is countable. But clearly $\cup\mathcal{F} = \cup\mathcal{F}'$, so $\cup\mathcal{F}$ is countable too. □

Another operation that preserves countability is the formation of finite sequences. Suppose A is a set and $a_1, a_2, \ldots, a_n$ is a list of elements of A. We might specify the terms in this list with a function $f : I_n \to A$, where for each i, $f(i) = a_i =$ the ith term in the list. Such a function is called a *finite sequence* of elements of A.

Definition 7.2.3. Suppose A is a set. A function $f : I_n \to A$, where n is a natural number, is called a *finite sequence* of elements of A, and n is called the *length* of the sequence.

Theorem 7.2.4. *Suppose A is a countable set. Then the set of all finite sequences of elements of A is also countable.*

Proof. For each $n \in \mathbb{N}$, let S_n be the set of all sequences of length n of elements of A. We first show that for every $n \in \mathbb{N}$, S_n is countable. We proceed by induction on n.

In the base case we assume $n = 0$. Note that $I_0 = \varnothing$, so a sequence of length 0 is a function $f : \varnothing \to A$, and the only such function is $\varnothing$. Thus, $S_0 = \{\varnothing\}$, which is clearly a countable set.

For the induction step, suppose n is a natural number and S_n is countable. We must show that S_{n+1} is countable. Consider the function $F : S_n \times A \to S_{n+1}$ defined as follows:

$$F(f, a) = f \cup \{(n + 1, a)\} \ .$$

In other words, for any sequence $f \in S_n$ and any element $a \in A$, $F(f, a)$ is the sequence you get by starting with f, which is a sequence a length n, and then tacking on a as term number $n + 1$. You are asked in exercise 2 to verify that F

is one-to-one and onto. Thus, $S_n \times A \sim S_{n+1}$. But S_n and A are both countable, so by Theorem 7.2.1 $S_n \times A$ is countable, and therefore S_{n+1} is countable.

This completes the inductive proof that for every $n \in \mathbb{N}$, S_n is countable. Finally, note that the set of all finite sequences of elements of A is $\cup_{n \in \mathbb{N}} S_n$, and this is countable by Theorem 7.2.2. □

As an example of the use of Theorem 7.2.4, you should be able to show that the set of all grammatical sentences of English is a denumerable set. (See exercise 10.)

By now you may be wondering if perhaps *all* sets are countable! Is there any set-theoretic operation that can be used to produce uncountable sets? We'll see in our next theorem that the answer is yes, the power set operation. This fact was discovered by the German mathematician Georg Cantor (1845–1918) by means of a famous and ingenious proof. In fact, it was Cantor who first conceived of the idea of comparing the sizes of infinite sets. Important mathematical theorems are often named after their discoverers, so we have identified Theorem 7.2.5 as *Cantor's theorem*. Cantor's proof is somewhat harder than the previous proofs in this chapter, so we'll discuss the strategy behind the proof before presenting the proof itself.

Theorem 7.2.5. *(Cantor's theorem) $\mathscr{P}(\mathbb{Z}^+)$ is uncountable.*

Scratch work

The proof is based on statement 2 of Theorem 7.1.5. We'll show that there is no function $f : \mathbb{Z}^+ \to \mathscr{P}(\mathbb{Z}^+)$ that is onto. Clearly $\mathscr{P}(\mathbb{Z}^+) \neq \varnothing$, so by Theorem 7.1.5 this shows that $\mathscr{P}(\mathbb{Z}^+)$ is not countable.

Our strategy will be to let $f : \mathbb{Z}^+ \to \mathscr{P}(\mathbb{Z}^+)$ be an arbitrary function and prove that f is not onto. Reexpressing this negative goal as a positive statement, we must show that $\exists D[D \in \mathscr{P}(\mathbb{Z}^+) \wedge \forall n \in \mathbb{Z}^+(D \neq f(n))]$. This suggests that we should try to find a particular set D for which we can prove both $D \in \mathscr{P}(\mathbb{Z}^+)$ and $\forall n \in \mathbb{Z}^+(D \neq f(n))$. This is the most difficult step in figuring out the proof. There is a set D that makes the proof work, but it will take some cleverness to come up with it.

We want to make sure that $D \in \mathscr{P}(\mathbb{Z}^+)$, or in other words $D \subseteq \mathbb{Z}^+$, so we know that we need only consider positive integers when deciding what the elements of D should be. But this still leaves us infinitely many decisions to make: For each positive integer n, we must decide whether or not we want n to be an element of D. We also need to make sure that $\forall n \in \mathbb{Z}^+(D \neq f(n))$. This imposes infinitely many restrictions on our choice of D: For each positive integer n, we must make sure that $D \neq f(n)$. Why not make each of our infinitely many decisions in such a way that it guarantees that the corresponding restriction is satisfied? In other words, for each positive integer n, we'll make our decision about whether or not n is an element of D in such a way that it will guarantee that $D \neq f(n)$. This isn't hard to do. We can let n be an element of D if $n \notin f(n)$, and leave n out of D if $n \in f(n)$. This will guarantee that $D \neq f(n)$, because one of these sets will contain n as an element and the other won't. This suggests that we should let $D = \{n \in \mathbb{Z}^+ \mid n \notin f(n)\}$.

Figure 1 may help you understand the definition of the set D. For each $m \in \mathbb{Z}^+$, $f(m)$ is a subset of $\mathbb{Z}^+$, which can be specified by saying, for each positive integer n, whether or not $n \in f(m)$. The answers to these questions can be arranged in a table as shown in Figure 1. Each row of the table gives the answers needed to specify the set $f(m)$ for a particular value of m. The set D can also be specified with a row of yesses and noes, as shown at the bottom of Figure 1. For each $n \in \mathbb{Z}^+$ we've decided to determine whether or not $n \in D$ by asking whether or not $n \in f(n)$, and the answers to these questions are the ones surrounded by boxes in Figure 1. Because $n \in D$ iff $n \notin f(n)$, the row of yesses and noes that specifies D can be found by reading the boxed answers along the diagonal of Figure 1, and reversing all the answers. This is guaranteed to be different from every row of the table in Figure 1, because for each $n \in \mathbb{Z}^+$ it differs from row n in the nth position.

Is $n \in f(m)$?		n = 1	2	3	4	5	
	1	[yes]	no	no	yes	yes	
	2	yes	[yes]	no	no	yes	
m	3	no	no	[no]	yes	no	$\cdots$
	4	yes	yes	no	[yes]	no	
	5	no	yes	yes	no	[no]	
				$\vdots$			
Is $n \in D$?		no	no	yes	no	yes	$\cdots$

Figure 1

If you found this reasoning difficult to follow, don't worry about it. Remember, the reasoning used in choosing the set D won't be part of the proof anyway! After you finish reading the proof, you can go back and try reading the last two paragraphs again.

It should be clear that the set D we have chosen is a subset of $\mathbb{Z}^+$, so $D \in \mathscr{P}(\mathbb{Z}^+)$. Our other goal is to prove that $\forall n \in \mathbb{Z}^+(D \neq f(n))$, so we let n

be an arbitrary positive integer and prove $D \neq f(n)$. Now recall that we chose D carefully so that we would be able to prove $D \neq f(n)$, and the reasoning behind this choice hinged on whether or not $n \in f(n)$. Perhaps the easiest way to write the proof is to consider the two cases $n \in f(n)$ and $n \notin f(n)$ separately. In each case, applying the definition of D easily leads to the conclusion that $D \neq f(n)$.

Proof. Suppose $f : \mathbb{Z}^+ \to \mathscr{P}(\mathbb{Z}^+)$. We will show that f cannot be onto by finding a set $D \in \mathscr{P}(\mathbb{Z}^+)$ such that $D \notin \text{Ran}(f)$. Let $D = \{n \in \mathbb{Z}^+ \mid n \notin f(n)\}$. Clearly $D \subseteq \mathbb{Z}^+$, so $D \in \mathscr{P}(\mathbb{Z}^+)$. Now let n be an arbitrary positive integer. We consider two cases.

Case 1. $n \in f(n)$. Since $D = \{n \in \mathbb{Z}^+ \mid n \notin f(n)\}$, we can conclude that $n \notin D$. But then since $n \in f(n)$ and $n \notin D$, it follows that $D \neq f(n)$.

Case 2. $n \notin f(n)$. Then by the definition of D, $n \in D$. Since $n \in D$ and $n \notin f(n)$, $D \neq f(n)$.

Since these cases are exhaustive, this shows that $\forall n \in \mathbb{Z}^+(D \neq f(n))$, so $D \notin \text{Ran}(f)$. Since f was arbitrary, this shows that there is no onto function $f : \mathbb{Z}^+ \to \mathscr{P}(\mathbb{Z}^+)$. Clearly $\mathscr{P}(\mathbb{Z}^+) \neq \varnothing$, so by Theorem 7.1.5, $\mathscr{P}(\mathbb{Z}^+)$ is uncountable. □

The method used in the proof of Theorem 7.2.5 is called *diagonalization* because of the diagonal arrangement of the boxed answers in Figure 1. Diagonalization is a powerful technique that can be used to prove many theorems, including our next theorem. However, rather than doing another diagonalization argument, we'll simply apply Theorem 7.2.5 to prove the next theorem.

Theorem 7.2.6. $\mathbb{R}$ *is uncountable.*

Proof. We will define a function $f : \mathscr{P}(\mathbb{Z}^+) \to \mathbb{R}$ and show that f is one-to-one. If $\mathbb{R}$ were countable, then there would be a one-to-one function $g : \mathbb{R} \to \mathbb{Z}^+$. But then $g \circ f$ would be a one-to-one function from $\mathscr{P}(\mathbb{Z}^+)$ to $\mathbb{Z}^+$ and therefore $\mathscr{P}(\mathbb{Z}^+)$ would be countable, contradicting Cantor's theorem. Thus, this will show that $\mathbb{R}$ is uncountable.

To define f, suppose $A \in \mathscr{P}(\mathbb{Z}^+)$. Then $f(A)$ will be a real number between 0 and 1 that we will specify by giving its decimal expansion. For each positive integer n, the nth digit of $f(A)$ will be the number d_n defined as follows:

$$d_n = \begin{cases} 3 & \text{if } n \notin A \\ 7 & \text{if } n \in A. \end{cases}$$

In other words, in decimal notation we have $f(A) = 0.d_1d_2d_3\ldots$. For example, if E is the set of all positive even integers, then $f(E) = 0.37373737\ldots$. If P is the set of all prime numbers, then $f(P) = 0.37737373337\ldots$.

To see that f is one-to-one, suppose that $A \in \mathscr{P}(\mathbb{Z}^+)$, $B \in \mathscr{P}(\mathbb{Z}^+)$, and $A \neq B$. Then there is some $n \in \mathbb{Z}^+$ such that either $n \in A$ and $n \notin B$, or $n \in B$ and $n \notin A$. But then $f(A)$ and $f(B)$ cannot be equal, since their decimal expansions differ in the nth digit. Thus, f is one-to-one.

Exercises

*1. (a) Prove that the set of all irrational numbers, $\mathbb{R} \setminus \mathbb{Q}$, is uncountable.
 (b) Prove that $\mathbb{R} \setminus \mathbb{Q} \sim \mathbb{R}$.

2. Let $F : S_n \times A \to S_{n+1}$ be the function defined in the proof of Theorem 7.2.4. Show that F is one-to-one and onto.

3. Let $P = \{X \in \mathscr{P}(\mathbb{Z}^+) \mid X \text{ is finite}\}$. Prove that P is denumerable.

*4. Prove the following more general form of Cantor's theorem: For any set A, $A \not\sim \mathscr{P}(A)$. (Hint: Imitate the proof of Theorem 7.2.5.)

5. For the meaning of the notation used in this exercise, see exercise 21 of Section 7.1.
 (a) Prove that for any sets A, B, and C, ${}^A(B \times C) \sim {}^AB \times {}^AC$.
 (b) Prove that for any sets A, B, and C, ${}^{(A\times B)}C \sim {}^A({}^BC)$.
 (c) Prove that for any set A, $\mathscr{P}(A) \sim {}^A\{\text{yes, no}\}$. (Note that if A is finite and $|A| = n$ then, by exercise 21(c) of Section 7.1, it follows that $|\mathscr{P}(A)| = |\{\text{yes, no}\}|^{|A|} = 2^n$. Of course, you already proved this, by a different method, in exercise 10 of Section 6.2.)
 (d) Prove that ${}^{\mathbb{Z}^+}\mathscr{P}(\mathbb{Z}^+) \sim \mathscr{P}(\mathbb{Z}^+)$.

6. Suppose A is denumerable. Prove that there is a partition P of A such that P is denumerable and for every $X \in P$, X is denumerable.

*7. Prove that if A and B are disjoint sets, then $\mathscr{P}(A \cup B) \sim \mathscr{P}(A) \times \mathscr{P}(B)$.

8. Suppose $A \subseteq \mathbb{R}^+$, $b \in \mathbb{R}^+$, and for every list $a_1, a_2, \ldots, a_n$ of finitely many distinct elements of A, $a_1 + a_2 + \cdots + a_n \leq b$. Prove that A is countable. (Hint: For each positive integer n, let $A_n = \{x \in A \mid x \geq 1/n\}$. What can you say about the number of elements in A_n?)

*9. Suppose $\mathcal{F} \subseteq \{f \mid f : \mathbb{Z}^+ \to \mathbb{R}\}$ and $\mathcal{F}$ is countable. Prove that there is a function $g : \mathbb{Z}^+ \to \mathbb{R}$ such that $\mathcal{F} \subseteq O(g)$. (See exercise 16 of Section 5.1 for the meaning of the notation used here.)

10. Prove that the set of all grammatical sentences of English is denumerable. (Hint: Every grammatical sentence of English is a finite sequence

of English words. First show that the set of all grammatical sentences is countable, and then show that it is infinite.)

11. Some real numbers can be defined by a phrase in the English language. For example, the phrase "the ratio of the circumference of a circle to its diameter" defines the number π.
    (a) Prove that the set of numbers that can be defined by an English phrase is denumerable. (Hint: See exercise 10.)
    (b) Prove that there are real numbers that cannot be defined by English phrases.

7.3. The Cantor–Schröder–Bernstein Theorem

Suppose A and B are sets and f is a one-to-one function from A to B. Then f shows that $A \sim \text{Ran}(f) \subseteq B$, so it is natural to think of B as being *at least as large as* A. This suggests the following notation:

Definition 7.3.1. If A and B are sets, then we will say that B *dominates* A, and write $A \precsim B$, if there is a function $f : A \to B$ that is one-to-one. If $A \precsim B$ and $A \nsim B$, then we say that B *strictly dominates* A, and write $A \prec B$.

For example, in the proof of Theorem 7.2.6 we gave a one-to-one function $f : \mathscr{P}(\mathbb{Z}^+) \to \mathbb{R}$, so $\mathscr{P}(\mathbb{Z}^+) \precsim \mathbb{R}$. Of course, for any sets A and B, if $A \sim B$ then also $A \precsim B$. It should also be clear that if $A \subseteq B$ then $A \precsim B$. For example, $\mathbb{Z}^+ \precsim \mathbb{R}$. In fact, by Theorem 7.2.6 we also know that $\mathbb{Z}^+ \nsim \mathbb{R}$, so we can say that $\mathbb{Z}^+ \prec \mathbb{R}$.

You might think that $\precsim$ would be a partial order, but it turns out that it isn't. You're asked in exercise 1 to check that $\precsim$ is reflexive and transitive, but it is not antisymmetric. (In the terminology of exercise 24 of Section 4.6, $\precsim$ is a preorder.) For example, $\mathbb{Z}^+ \sim \mathbb{Q}$, so $\mathbb{Z}^+ \precsim \mathbb{Q}$ and $\mathbb{Q} \precsim \mathbb{Z}^+$, but of course $\mathbb{Z}^+ \neq \mathbb{Q}$. But this suggests an interesting question: If $A \precsim B$ and $B \precsim A$, then A and B might not be equal, but must they be equinumerous?

The answer, it turns out, is yes, as we'll prove in our next theorem. Several mathematicians' names are usually associated with this theorem. Cantor proved a limited version of the theorem, and later Ernst Schröder (1841–1902) and Felix Bernstein (1878–1956) discovered proofs independently.

Theorem 7.3.2. (Cantor–Schröder–Bernstein theorem) *Suppose A and B are sets. If $A \precsim B$ and $B \precsim A$, then $A \sim B$.*

Scratch work

We start by assuming that $A \precsim B$ and $B \precsim A$, which means that we can choose one-to-one functions $f : A \to B$ and $g : B \to A$. To prove that $A \to B$ we need to find a one-to-one, onto function $h : A \to B$.

At this point, we don't know much about A and B. The only tools we have to help us match up the elements of A and B are the functions f and g. If f is onto, then of course we can let $h = f$; and if g is onto, then we can let $h = g^{-1}$. But it may turn out that neither f nor g is onto. How can we come up with the required function h in this case?

Our solution will be to combine parts of f and g^{-1} to get h. To do this, we'll split A into two pieces X and Y, and B into two pieces W and Z, in such a way that X and W can be matched up by f, and Y and Z can be matched up by g. More precisely, we'll have $W = f(X) = \{f(x) \mid x \in X\}$ and $Y = g(Z) = \{g(z) \mid z \in Z\}$. The situation is illustrated in Figure 1. Once we have this, we'll be able to define h by letting $h(a) = f(a)$ for $a \in X$, and $h(a) = g^{-1}(a)$ for $a \in Y$.

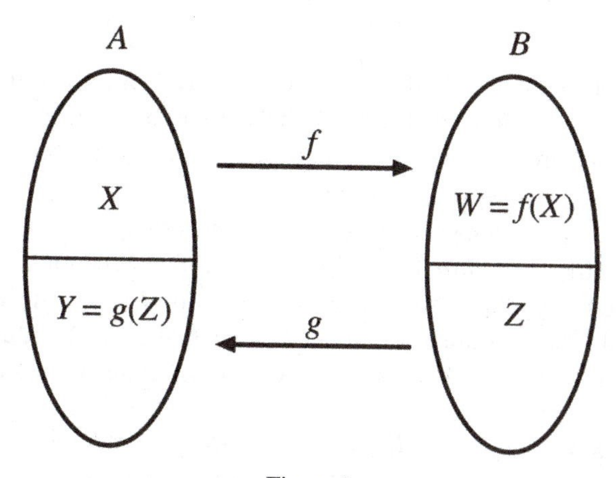

Figure 1

How can we choose the sets X, Y, W, and Z? First of all, note that every element of Y must be in $\text{Ran}(g)$, so any element of A that is not in $\text{Ran}(g)$ must be in X. In other words, if we let $A_1 = A \setminus \text{Ran}(g)$, then we must have $A_1 \subseteq X$. But now consider any $a \in A_1$. We know that we must have $a \in X$, and therefore $f(a) \in W$. But now note that since g is one-to-one, $g(f(a))$ will be different from $g(z)$ for every $z \in Z$, and therefore $g(f(a)) \notin g(Z) = Y$. Thus, we must have $g(f(a)) \in X$. Since a was an arbitrary element of A_1, this shows that if we let $A_2 = g(f(A_1)) = \{g(f(a)) \mid a \in A_1\}$, then we must have $A_2 \subseteq X$. Similarly, if we let $A_3 = g(f(A_2))$, then it will turn out that we must have $A_3 \subseteq X$. Continuing in this way we can define sets A_n for every positive integer n, and for every n we must have $A_n \subseteq X$. As you will see, letting $X = \bigcup_{n \in \mathbb{Z}^+} A_n$ works. In the following proof, we actually do not mention the sets W and Z.

Proof. Suppose $A \precsim B$ and $B \precsim A$. Then we can choose one-to-one functions $f : A \to B$ and $g : B \to A$. Let $R = \text{Ran}(g) \subseteq A$. Then if we think of g as a function from B to R, it is one-to-one and onto, so by Theorem 5.3.4, $g^{-1} : R \to B$.

We now define a sequence of sets $A_1, A_2, A_3, \ldots$ by recursion as follows:

$$A_1 = A \setminus R;$$
$$\text{for every } n \in \mathbb{Z}^+,\ A_{n+1} = g(f(A_n)) = \{g(f(a)) \mid a \in A_n\}.$$

Let $X = \bigcup_{n \in \mathbb{Z}^+} A_n$ and $Y = A \setminus X$. Of course, every element of A is in either X or Y, but not both. Now define $h : A \to B$ as follows:

$$h(a) = \begin{cases} f(a) & \text{if } a \in X \\ g^{-1}(a) & \text{if } a \in Y. \end{cases}$$

Note that for every $a \in A$, if $a \notin R$ then $a \in A_1 \subseteq X$. Thus, if $a \in Y$ then $a \in R$, so $g^{-1}(a)$ is defined. Therefore this definition makes sense.

We will show that h is one-to-one and onto, which will establish that $A \sim B$. To see that h is one-to-one, suppose $a_1 \in A$, $a_2 \in A$, and $h(a_1) = h(a_2)$.

Case 1. $a_1 \in X$. Suppose $a_2 \in Y$. Then according to the definition of h, $h(a_1) = f(a_1)$ and $h(a_2) = g^{-1}(a_2)$. Thus, the equation $h(a_1) = h(a_2)$ means $f(a_1) = g^{-1}(a_2)$, so $g(f(a_1)) = g(g^{-1}(a_2)) = a_2$. Since $a_1 \in X = \bigcup_{n \in \mathbb{Z}^+} A_n$, we can choose some $n \in \mathbb{Z}^+$ such that $a_1 \in A_n$. But then $a_2 = g(f(a_1)) \in g(f(A_n)) = A_{n+1}$, so $a_2 \in X$, contradicting our assumption that $a_2 \in Y$.

Thus, $a_2 \notin Y$, so $a_2 \in X$. This means that $h(a_2) = f(a_2)$, so from the equation $h(a_1) = h(a_2)$ we get $f(a_1) = f(a_2)$. But f is one-to-one, so it follows that $a_1 = a_2$.

Case 2. $a_1 \in Y$. As in case 1, if $a_2 \in X$, then we can derive a contradiction, so we must have $a_2 \in Y$. Thus, the equation $h(a_1) = h(a_2)$ means $g^{-1}(a_1) = g^{-1}(a_2)$. Therefore, $a_1 = g(g^{-1}(a_1)) = g(g^{-1}(a_2)) = a_2$.

In both cases we have $a_1 = a_2$, so h is one-to-one.

To see that h is onto, suppose $b \in B$. Then $g(b) \in A$, so either $g(b) \in X$ or $g(b) \in Y$.

Case 1. $g(b) \in X$. Choose n such that $g(b) \in A_n$. Note that $g(b) \in \text{Ran}(g) = R$ and $A_1 = A \setminus R$, so $g(b) \notin A_1$. Thus, $n > 1$, so $A_n = g(f(A_{n-1}))$, and therefore we can choose some $a \in A_{n-1}$ such that $g(f(a)) = g(b)$. But then since

g is one-to-one, $f(a) = b$. Since $a \in A_{n-1}$, $a \in X$, so $h(a) = f(a) = b$. Thus, $b \in \text{Ran}(h)$.

Case 2. $g(b) \in Y$. Then $h(g(b)) = g^{-1}(g(b)) = b$, so $b \in \text{Ran}(h)$.

In both cases we have $b \in \text{Ran}(h)$, so h is onto. □

The Cantor–Schröder–Bernstein theorem is often useful for showing that sets are equinumerous. For example, in exercise 3 of Section 7.1 you were asked to show that $(0, 1] \sim (0, 1)$, where

$$(0, 1] = \{x \in \mathbb{R} \mid 0 < x \leq 1\}$$

and

$$(0, 1) = \{x \in \mathbb{R} \mid 0 < x < 1\}.$$

It is surprisingly difficult to find a one-to-one correspondence between these two sets, but it is easy to show that they are equinumerous using the Cantor–Schröder–Bernstein theorem. Of course, $(0, 1) \subseteq (0, 1]$, so clearly $(0, 1) \precsim (0, 1]$. For the other direction, define $f : (0, 1] \to (0, 1)$ by the formula

$$f(x) = \frac{x}{2}.$$

It is easy to check that this function is one-to-one (although it is not onto), so $(0, 1] \precsim (0, 1)$. Thus, by the Cantor–Schröder–Bernstein theorem, $(0, 1] \sim (0, 1)$. For more on this example see exercise 9.

Our next theorem gives a more surprising consequence of the Cantor–Schröder–Bernstein theorem.

Theorem 7.3.3. $\mathbb{R} \sim \mathscr{P}(\mathbb{Z}^+)$.

It is quite difficult to prove Theorem 7.3.3 directly by giving an example of a one-to-one, onto function from $\mathbb{R}$ to $\mathscr{P}(\mathbb{Z}^+)$. In our proof we'll use the Cantor–Schröder–Bernstein theorem and the following lemma.

Lemma 7.3.4. *Suppose x and y are real numbers and $x < y$. Then there is a rational number q such that $x < q < y$.*

Proof. Let k be a positive integer larger than $\frac{1}{y-x}$. Then $\frac{1}{k} < y - x$. We will show that there is a fraction with denominator k that is between x and y.

Let m and n be integers such that $m < x < n$, and let $S = \{j \in \mathbb{N} \mid m + \frac{j}{k} > x\}$. Note that $m + \frac{k(n-m)}{k} = n > x$, and therefore $k(n - m) \in S$. Thus $S \neq \varnothing$, so by the well-ordering principle it has a smallest element. Let j be the smallest element of S. Note also that $m + \frac{0}{k} = m < x$, so $0 \notin S$, and therefore $j > 0$.

Thus, $j-1$ is a natural number, but since j is the smallest element of S, $j-1 \notin S$. It follows that $m + \frac{j-1}{k} \leq x$.

Let $q = m + \frac{j}{k}$. Clearly q is a rational number, and since $j \in S$, $q = m + \frac{j}{k} > x$. Also, combining the observations that $m + \frac{j-1}{k} \leq x$ and $\frac{1}{k} < y - x$, we have

$$q = m + \frac{j}{k} = m + \frac{j-1}{k} + \frac{1}{k} < x + (y - x) = y.$$

Thus, we have $x < q < y$, as required. □

Proof of Theorem 7.3.3. As we observed earlier, we already know that $\mathscr{P}(\mathbb{Z}^+) \precsim \mathbb{R}$. But now consider the function $f : \mathbb{R} \to \mathscr{P}(\mathbb{Q})$ defined as follows:

$$f(x) = \{q \in \mathbb{Q} \mid q < x\}.$$

We claim that f is one-to-one. To see why, suppose $x \in \mathbb{R}$, $y \in \mathbb{R}$, and $x \neq y$. Then either $x < y$ or $y < x$. Suppose first that $x < y$. By Lemma 7.3.4, we can choose a rational number q such that $x < q < y$. But then $q \in f(y)$ and $q \notin f(x)$, so $f(x) \neq f(y)$. A similar argument shows that if $y < x$ then $f(x) \neq f(y)$, so f is one-to-one.

Since f is one-to-one, we have shown that $\mathbb{R} \precsim \mathscr{P}(\mathbb{Q})$. But we also know that $\mathbb{Q} \sim \mathbb{Z}^+$, so by exercise 5 in Section 7.1 it follows that $\mathscr{P}(\mathbb{Q}) \sim \mathscr{P}(\mathbb{Z}^+)$. Thus, $\mathbb{R} \precsim \mathscr{P}(\mathbb{Q}) \precsim \mathscr{P}(\mathbb{Z}^+)$, so by transitivity of $\precsim$ we have $\mathbb{R} \precsim \mathscr{P}(\mathbb{Z}^+)$. Combining this with the fact that $\mathscr{P}(\mathbb{Z}^+) \precsim \mathbb{R}$ and applying the Cantor–Schröder–Bernstein theorem, we conclude that $\mathbb{R} \sim \mathscr{P}(\mathbb{Z}^+)$. □

We said at the beginning of this chapter that we would show that infinity comes in different sizes. We now see that, so far, we have found only two sizes of infinity. One size is represented by the denumerable sets, which are all equinumerous with each other. The only examples of nondenumerable infinite sets we have given so far are $\mathscr{P}(\mathbb{Z}^+)$ and $\mathbb{R}$, which we now know are equinumerous. In fact, there are many more sizes of infinity. For example, $\mathscr{P}(\mathbb{R})$ is an infinite set that is neither denumerable nor equinumerous with $\mathbb{R}$. Thus, it represents a third size of infinity. For more on this see exercise 8.

Because $\mathbb{Z}^+ \prec \mathbb{R}$, it is natural to think of the set of real numbers as *larger* than the set of positive integers. In 1878, Cantor asked whether there was a size of infinity between these two sizes. More precisely, is there a set X such that $\mathbb{Z}^+ \prec X \prec \mathbb{R}$? Cantor conjectured that the answer was no, but he was unable to prove it. His conjecture is known as the *continuum hypothesis*. At the Second International Congress of Mathematicians in 1900, David Hilbert

(1862–1943) gave a famous lecture in which he listed what he believed to be the most important unsolved mathematical problems of the time, and the proof or disproof of the continuum hypothesis was number one on his list.

The status of the continuum hypothesis was "resolved" in a remarkable way by the work of Kurt Gödel (1906–1978) in 1939 and Paul Cohen (1934–) in 1963. The resolution turns out to require even more careful analyses than we have given in this book of both the notion of proof and the basic assumptions underlying set theory. Once such analyses have been given, it is possible to prove theorems about what can be proven and what cannot be proven. What Gödel and Cohen proved was that, using the methods of mathematical proof and set-theoretic assumptions accepted by most mathematicians today, it is impossible to prove the continuum hypothesis, and it is also impossible to disprove it!

Exercises

*1. Prove that $\precsim$ is reflexive and transitive. In other words:
   (a) For every set A, $A \precsim A$.
   (b) For all sets A, B, and C, if $A \precsim B$ and $B \precsim C$ then $A \precsim C$.

2. Prove that $\prec$ is irreflexive and transitive. In other words:
   (a) For every set A, $A \not\prec A$.
   (b) For all sets A, B, and C, if $A \prec B$ and $B \prec C$ then $A \prec C$.

3. Suppose $A \subseteq B \subseteq C$ and $A \sim C$. Prove that $B \sim C$.

4. Suppose $A \precsim B$ and $C \precsim D$.
   (a) Prove that $A \times C \precsim B \times D$.
   (b) Prove that if A and C are disjoint and B and D are disjoint, then $A \cup C \precsim B \cup D$.
   (c) Prove that $\mathscr{P}(A) \precsim \mathscr{P}(B)$.

*5. For the meaning of the notation used in this exercise, see exercise 21 of Section 7.1. Suppose $A \precsim B$ and $C \precsim D$.
   (a) Prove that if $A \neq \varnothing$ then ${}^{A}C \precsim {}^{B}D$.
   (b) Is the assumption that $A \neq \varnothing$ needed in part (a)?

6. (a) Prove that if $A \precsim B$ and B is finite, then A is finite and $|A| \leq |B|$.
   (b) Prove that if $A \prec B$ and B is finite, then A is finite and $|A| < |B|$.

7. Prove that for every set A, $A \prec \mathscr{P}(A)$. (Hint: See exercise 4 of Section 7.2. Note that in particular, if A is finite and $|A| = n$ then, by exercise 10 of Section 6.2, $|\mathscr{P}(A)| = 2^n$. It follows, by exercise 6(b), that $2^n > n$. Of course, you already proved this, by a different method, in exercise 12(a) of Section 6.3.)

*8. Let $A_1 = \mathbb{Z}^+$, and for all $n \in \mathbb{Z}^+$ let $A_{n+1} = \mathscr{P}(A_n)$.
   (a) Prove that for all $n \in \mathbb{Z}^+$ and $m \in \mathbb{Z}^+$, if $n < m$ then $A_n \prec A_m$.
   (b) The sets A_n, for $n \in \mathbb{Z}^+$, represent infinitely many sizes of infinity. Are there any more sizes of infinity? In other words, can you think of an infinite set that is not equinumerous with A_n for any $n \in \mathbb{Z}^+$?

9. The proof of the Cantor–Schröder–Bernstein theorem gives a method for constructing a one-to-one and onto function $h : A \to B$ from one-to-one functions $f : A \to B$ and $g : B \to A$. Use this method to find a one-to-one, onto function $h : (0, 1] \to (0, 1)$. Start with the functions $f : (0, 1] \to (0, 1)$ and $g : (0, 1) \to (0, 1]$ given by the formulas:

$$f(x) = \frac{x}{2}, \qquad g(x) = x.$$

*10. Let $\mathcal{E} = \{R \mid R \text{ is an equivalence relation on } \mathbb{Z}^+\}$.
   (a) Prove that $\mathcal{E} \precsim \mathscr{P}(\mathbb{Z}^+)$.
   (b) Let $A = \mathbb{Z}^+ \setminus \{1, 2\}$ and let $\mathcal{P}$ be the set of all partitions of $\mathbb{Z}^+$. Define $f : \mathscr{P}(A) \to \mathcal{P}$ by the formula $f(X) = \{X \cup \{1\}, (A \setminus X) \cup \{2\}\}$. Prove that f is one-to-one.
   (c) Prove that $\mathcal{E} \sim \mathscr{P}(\mathbb{Z}^+)$.

11. Let $\mathcal{T} = \{R \mid R \text{ is a total order on } \mathbb{Z}^+\}$. Prove that $\mathcal{T} \sim \mathscr{P}(\mathbb{Z}^+)$. (Hint: Imitate the solution to exercise 10.)

12. (a) Prove that if A has at least two elements and $A \times A \sim A$ then $\mathscr{P}(A) \times \mathscr{P}(A) \sim \mathscr{P}(A)$. (Hint: Use exercise 7 of Section 7.2.)
   (b) Prove that $\mathbb{R} \times \mathbb{R} \sim \mathbb{R}$.

13. An *interval* is a set $I \subseteq \mathbb{R}$ with the property that for all real numbers x, y, and z, if $x \in I$, $z \in I$, and $x < y < z$, then $y \in I$. An interval is *nondegenerate* if it contains at least two different real numbers. Suppose $\mathcal{F}$ is a set of nondegenerate intervals and $\mathcal{F}$ is pairwise disjoint. Prove that $\mathcal{F}$ is countable. (Hint: By Lemma 7.3.4, every nondegenerate interval contains a rational number.)

*14. For the meaning of the notation used in this exercise, see exercise 21 of Section 7.1.
   (a) Prove that ${}^{\mathbb{R}}\mathbb{R} \sim \mathscr{P}(\mathbb{R})$.
   (b) Prove that ${}^{\mathbb{Q}}\mathbb{R} \sim \mathbb{R}$.
   (c) (For students who have studied calculus) Let $\mathcal{C} = \{f \in {}^{\mathbb{R}}\mathbb{R} \mid f \text{ is continuous}\}$. Prove that $\mathcal{C} \sim \mathbb{R}$. (Hint: Show that if f and g are continuous functions and $\forall x \in \mathbb{Q}(f(x) = g(x))$, then $f = g$.)

Appendix 1

Solutions to Selected Exercises

Introduction

1. (a) One possible answer is $32{,}767 = 31 \cdot 1057$.
   (b) One possible answer is $x = 2^{31} - 1 = 2{,}147{,}483{,}647$.
3. (a) The method yields the prime number 211.
   (b) The method yields two primes, 3 and 37.

Chapter 1

Section 1.1

1. (a) $(R \vee H) \wedge \neg(H \wedge T)$, where R stands for the statement "We'll have a reading assignment," H stands for "We'll have homework problems," and T stands for "We'll have a test."
   (b) $\neg G \vee (G \wedge \neg S)$, where G stands for "You'll go skiing," and S stands for "There will be snow."
   (c) $\neg[(\sqrt{7} < 2) \vee (\sqrt{7} = 2)]$.
5. (a) I won't buy the pants without the shirt.
   (b) I won't buy the pants and I won't buy the shirt.
   (c) Either I won't buy the pants or I won't buy the shirt.

Section 1.2

1. (a)

P	Q	$\neg P \vee Q$
F	F	T
F	T	T
T	F	F
T	T	T

(b)

S	G	$(S \lor G) \land (\neg S \lor \neg G)$
F	F	F
F	T	T
T	F	T
T	T	F

5. (a)

P	Q	$P \downarrow Q$
F	F	T
F	T	F
T	F	F
T	T	F

   (b) $\neg(P \lor Q)$.
   (c) $\neg P$ is equivalent to $P \downarrow P$, $P \lor Q$ is equivalent to $(P \downarrow Q) \downarrow (P \downarrow Q)$, and $P \land Q$ is equivalent to $(P \downarrow P) \downarrow (Q \downarrow Q)$.

7. (a) and (c) are valid; (b) and (d) are invalid.
9. (a) is neither a contradiction nor a tautology; (b) is a contradiction; (c) and (d) are tautologies.
11. (a) $P \lor Q$.
    (b) P.
    (c) $\neg P \lor Q$.
14. We use the associative law for $\land$ twice:

$$[P \land (Q \land R)] \land S \text{ is equivalent to } [(P \land Q) \land R] \land S$$
$$\text{which is equivalent to } (P \land Q) \land (R \land S)$$

16. $P \lor \neg Q$.

Section 1.3

1. (a) $D(6) \land D(9) \land D(15)$, where $D(x)$ means "x is divisible by 3."
   (b) $D(x, 2) \land D(x, 3) \land \neg D(x, 4)$, where $D(x, y)$ means "x is divisible by y."
   (c) $N(x) \land N(y) \land [(P(x) \land \neg P(y)) \lor (P(y) \land \neg P(x))]$, where $N(x)$ means "x is a natural number" and $P(x)$ means "x is prime."
3. (a) $\{x \mid x \text{ is a planet}\}$.
   (b) $\{x \mid x \text{ is an Ivy League school}\}$.
   (c) $\{x \mid x \text{ is a state in the United States}\}$.
   (d) $\{x \mid x \text{ is a province or territory in Canada}\}$.

5. (a) $(-3 \in \mathbb{R}) \wedge (13 - 2(-3) > 1)$. Bound variables: x; no free variables. This statement is true.
   (b) $(4 \in \mathbb{R}) \wedge (4 < 0) \wedge (13 - 2(4) > 1)$. Bound variables: x; no free variables. This statement is false.
   (c) $\neg[(5 \in \mathbb{R}) \wedge (13 - 2(5) > c)]$. Bound variables: x; free variables: c.
7. (a) $\{x \mid$ Elizabeth Taylor was once married to $x\} = \{$Conrad Hilton Jr., Michael Wilding, Michael Todd, Eddie Fisher, Richard Burton, John Warner, Larry Fortensky$\}$.
   (b) $\{x \mid x$ is a logical connective studied in Section 1.1$\} = \{\wedge, \vee, \neg\}$.
   (c) $\{x \mid x$ is the author of this book$\} = \{$Daniel J. Velleman$\}$.

Section 1.4

1. (a) $\{3, 12\}$.
   (b) $\{1, 12, 20, 35\}$.
   (c) $\{1, 3, 12, 20, 35\}$.
   The sets in parts (a) and (b) are both subsets of the set in part (c).
4. (a) Both Venn diagrams look like this:

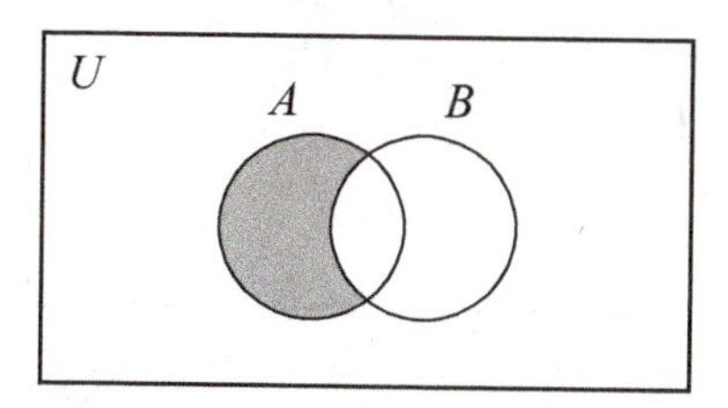

   (b) Both Venn diagrams look like this:

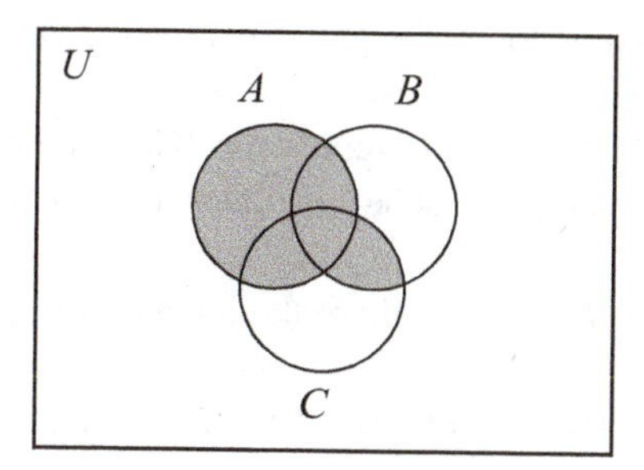

8. Sets (a), (d), and (e) are equal, and sets (b) and (c) are equal.

10. (a) There is no region corresponding to the set $(A \cap D) \setminus (B \cup C)$, but this set could have elements.

    (b) Here is one possibility:

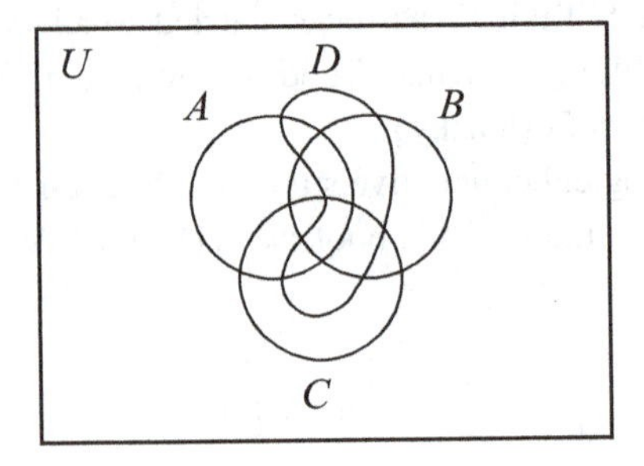

12. The Venn diagrams for both sets look like this:

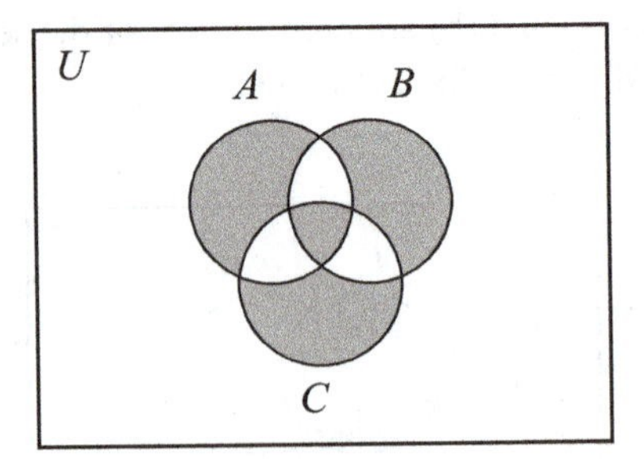

Section 1.5

1. (a) $(S \vee \neg E) \rightarrow \neg H$, where S stands for "This gas has an unpleasant smell," E stands for "This gas is explosive," and H stands for "This gas is hydrogen."

   (b) $(F \wedge H) \rightarrow D$, where F stands for "George has a fever," H stands for "George has a headache," and D stands for "George will go to the doctor."

   (c) $(F \rightarrow D) \wedge (H \rightarrow D)$, where the letters have the same meanings as in part (b).

   (d) $(x \neq 2) \rightarrow (P(x) \rightarrow O(x))$, where $P(x)$ stands for "x is prime" and $O(x)$ stands for "x is odd."

4. (a) and (b) are valid, but (c) is invalid.

6. (a) Either make a truth table, or reason as follows:

$(P \to R) \wedge (Q \to R)$ is equivalent to $(\neg P \vee R) \wedge (\neg Q \vee R)$
which is equivalent to $(\neg P \wedge \neg Q) \vee R$
which is equivalent to $\neg(P \vee Q) \vee R$
which is equivalent to $(P \vee Q) \to R$

(b) $(P \to R) \vee (Q \to R)$ is equivalent to $(P \wedge Q) \to R$.

8. $\neg(P \to \neg Q)$.

Chapter 2

Section 2.1

1. (a) $\forall x[\exists y F(x, y) \to S(x)]$, where $F(x, y)$ stands for "x has forgiven y," and $S(x)$ stands for "x is a saint."
   (b) $\neg\exists x[C(x) \wedge \forall y(D(y) \to S(x, y))]$, where $C(x)$ stands for "x is in the calculus class," $D(y)$ stands for "y is in the discrete math class," and $S(x, y)$ stands for "x is smarter than y."
   (c) $\forall x(\neg(x = m) \to L(x, m))$, where $L(x, y)$ stands for "x likes y," and m stands for Mary.
   (d) $\exists x(P(x) \wedge S(j, x)) \wedge \exists y(P(y) \wedge S(r, y))$, where $P(x)$ stands for "x is a police officer," $S(x, y)$ stands for "x saw y," j stands for Jane, and r stands for Roger.
   (e) $\exists x(P(x) \wedge S(j, x) \wedge S(r, x))$, where the letters have the same meanings as in part (d).
4. (a) All unmarried men are unhappy.
   (b) y is a sister of one of x's parents; i.e., y is x's blood aunt.
7. (a), (d), and (e) are true; (b), (c), and (f) are false.

Section 2.2

1. (a) $\exists x[M(x) \wedge \forall y(F(x, y) \to \neg H(y))]$, where $M(x)$ stands for "x is majoring in math," $F(x, y)$ stands for "x and y are friends," and $H(y)$ stands for "y needs help with his homework." In English: There is a math major all of whose friends don't need help with their homework.
   (b) $\exists x \forall y(R(x, y) \to \exists z L(y, z))$, where $R(x, y)$ stands for "x and y are roommates" and $L(y, z)$ stands for "y likes z." In English: There is someone all of whose roommates like at least one person.
   (c) $\exists x[(x \in A \vee x \in B) \wedge (x \notin C \vee x \in D)]$.
   (d) $\forall x \exists y[y > x \wedge \forall z(z^2 + 5z \neq y)]$.
4. Hint: Begin by replacing $P(x)$ with $\neg P(x)$ in the first quantifier negation law, to get the fact that $\neg\exists x \neg P(x)$ is equivalent to $\forall x \neg\neg P(x)$.

6. Hint: Begin by showing that $\exists x(P(x) \vee Q(x))$ is equivalent to $\neg\forall x \neg(P(x) \vee Q(x))$.
8. $(\forall x \in A P(x)) \wedge (\forall x \in B P(x))$
is equivalent to $\forall x(x \in A \to P(x)) \wedge \forall x(x \in B \to P(x))$
which is equivalent to $\forall x[(x \in A \to P(x)) \wedge (x \in B \to P(x))]$
which is equivalent to $\forall x[(x \notin A \vee P(x)) \wedge (x \notin B \vee P(x))]$
which is equivalent to $\forall x[(x \notin A \wedge x \notin B) \vee P(x)]$
which is equivalent to $\forall x[\neg(x \in A \vee x \in B) \vee P(x)]$
which is equivalent to $\forall x[x \in (A \cup B) \to P(x)]$
which is equivalent to $\forall x \in (A \cup B) P(x)$.
11. $A \setminus B = \varnothing$ is equivalent to $\neg\exists x(x \in A \wedge x \notin B)$
which is equivalent to $\forall x \neg(x \in A \wedge x \notin B)$
which is equivalent to $\forall x(x \notin A \vee x \in B)$
which is equivalent to $\forall x(x \in A \to x \in B)$
which is equivalent to $A \subseteq B$.

Section 2.3

1. (a) $\forall x(x \in \mathcal{F} \to \forall y(y \in x \to y \in A))$.
   (b) $\forall x(x \in A \to \exists n \in \mathbb{N}(x = 2n + 1))$.
   (c) $\forall n \in \mathbb{N} \exists m \in \mathbb{N}(n^2 + n + 1 = 2m + 1)$.
   (d) $\exists x(\forall y(y \in x \to \exists i \in I(y \in A_i)) \wedge \forall i \in I \exists y(y \in x \wedge y \notin A_i))$.
4. $\cap\mathcal{F} = \{\text{red, blue}\}$ and $\cup\mathcal{F} = \{\text{red, green, blue, orange, purple}\}$.
8. (a) $A_2 = \{2, 4\}$, $A_3 = \{3, 6\}$, $B_2 = \{2, 3\}$, $B_3 = \{3, 4\}$.
   (b) $\bigcap_{i \in I}(A_i \cup B_i) = \{3, 4\}$ and $(\bigcap_{i \in I} A_i) \cup (\bigcap_{i \in I} B_i) = \{3\}$.
   (c) They are not equivalent.
11. One example is $A = \{1, 2\}$ and $B = \{2, 3\}$.
13. (a) $B_3 = \{1, 2, 3, 4, 5\}$ and $B_4 = \{1, 2, 4, 5, 6\}$.
   (b) $\bigcap_{j \in J} B_j = \{1, 2, 4, 5\}$.
   (c) $\bigcup_{i \in I}(\bigcap_{j \in J} A_{i,j}) = \{1, 2, 4\}$.
   (d) $x \in \bigcap_{j \in J}(\bigcup_{i \in I} A_{i,j})$ means $\forall j \in J \exists i \in I(x \in A_{i,j})$ and $x \in \bigcup_{i \in I}(\bigcap_{j \in J} A_{i,j})$ means $\exists i \in I \forall j \in J(x \in A_{i,j})$. They are not equivalent.

Chapter 3

Section 3.1

1. (a) Hypotheses: n is an integer larger than 1 and n is not prime. Conclusion: $2^n - 1$ is not prime. The hypotheses are true when $n = 6$, so

the theorem tells us that $2^6 - 1$ is not prime. This is correct, since $2^6 - 1 = 63 = 9 \cdot 7$.

(b) We can conclude that 32767 is not prime. This is correct, since $32767 = 151 \cdot 217$.

(c) The theorem tells us nothing; 11 is prime, so the hypotheses are not satisfied.

4. Suppose $0 < a < b$. Then $b - a > 0$. Multiplying both sides by the positive number $b + a$, we get $(b + a) \cdot (b - a) > (b + a) \cdot 0$, or in other words $b^2 - a^2 > 0$. Since $b^2 - a^2 > 0$, it follows that $a^2 < b^2$. Therefore if $0 < a < b$ then $a^2 < b^2$.

9. Hint: Add b to both sides of the inequality $a < b$.

11. We will prove the contrapositive. Suppose $c \leq d$. Multiplying both sides of this inequality by the positive number a, we get $ac \leq ad$. Also, multiplying both sides of the given inequality $a < b$ by the positive number d gives us $ad < bd$. Combining $ac \leq ad$ and $ad < bd$, we can conclude that $ac < bd$. Thus, if $ac \geq bd$ then $c > d$.

14. Since $x > 3 > 0$, by the theorem in Example 3.1.2, $x^2 > 9$. Also, multiplying both sides of the given inequality $y < 2$ by -2 (and reversing the direction of the inequality, since -2 is negative) we get $-2y > -4$. Finally, adding the inequalities $x^2 > 9$ and $-2y > -4$ gives us $x^2 - 2y > 5$.

Section 3.2

1. (a) Suppose P. Since $P \to Q$, it follows that Q. But then, since $Q \to R$, we can conclude R. Thus, $P \to R$.

   (b) Suppose P. To prove that $Q \to R$, we will prove the contrapositive, so suppose $\neg R$. Since $\neg R \to (P \to \neg Q)$, it follows that $P \to \neg Q$, and since we know P, we can conclude $\neg Q$. Thus, $Q \to R$, so $P \to (Q \to R)$.

5. Suppose $a \in A \setminus B$. This means that $a \in A$ and $a \notin B$. Since $a \in A$ and $a \in C$, $a \in A \cap C$. But then since $A \cap C \subseteq B$, it follows that $a \in B$, and this contradicts the fact that $a \notin B$. Thus, $a \notin A \setminus B$.

8. Hint: Assume $a < 1/a < b < 1/b$. Now prove that $a < 0$, and then use this fact to prove that $a < -1$.

11. (a) The sentence "Then $x = 3$ and $y = 8$" is incorrect. (Why?)

    (b) One counterexample is $x = 3$, $y = 7$.

14.

P	Q	R	$P \to (Q \to R)$	$\neg R \to (P \to \neg Q)$
F	F	F	T	T
F	F	T	T	T
F	T	F	T	T
F	T	T	T	T
T	F	F	T	T
T	F	T	T	T
T	T	F	F	F
T	T	T	T	T

Section 3.3

1. Suppose $\exists x(P(x) \to Q(x))$. Then we can choose some x_0 such that $P(x_0) \to Q(x_0)$. Now suppose that $\forall x P(x)$. Then in particular, $P(x_0)$, and since $P(x_0) \to Q(x_0)$, it follows that $Q(x_0)$. Since we have found a particular value of x for which $Q(x)$ holds, we can conclude that $\exists x Q(x)$. Thus $\forall x P(x) \to \exists x Q(x)$.
3. Suppose that $A \subseteq B \setminus C$, but A and C are not disjoint. Then we can choose some x such that $x \in A$ and $x \in C$. Since $x \in A$ and $A \subseteq B \setminus C$, it follows that $x \in B \setminus C$, which means that $x \in B$ and $x \notin C$. But now we have both $x \in C$ and $x \notin C$, which is a contradiction. Thus, if $A \subseteq B \setminus C$ then A and C are disjoint.
7. Suppose $x > 2$. Let $y = (x + \sqrt{x^2 - 4})/2$, which is defined since $x^2 - 4 > 0$. Then
$$y + \frac{1}{y} = \frac{x + \sqrt{x^2 - 4}}{2} + \frac{2}{x + \sqrt{x^2 - 4}} = \frac{2x^2 + 2x\sqrt{x^2 - 4}}{2(x + \sqrt{x^2 - 4})} = x.$$
9. Suppose $\mathcal{F}$ is a family of sets and $A \in \mathcal{F}$. Suppose $x \in \cap\mathcal{F}$. Then by the definition of $\cap\mathcal{F}$, since $x \in \cap\mathcal{F}$ and $A \in \mathcal{F}$, $x \in A$. But x was an arbitrary element of $\cap\mathcal{F}$, so it follows that $\cap\mathcal{F} \subseteq A$.
12. Hint: Assume $\mathcal{F} \subseteq \mathcal{G}$ and let x be an arbitrary element of $\cup\mathcal{F}$. You must prove that $x \in \cup\mathcal{G}$, which means $\exists A \in \mathcal{G}(x \in A)$, so you should try to find some $A \in \mathcal{G}$ such that $x \in A$. To do this, write out the givens in logical notation. You will find that one of them is a universal statement, and one is existential. Apply existential instantiation to the existential one.
14. Suppose $x \in \cup_{i \in I}\mathscr{P}(A_i)$. Then we can choose some $i \in I$ such that $x \in \mathscr{P}(A_i)$, or in other words $x \subseteq A_i$. Now let a be an arbitrary element of x. Then $a \in A_i$, and therefore $a \in \cup_{i \in I} A_i$. Since a was an arbitrary element of x, it follows that $x \subseteq U_{i \in I} A_i$, which means that $x \in \mathscr{P}(U_{i \in I} A_i)$. Thus $\cup_{i \in I}\mathscr{P}(A_i) \subseteq \mathscr{P}(\cup_{i \in I} A_i)$.

17. Hint: The last hypothesis means $\forall A \in \mathcal{F} \forall B \in \mathcal{G}(A \subseteq B)$, so if in the course of the proof you ever come across sets $A \in \mathcal{F}$ and $B \in \mathcal{G}$, you can conclude that $A \subseteq B$. Start the proof by letting x be arbitrary and assuming $x \in \cup\mathcal{F}$, and prove that $x \in \cap\mathcal{G}$. To see where to go from there, write these statements in logical symbols.
20. The sentence "Then for every real number x, $x^2 < 0$" is incorrect. (Why?)
22. Based on the logical form of the statement to be proven, the proof should have this outline:

    Let $x = \ldots$.
    Let y be an arbitrary real number.
    [Proof of $xy^2 = y - x$ goes here.]
    Since y was arbitrary, $\forall y \in \mathbb{R}(xy^2 = y - x)$.
    Thus, $\exists x \in \mathbb{R} \forall y \in \mathbb{R}(xy^2 = y - x)$.

    This outline makes it clear that y should be introduced into the proof *after* x. Therefore, x cannot be defined in terms of y, because y will not yet have been introduced into the proof when x is being defined. But in the given proof, x is defined in terms of y in the first sentence. (The mistake has been disguised by the fact that the sentence "Let y be an arbitrary real number" has been left out of the proof. If you try to add this sentence to the proof, you will find that there is nowhere it could be added that would lead to a correct proof of the incorrect theorem.)
25. Here is the beginning of the proof: Let x be an arbitrary real number. Let $y = 2x$. Now let z be an arbitrary real number. Then....

Section 3.4

1. ($\rightarrow$) Suppose $\forall x(P(x) \wedge Q(x))$. Let y be arbitrary. Then since $\forall x(P(x) \wedge Q(x))$, $P(y) \wedge Q(y)$, and so in particular $P(y)$. Since y was arbitrary, this shows that $\forall x P(x)$. A similar argument proves $\forall x Q(x)$: for arbitrary y, $P(y) \wedge Q(y)$, and therefore $Q(y)$. Thus, $\forall x P(x) \wedge \forall x Q(x)$.

    ($\leftarrow$) Suppose $\forall x P(x) \wedge \forall x Q(x)$. Let y be arbitrary. Then since $\forall x P(x)$, $P(y)$, and similarly since $\forall x Q(x)$, $Q(y)$. Thus, $P(y) \wedge Q(y)$, and since y was arbitrary, it follows that $\forall x(P(x) \wedge Q(x))$.
4. Suppose that $A \subseteq B$ and $A \not\subseteq C$. Since $A \not\subseteq C$, we can choose some $a \in A$ such that $a \notin C$. Since $a \in A$ and $A \subseteq B$, $a \in B$. Since $a \in B$ and $a \notin C$, $B \not\subseteq C$.
7. Let A and B be arbitrary sets. Let x be arbitrary, and suppose that $x \in \mathscr{P}(A \cap B)$. Then $x \subseteq A \cap B$. Now let y be all arbitrary element of x. Then since $x \subseteq A \cap B$, $y \in A \cap B$, and therefore $y \in A$. Since y was

arbitrary, this shows that $x \subseteq A$, so $x \in \mathscr{P}(A)$. A similar argument shows that $x \subseteq B$, and therefore $x \in \mathscr{P}(B)$. Thus, $x \in \mathscr{P}(A) \cap \mathscr{P}(B)$.

Now suppose that $x \in \mathscr{P}(A) \cap \mathscr{P}(B)$. Then $x \in \mathscr{P}(A)$ and $x \in \mathscr{P}(B)$, so $x \subseteq A$ and $x \subseteq B$. Suppose that $y \in x$. Then since $x \subseteq A$ and $x \subseteq B$, $y \in A$ and $y \in B$, so $y \in A \cap B$. Thus, $x \subseteq A \cap B$, so $x \in \mathscr{P}(A \cap B)$.

9. Suppose that x and y are odd. Then we can choose integers j and k such that $x = 2j + 1$ and $y = 2k + 1$. Therefore $xy = (2j + 1)(2k + 1) = 4jk + 2j + 2k + 1 = 2(2jk + j + k) + 1$. Since $2jk + j + k$ is an integer, it follows that xy is odd.

12. Hint: Let $x \in \mathbb{R}$ be arbitrary, and prove both directions of the biconditional separately. For the "$\rightarrow$" direction, use existential instantiation and proof by contradiction. For the "$\leftarrow$" direction, assume that $x \neq 1$ and then solve the equation $x + y = xy$ for y in order to decide what value to choose for y.

15. Suppose that $\cup\mathcal{F}$ and $\cap\mathcal{G}$ are not disjoint. Then we can choose some x such that $x \in \cup\mathcal{F}$ and $x \in \cap\mathcal{G}$. Since $x \in \cup\mathcal{F}$, we can choose some $A \in \mathcal{F}$ such that $x \in A$. Since we are given that every element of $\mathcal{F}$ is disjoint from some element of $\mathcal{G}$, there must be some $B \in \mathcal{G}$ such that $A \cap B = \varnothing$. Since $x \in A$, it follows that $x \notin B$. But we also have $x \in \cap\mathcal{G}$ and $B \in \mathcal{G}$, from which it follows that $x \in B$, which is a contradiction. Thus, $\cup\mathcal{F}$ and $\cap\mathcal{G}$ must be disjoint.

17. (a) Suppose $x \in \cup(\mathcal{F} \cap \mathcal{G})$. Then we can choose some $A \in \mathcal{F} \cap \mathcal{G}$ such that $x \in A$. Since $x \in A$ and $A \in \mathcal{F}$, $x \in \cup\mathcal{F}$, and similarly since $x \in A$ and $A \in \mathcal{G}$, $x \in \cup\mathcal{G}$. Therefore, $x \in (\cup\mathcal{F}) \cap (\cup\mathcal{G})$. Since x was arbitrary, this shows that $\cup(\mathcal{F} \cap \mathcal{G}) \subseteq (\cup\mathcal{F}) \cap (\cup\mathcal{G})$.
    (b) The sentence "Thus, we can choose a set A such that $A \in \mathcal{F}$, $A \in \mathcal{G}$, and $x \in A$" is incorrect. (Why?)
    (c) One example is $\mathcal{F} = \{\{1\}, \{2\}\}$, $\mathcal{G} = \{\{1\}, \{1, 2\}\}$.

21. Suppose that $\cup\mathcal{F} \not\subseteq \cup\mathcal{G}$. Then there is some $x \in \cup\mathcal{F}$ such that $x \notin \cup\mathcal{G}$. Since $x \in \cup\mathcal{F}$, we can choose some $A \in \mathcal{F}$ such that $x \in A$. Now let $B \in \mathcal{G}$ be arbitrary. If $A \subseteq B$, then since $x \in A$, $x \in B$. But then since $x \in B$ and $B \in \mathcal{G}$, $x \in \cup\mathcal{G}$, which we already know is false. Therefore $A \not\subseteq B$. Since B was arbitrary, this shows that for all $B \in \mathcal{G}$, $A \not\subseteq B$. Thus, we have shown that there is some $A \in \mathcal{F}$ such that for all $B \in \mathcal{G}$, $A \not\subseteq B$.

23. (a) Suppose $x \in \cup_{i \in I}(A_i \setminus B_i)$. Then we can choose some $i \in I$ such that $x \in A_i \setminus B_i$, which means $x \in A_i$ and $x \notin B_i$. Since $x \in A_i$, $x \in \cup_{i \in I} A_i$, and since $x \notin B_i$, $x \notin \cap_{i \in I} B_i$. Thus, $x \in (\cup_{i \in I} A_i) \setminus (\cap_{i \in I} B_i)$.
    (b) One example is $I = \{1, 2\}$, $A_1 = B_1 = \{1\}$, $A_2 = B_2 = \{2\}$.

Section 3.5

1. Suppose $x \in A \cap (B \cup C)$. Then $x \in A$, and either $x \in B$ or $x \in C$.
   Case 1. $x \in B$. Then since $x \in A$, $x \in A \cap B$, so $x \in (A \cap B) \cup C$.
   Case 2. $x \in C$. Then clearly $x \in (A \cap B) \cup C$.
   Since x was arbitrary, we can conclude that $A \cap (B \cup C) \subseteq (A \cap B) \cup C$.
4. Suppose $x \in A$. We now consider two cases:
   Case 1. $x \in C$. Then $x \in A \cap C$, so since $A \cap C \subseteq B \cap C$, $x \in B \cap C$, and therefore $x \in B$.
   Case 2. $x \notin C$. Since $x \in A$, $x \in A \cup C$, so since $A \cup C \subseteq B \cup C$, $x \in B \cup C$. But $x \notin C$, so we must have $x \in B$.
   Thus, $x \in B$, and since x was arbitrary, $A \subseteq B$.
7. Hint: Assume $x \in \mathscr{P}(A) \cup \mathscr{P}(B)$, which means that either $x \in \mathscr{P}(A)$ or $x \in \mathscr{P}(B)$. Treat these as two separate cases. In case 1, assume $x \in \mathscr{P}(A)$, which means $x \subseteq A$, and prove $x \in \mathscr{P}(A \cup B)$, which means $x \subseteq A \cup B$. Case 2 is similar.
11. Let x be an arbitrary real number.
   ($\leftarrow$) Suppose $|x - 4| > 2$.
   Case 1. $x - 4 \geq 0$. Then $|x - 4| = x - 4$, so we have $x - 4 > 2$, and therefore $x > 6$. Adding x to both sides gives us $2x > 6 + x$, so $2x - 6 > x$. Since $x > 6$, this implies that $2x - 6$ is positive, so $|2x - 6| = 2x - 6 > x$.
   Case 2. $x - 4 < 0$. Then $|x - 4| = 4 - x$, so we have $4 - x > 2$, and therefore $x < 2$. Therefore $3x < 6$, and subtracting $2x$ from both sides we get $x < 6 - 2x$. Also, from $x < 2$ we get $2x < 4$, so $2x - 6 < -2$. Therefore $2x - 6$ is negative, so $|2x - 6| = 6 - 2x > x$.
   ($\rightarrow$) Hint: Imitate the "$\leftarrow$" direction, using the cases $2x - 6 \geq 0$ and $2x - 6 < 0$.
15. (a) Suppose $x \in \cup(\mathcal{F} \cup \mathcal{G})$. Then we can choose some $A \in \mathcal{F} \cup \mathcal{G}$ such that $x \in A$. Since $A \in \mathcal{F} \cup \mathcal{G}$, either $A \in \mathcal{F}$ or $A \in \mathcal{G}$. If $A \in \mathcal{F}$ then, since $x \in A$, it follows that $x \in \cup\mathcal{F}$. Similarly, if $A \in \mathcal{G}$ then $x \in \cup\mathcal{G}$. Thus either $x \in \cup\mathcal{F}$ or $x \in \cup\mathcal{G}$, so $x \in (\cup\mathcal{F}) \cup (\cup\mathcal{G})$.
   Now suppose that $x \in (\cup\mathcal{F}) \cup (\cup\mathcal{G})$. Then either $x \in \cup\mathcal{F}$ or $x \in \cup\mathcal{G}$. If $x \in \cup\mathcal{F}$, then we can choose some $A \in \mathcal{F}$ such that $x \in A$. Since $A \in \mathcal{F}$, $A \in \mathcal{F} \cup \mathcal{G}$, so since $x \in A$, it follows that $x \in \cup(\mathcal{F} \cup \mathcal{G})$. A similar argument shows that if $x \in \cup\mathcal{G}$ then $x \in \cup(\mathcal{F} \cup \mathcal{G})$.
   (b) The theorem is: $\cap(\mathcal{F} \cup \mathcal{G}) = (\cap\mathcal{F}) \cap (\cap\mathcal{G})$.
19. ($\rightarrow$) Suppose that $A \mathbin{\triangle} B$ and C are disjoint. Let x be an arbitrary element of $A \cap C$. Then $x \in A$ and $x \in C$. If $x \notin B$, then since $x \in A$, $x \in A \setminus B$, and therefore $x \in A \mathbin{\triangle} B$. But also $x \in C$, so this contradicts our assumption

that $A \triangle B$ and C are disjoint. Therefore $x \in B$. Since we also know $x \in C$, we have $x \in B \cap C$. Since x was an arbitrary element of $A \cap C$, this shows that $A \cap C \subseteq B \cap C$. A similar argument shows that $B \cap C \subseteq A \cap C$.

($\leftarrow$) Suppose that $A \cap C = B \cap C$. Suppose that $A \triangle B$ and C are not disjoint. Then we can choose some x such that $x \in A \triangle B$ and $x \in C$. Since $x \in A \triangle B$, either $x \in A \setminus B$ or $x \in B \setminus A$.

Case 1. $x \in A \setminus B$. Then $x \in A$ and $x \notin B$. Since we also know $x \in C$, we can conclude that $x \in A \cap C$ but $x \notin B \cap C$. This contradicts the fact that $A \cap C = B \cap C$.

Case 2. $x \in B \setminus A$. Similarly, this leads to a contradiction.

Thus we can conclude that $A \triangle B$ and C are disjoint.

22. (a) Hint: Suppose $x \in A \setminus C$, and then break the proof into cases, depending on whether or not $x \in B$.
    (b) Hint: Apply part (a).
23. (a) Suppose $x \in (A \cup B) \triangle C$. Then either $x \in (A \cup B) \setminus C$ or $x \in C \setminus (A \cup B)$.

    Case 1. $x \in (A \cup B) \setminus C$. Then either $x \in A$ or $x \in B$, and $x \notin C$. We now break case 1 into two subcases, depending on whether $x \in A$ or $x \in B$:

    Case 1a. $x \in A$. Then $x \in A \setminus C$, so $x \in A \triangle C$, so $x \in (A \triangle C) \cup (B \triangle C)$.

    Case 1b. $x \in B$. Similarly, $x \in B \triangle C$, so $x \in (A \triangle C) \cup (B \triangle C)$.

    Case 2. $x \in C \setminus (A \cup B)$. Then $x \in C, x \notin A$, and $x \notin B$. It follows that $x \in A \triangle C$ and $x \in B \triangle C$, so certainly $x \in (A \triangle C) \cup (B \triangle C)$.

    (b) Here is one example: $A = \{1\}$, $B = \{2\}$, $C = \{1, 2\}$.
26. The proof is incorrect, because it only establishes that either $0 < x$ or $x < 6$, but what must be proven is that $0 < x$ and $x < 6$. However, it can be fixed.
28. The proof is correct.
30. Hint: Here is a counterexample to the theorem: $A = \{1, 2\}$, $B = \{1\}$, $C = \{2\}$.

Section 3.6

1. Let x be an arbitrary real number. Let $y = x/(x^2 + 1)$. Then

$$x - y = x - \frac{x}{x^2+1} = \frac{x^3+x}{x^2+1} - \frac{x}{x^2+1} = \frac{x^3}{x^2+1} = x^2\frac{x}{x^2+1} = x^2 y.$$

To see that y is unique, suppose that $x^2 z = x - z$. Then $z(x^2 + 1) = x$, and since $x^2 + 1 \neq 0$, we can divide both sides by $x^2 + 1$ to conclude that $z = x/(x^2 + 1) = y$.

4. Suppose $x \neq 0$. Let $y = 1/x$. Now let z be an arbitrary real number. Then $zy = z(1/x) = z/x$, as required.

   To see that y is unique, suppose that y' is a number with the property that $\forall z \in \mathbb{R}(zy' = z/x)$. Then in particular, taking $z = 1$, we have $y' = 1/x$, so $y' = y$.

6. (a) Let $A = \varnothing \in \mathscr{P}(U)$. Then clearly for any $B \in \mathscr{P}(U)$, $A \cup B = \varnothing \cup B = B$.

   To see that A is unique, suppose that $A' \in \mathscr{P}(U)$ and for all $B \in \mathscr{P}(U)$, $A' \cup B = B$. Then in particular, taking $B = \varnothing$, we can conclude that $A' \cup \varnothing = \varnothing$. But clearly $A' \cup \varnothing = A'$, so we have $A' = \varnothing = A$.

   (b) Hint: Let $A = U$.

11. Existence: We are given that for every $\mathcal{G} \subseteq \mathcal{F}$, $\cup\mathcal{G} \in \mathcal{F}$, so in particular, since $\mathcal{F} \subseteq \mathcal{F}$, $\cup\mathcal{F} \in \mathcal{F}$. Let $A = \cup\mathcal{F}$. Now suppose $B \in \mathcal{F}$. Then by exercise 8 of Section 3.3, $B \subseteq \cup\mathcal{F} = A$, as required.

    Uniqueness: Suppose that $A_1 \in \mathcal{F}$, $A_2 \in \mathcal{F}$, $\forall B \in \mathcal{F}(B \subseteq A_1)$, and $\forall B \in \mathcal{F}(B \subseteq A_2)$. Applying this last fact with $B = A_1$ we can conclude that $A_1 \subseteq A_2$, and similarly the previous fact implies that $A_2 \subseteq A_1$. Thus $A_1 = A_2$.

Section 3.7

1. Hint: Comparing (b) to exercise 16 of Section 3.3 may give you an idea of what to use for A.

4. Suppose $\mathscr{P}(\cup_{i \in I} A_i) \subseteq \cup_{i \in I} \mathscr{P}(A_i)$. Clearly $\cup_{i \in I} A_i \subseteq \cup_{i \in I} A_i$, so $\cup_{i \in I} A_i \in \mathscr{P}(\cup_{i \in I} A_i)$ and therefore $\cup_{i \in I} A_i \in \cup_{i \in I} \mathscr{P}(A_i)$. By the definition of the union of a family, this means that there is some $i \in I$ such that $\cup_{i \in I} A_i \subseteq A_i$. Now let $j \in I$ be arbitrary. Then it is not hard to see that $A_j \subseteq \cup_{i \in I} A_i$, so $A_j \subseteq A_i$.

7. Suppose that $\lim_{x \to c} f(x) = L > 0$. Let $\epsilon = L$. Then by the definition of limit, there is some $\delta > 0$ such that for all x, if $0 < |x - c| < \delta$ then $|f(x) - L| < \epsilon = L$. But if $|f(x) - L| < L$ then $-L < f(x) - L < L$, so $0 < f(x) < 2L$. Therefore, if $0 < |x - c| < \delta$ then $f(x) > 0$.

9. The proof is correct.

Chapter 4

Section 4.1

1. (a) $\{(x, y) \in P \times P \mid x \text{ is a parent of } y\} = \{$(George H. W. Bush, George W. Bush), (Goldie Hawn, Kate Hudson), . . .$\}$.
   (b) $\{(x, y) \in C \times U \mid$ there is someone who lives in x and attends $y\}$. If you are a university student, then let x be the city you live in, and let y be the university you attend; (x, y) will then be an element of this truth set.
4. $A \times (B \cap C) = (A \times B) \cap (A \times C) = \{(1, 4), (2, 4), (3, 4)\}$,
   $A \times (B \cup C) = (A \times B) \cup (A \times C) = \{(1, 1), (2, 1), (3, 1), (1, 3), (2, 3), (3, 3), (1, 4), (2, 4), (3, 4)\}$,
   $(A \times B) \cap (C \times D) = (A \cap C) \times (B \cap D) = \varnothing$,
   $(A \times B) \cup (C \times D) = \{(1, 1), (2, 1), (3, 1), (1, 4), (2, 4), (3, 4), (3, 5), (4, 5)\}$,
   $(A \cup C) \times (B \cup D) = \{(1, 1), (2, 1), (3, 1), (4, 1), (1, 4), (2, 4), (3, 4), (4, 4), (1, 5), (2, 5), (3, 5), (4, 5)\}$.
6. The cases are not exhaustive.
8. True.
12. The theorem is incorrect. Counterexample: $A = \{1\}$, $B = C = D = \varnothing$. Notice that $A \not\subseteq C$. Where is the mistake in the proof that $A \subseteq C$?

Section 4.2

1. (a) Domain $= \{p \in P \mid p$ has a living child$\}$; Range $= \{p \in P \mid p$ has a living parent$\}$.
   (b) Domain $= \mathbb{R}$; Range $= \mathbb{R}^+$.
4. (a) $\{(1, 4), (1, 5), (1, 6), (2, 4), (3, 6)\}$.
   (b) $\{(4, 4), (5, 5), (5, 6), (6, 5), (6, 6)\}$.
7. $E \circ E \subseteq F$.
10. We prove the contrapositives of both directions.

    ($\rightarrow$) Suppose Ran(R) and Dom(S) are not disjoint. Then we can choose some $b \in \text{Ran}(R) \cap \text{Dom}(S)$. Since $b \in \text{Ran}(R)$, we can choose some $a \in A$ such that $(a, b) \in R$. Similarly, since $b \in \text{Dom}(S)$, we can choose some $c \in C$ such that $(b, c) \in S$. But then $(a, c) \in S \circ R$, so $S \circ R \neq \varnothing$.

    ($\leftarrow$) Suppose $S \circ R \neq \varnothing$. Then we can choose some $(a, c) \in S \circ R$. By definition of $S \circ R$, this means that we can choose some $b \in B$ such that $(a, b) \in R$ and $(b, c) \in S$. But then $b \in \text{Ran}(R)$ and $b \in \text{Dom}(S)$, so Ran(R) and Dom(S) are not disjoint.

Section 4.3

1.

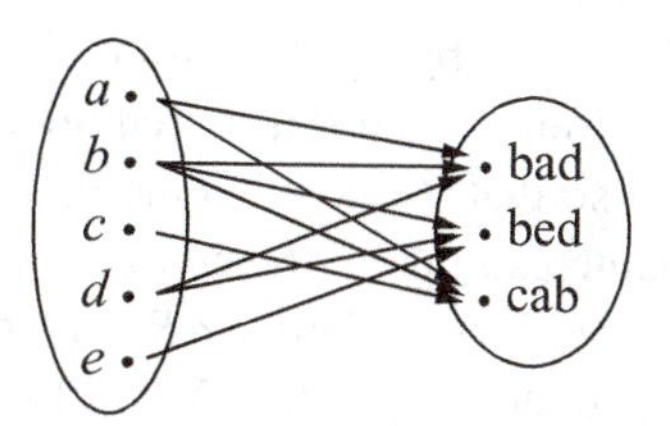

3.

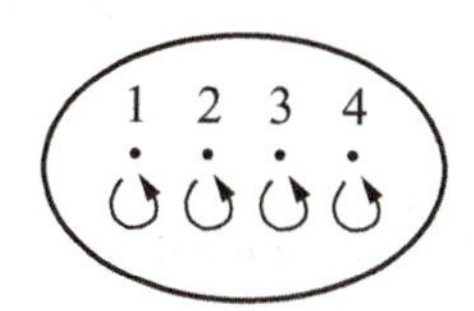

5. $S \circ R = \{(a, y), (a, z), (b, x), (c, y), (c, z)\}$.

7. $(\rightarrow)$ Suppose R is reflexive. Let (x, y) be arbitrary element of i_A. Then by the definition of i_A, $x = y \in A$. Since R is reflexive, $(x, y) = (x, x) \in R$. Since (x, y) was arbitrary, this shows that $i_A \subseteq R$.

   $(\leftarrow)$ Suppose $i_A \subseteq R$. Let $x \in A$ be arbitrary. Then $(x, x) \in i_A$, so since $i_A \subseteq R$, $(x, x) \in R$. Since x was arbitrary, this shows that R is reflexive.

10. Suppose $(x, y) \in i_D$. Then $x = y \in D = \text{Dom}(S)$, so there is some $z \in A$ such that $(x, z) \in S$. Therefore $(z, x) \in S^{-1}$, so $(x, y) = (x, x) \in S^{-1} \circ S$. Thus, $i_D \subseteq S^{-1} \circ S$. The proof of the other statement is similar.

13. (a) Yes. To prove it, suppose R_1 and R_2 are reflexive, and suppose $a \in A$. Since R_1 is reflexive, $(a, a) \in R_1$, so $(a, a) \in R_1 \cup R_2$.
    (b) Yes. To prove it, suppose R_1 and R_2 are symmetric, and suppose $(x, y) \in R_1 \cup R_2$. Then either $(x, y) \in R_1$ or $(x, y) \in R_2$. If $(x, y) \in R_1$ then since R_1 is symmetric, $(y, x) \in R_1$, so $(y, x) \in R_1 \cup R_2$. Similar reasoning shows that if $(x, y) \in R_2$ then $(y, x) \in R_1 \cup R_2$.
    (c) No. Counterexample: $A = \{1, 2, 3\}$, $R_1 = \{(1, 2)\}$, $R_2 = \{(2, 3)\}$.

17. First note that by part 2 of Theorem 4.3.4, since R and S are symmetric, $R = R^{-1}$ and $S = S^{-1}$. Therefore

$$
\begin{aligned}
R \circ S \text{ is symmetric} \quad &\text{iff } R \circ S = (R \circ S)^{-1} && \text{(Theorem 4.3.4, part 2)} \\
&\text{iff } R \circ S = S^{-1} \circ R^{-1} && \text{(Theorem 4.2.5, part 5)} \\
&\text{iff } R \circ S = S \circ R.
\end{aligned}
$$

20. Suppose R is transitive, and suppose $(X, Y) \in S$ and $(Y, Z) \in S$. To prove that $(X, Z) \in S$ we must show that $\forall x \in X \forall z \in Z(xRz)$, so let $x \in X$ and $z \in Z$ be arbitrary. Since $Y \in B$, $Y \neq \varnothing$, so we can choose $y \in Y$. Since $(X, Y) \in S$ and $(Y, Z) \in S$, by the definition of S we have xRy and yRz. But then since R is transitive, xRz, as required. The empty set had to be excluded from B so that we could come up with $y \in Y$ in this proof. (Can you find a counterexample if the empty set is not excluded?)
23. Hint: Suppose aRb and bRc. To prove aRc, suppose that $X \subseteq A \setminus \{a, c\}$ and $X \cup \{a\} \in \mathcal{F}$; you must prove that $X \cup \{c\} \in \mathcal{F}$. To do this, you may find it helpful to consider two cases: $b \notin X$ or $b \in X$. In the second of these cases, try working with the sets $X' = (X \cup \{a\}) \setminus \{b\}$ and $X'' = (X \cup \{c\}) \setminus \{b\}$.

Section 4.4

1. (a) Partial order, but not total order.
   (b) Not a partial order.
   (c) Partial order, but not total order.
4. ($\rightarrow$) Suppose that R is both antisymmetric and symmetric. Suppose that $(x, y) \in R$. Then since R is symmetric, $(y, x) \in R$, and since R is antisymmetric, it follows that $x = y$. Therefore $(x, y) \in i_A$. Since (x, y) was arbitrary, this shows that $R \subseteq i_A$.

   ($\leftarrow$) Suppose that $R \subseteq i_A$. Suppose $(x, y) \in R$. Then $(x, y) \in i_A$, so $x = y$, and therefore $(y, x) = (x, y) \in R$. This shows that R is symmetric. To see that R is antisymmetric, suppose that $(x, y) \in R$ and $(y, x) \in R$. Then $(x, y) \in i_A$, so $x = y$.
8. To see that T is reflexive, consider an arbitrary $(a, b) \in A \times B$. Since R and S are both reflexive, we have aRa and bSb. By the definition of T, it follows that $(a, b)T(a, b)$. To see that T is antisymmetric, suppose that $(a, b)T(a', b')$ and $(a', b')T(a, b)$. Then aRa' and $a'Ra$, so since R is antisymmetric, $a = a'$. Similarly, bSb' and $b'Sb$, so since S is antisymmetric, we also have $b = b'$. Thus $(a, b) = (a', b')$, as required. Finally, to see that T is transitive, suppose that $(a, b)T(a', b')$ and $(a', b')T(a'', b'')$. Then aRa' and $a'Ra''$, so since R is transitive, aRa''. Similarly, bSb' and $b'Sb''$, so bSb'', and therefore $(a, b)T(a'', b'')$.

   Even if both R and S are total orders, T need not be a total order.
11. The minimal elements of B are the prime numbers. B has no smallest element.

14. (a)

$$\begin{aligned} b \text{ is the } R\text{-largest element of } B \ \text{ iff } & b \in B \text{ and } \forall x \in B(xRb) \\ \text{iff } & b \in B \text{ and } \forall x \in B(bR^{-1}x) \\ \text{iff } & b \text{ is the } R^{-1}\text{-smallest element of } B. \end{aligned}$$

(b)

$$\begin{aligned} b \text{ is an } R\text{-maximal element of } B \ \text{ iff } & b \in B \text{ and } \neg\exists x \in B(bRx \wedge b \neq x) \\ \text{iff } & b \in B \text{ and } \neg\exists x \in B(xR^{-1}b \\ & \wedge x \neq b) \\ \text{iff } & b \text{ is an } R\text{-minimal element of B.} \end{aligned}$$

17. No. Let $A = \mathbb{R} \times \mathbb{R}$, and let $R = \{((x, y), (x', y')) \in A \times A \mid x \leq x'$ and $y \leq y'\}$. (You might want to compare this to exercise 8.) Let $B = \{(0, 0)\} \cup (\{1\} \times \mathbb{R})$. We will leave it to you to check that R is a partial order on A, and that $(0, 0)$ is the only minimal element of B, but it is not a smallest element.
21. (a) Suppose that $x \in U$ and xRy. To prove that $y \in U$, we must show that y is an upper bound for B, so suppose that $b \in B$. Since $x \in U$, x is an upper bound for B, so bRx. But we also have xRy, so by transitivity of R we can conclude that bRy. Since b was arbitrary, this shows that y is an upper bound for B.
    (b) Suppose $b \in B$. To prove that b is a lower bound for U, let x be an arbitrary element of U. Then by definition of U, x is an upper bound for B, so bRx. Since x was arbitrary, this shows that b is a lower bound for U.
    (c) Hint: Suppose x is the greatest lower bound of U. First use part (b) to show that x is an upper bound for B, and therefore $x \in U$. Then use the fact that x is a lower bound for U to show that x is the smallest element of U – in other words, it is the least upper bound of B.

Section 4.5

1. (a) Reflexive closure: $\{(a, a), (a, b), (b, c), (c, b), (b, b), (c, c)\}$.
    Symmetric closure: $\{(a, a), (a, b), (b, c), (c, b), (b, a)\}$.
    Transitive closure: $\{(a, a), (a, b), (b, c), (c, b), (a, c), (b, b), (c, c)\}$.
    (b) Reflexive closure: $\{(x, y) \in \mathbb{R} \times \mathbb{R} \mid x \leq y\}$.
    Symmetric closure: $\{(x, y) \in \mathbb{R} \times \mathbb{R} \mid x \neq y\}$.
    Transitive closure: R.

(c) Reflexive closure and symmetric closure are both D_r. Transitive closure is $\mathbb{R} \times \mathbb{R}$.

3. (a) Suppose that R is an asymmetric relation on A. Then the statement $\forall x \in A \forall y \in A((xRy \wedge yRx) \rightarrow x = y)$ is vacuously true, because $xRy \wedge yRx$ is always false.

   (b) Suppose that R is a strict partial order, and suppose that for some $x, y \in A$, $(x, y) \in R$ and $(y, x) \in R$. Then by transitivity of R, $(x, x) \in R$, which contradicts the fact that R is irreflexive. Therefore, R is asymmetric.

5. (a) Hint: Let $\mathcal{F} = \{T \subseteq A \times A \mid T \subseteq R$ and T is irreflexive $\}$. Then you must prove that $S \in \mathcal{F}$ and $\forall T \in \mathcal{F}(T \subseteq S)$. For the first of these, you must prove that $S \subseteq R$ and S is irreflexive. Both of these follow easily from the definition of S. For the second, let $T \in \mathcal{F}$ be arbitrary and prove $T \subseteq S$. Since $T \in \mathcal{F}$, you know that $T \subseteq R$ and T is irreflexive. Let (x, y) be an arbitrary element of T, and use these facts about T, together with the definition of S, to prove $(x, y) \in S$.

   (b) Suppose R is a partial order on A. We already showed in part (a) that S is irreflexive. To show that it is transitive, suppose $(x, y) \in S$ and $(y, z) \in S$. Then by the definition of S, $(x, y) \in R$ and $(y, z) \in R$, so since R is transitive, $(x, z) \in R$. If $x = z$ then we have $(x, y) \in R$ and $(y, x) \in R$, so by the antisymmetry of R, $x = y$. But then $(x, y) \in i_A$, which contradicts the fact that $(x, y) \in S = R \setminus i_A$. Therefore $x \neq z$, so $(x, z) \notin i_A$ and hence $(x, z) \in S$.

7. (a) Let S be the reflexive closure of R.

   ($\rightarrow$) Suppose R is reflexive. By clause 1 in the definition of reflexive closure (Definition 4.5.1), $R \subseteq S$, and by clause 3 (with $T = R$), $S \subseteq R$. Therefore $R = S$.

   ($\leftarrow$) Suppose $R = S$. By clause 2 in the definition of reflexive closure, R is reflexive.

   (b) Yes; the proofs are very similar.

9. Hint: Let $T = \{(x, y) \in S \mid x \in \mathrm{Dom}(R)$ and $y \in \mathrm{Ran}(R)\}$. Prove that $R \subseteq T$ and T is transitive.

12. (a) $S_1 \cup S_2 = (R_1 \cup i_A) \cup (R_2 \cup i_A) = (R_1 \cup R_2) \cup i_A = R \cup i_A = S$.

   (b) It is possible to give a proof that is similar to the proof in part (a), using formulas for S_1, S_2, and S. However, we will take a different approach. First, note that $R_1 \subseteq R$ and $R_2 \subseteq R$. It follows, by exercise 11, that $S_1 \subseteq S$ and $S_2 \subseteq S$, so $S_1 \cup S_2 \subseteq S$. For the other direction, note that $R = R_1 \cup R_2 \subseteq S_1 \cup S_2$, and by exercise 13(b) of Section 4.3, $S_1 \cup S_2$ is symmetric. Therefore, by definition of symmetric closure, $S \subseteq S_1 \cup S_2$.

(c) Imitating the first half of the proof in part (b), we can use exercise 11 to show that $S_1 \cup S_2 \subseteq S$. However, the answer to exercise 13(c) of Section 4.3 was no, so we can't imitate the second half of the proof. In fact, the example given in the solution to exercise 13(c) works as an example for which $S_1 \cup S_2 \neq S$.

15. Hint: Let $S = R \cup R^{-1} \cup i_A$.

18. (a) We have $R \subseteq Q$ and $Q \subseteq S$, so $R \subseteq S$. By definition of symmetric closure, Q is symmetric, and therefore, by exercise 17, S is symmetric. By definition of transitive closure, S is also transitive. Now suppose that $T \subseteq A \times A$, $R \subseteq T$, and T is both symmetric and transitive. Since Q is the *smallest* symmetric relation on A containing R, $Q \subseteq T$. But then since S is the *smallest* transitive relation on A containing Q, $S \subseteq T$.

(b) Since $R \subseteq Q$, Q' is the transitive closure of R, and S is the transitive closure of Q, by exercise 11, $Q' \subseteq S$. Since S is symmetric and S' is the *smallest* symmetric relation on A containing Q', $S' \subseteq S$.

(c) No. Counterexample: $A = \{1, 2, 3\}$, $R = \{(1, 2), (3, 2)\}$.

20. (a) One example is {(San Francisco, Chicago), (Chicago, Dallas), (Dallas, New York), (New York, Washington, D.C.), (Washington, D.C., San Francisco)}.

(b) No.

Section 4.6

1. Here is a list of all partitions:

$$\{\{1, 2, 3\}\}$$
$$\{\{1, 2\}, \{3\}\}$$
$$\{\{1, 3\}, \{2\}\}$$
$$\{\{2, 3\}, \{1\}\}$$
$$\{\{1\}, \{2\}, \{3\}\}$$

3. (a) R is an equivalence relation. There are 26 equivalence classes – one for each letter of the alphabet. The equivalence classes are: the set of all words that start with a, the set of all words that start with $b, \ldots,$ the set of all words that start with z.

(b) S is not an equivalence relation, because it is not transitive.

(c) T is an equivalence relation. The equivalence classes are: the set of all one-letter words, the set of all two-letter words, ..., the set of all n-letter words, where n is the length of the longest English word.

5. The assumption that is needed is that for every date d, someone was born on the date d. What would go wrong if, say, just by chance, no one was born on April 23? Where in the proof is this assumption used?
9. Since S is the equivalence relation determined by $\mathcal{F}$, the proof of Theorem 4.6.6 shows that $A/S = \mathcal{F} = A/R$. The desired conclusion now follows from exercise 8.
12. Suppose $a \equiv c \pmod{m}$ and $b \equiv d \pmod{m}$. Then $m \mid (a-c)$ and $m \mid (b-d)$. By exercise 18(a) of Section 3.3, it follows that $m \mid (a-c+b-d)$. But $a-c+b-d = (a+b)-(c+d)$, so $m \mid ((a+b)-(c+d))$, and therefore $a+b \equiv c+d \pmod{m}$.

   For the second half of the problem, you might find it useful to begin with the equation $ab - cd = (ab - ad) + (ad - cd)$.
15. By exercise 15(a) of Section 3.5, $\cup(\mathcal{F} \cup \mathcal{G}) = (\cup\mathcal{F}) \cup (\cup\mathcal{G}) = A \cup B$. To see that $\mathcal{F} \cup \mathcal{G}$ is pairwise disjoint, suppose that $X \in \mathcal{F} \cup \mathcal{G}$, $Y \in \mathcal{F} \cup \mathcal{G}$, and $X \cap Y \neq \varnothing$. If $X \in \mathcal{F}$ and $Y \in \mathcal{G}$ then $X \subseteq A$ and $Y \subseteq B$, and since A and B are disjoint it follows that X and Y are disjoint, which is a contradiction. Thus it cannot be the case that $X \in \mathcal{F}$ and $Y \in \mathcal{G}$, and a similar argument can be used to rule out the possibility that $X \in \mathcal{G}$ and $Y \in \mathcal{F}$. Thus, X and Y are either both elements of $\mathcal{F}$ or both elements of $\mathcal{G}$. If they are both in $\mathcal{F}$, then since $\mathcal{F}$ is pairwise disjoint, $X = Y$. A similar argument applies if they are both in $\mathcal{G}$. Finally, we have $\forall X \in \mathcal{F}(X \neq \varnothing)$ and $\forall X \in \mathcal{G}(X \neq \varnothing)$, and it follows by exercise 8 of Section 2.2 that $\forall X \in \mathcal{F} \cup \mathcal{G}(X \neq \varnothing)$.
19. (a) Here is the proof of transitivity: Suppose $(x, y) \in T$ and $(y, z) \in T$. Then since $T = R \cap S$, $(x, y) \in R$ and $(y, z) \in R$, so since R is transitive, $(x, z) \in R$. Similarly, $(x, z) \in S$, so $(x, z) \in R \cap S = T$.

    (b) Suppose $x \in A$. Then for all $y \in A$,

    $$y \in [x]_T \text{ iff } (y, x) \in T \text{ iff } (y, x) \in R \wedge (y, x) \in S$$
    $$\text{iff } y \in [x]_R \wedge y \in [x]_S \text{ iff } y \in [x]_R \cap [x]_S.$$

    (c) Suppose $X \in A/T$. Then since A/T is a partition, $X \neq \varnothing$. Also, for some $x \in A$, $X = [x]_T = [x]_R \cap [x]_S$, so since $[x]_R \in A/R$ and $[x]_S \in A/S$, $X \in (A/R) \cdot (A/S)$.

    Now suppose $X \in (A/R) \cdot (A/S)$. Then for some y and z in A, $X = [y]_R \cap [z]_S$. Also, $X \neq \varnothing$, so we can choose some $x \in X$. Therefore $x \in [y]_R$ and $x \in [z]_S$, and by part 2 of Lemma 4.6.5 it follows that $[x]_R = [y]_R$ and $[x]_S = [z]_S$. Therefore $X = [x]_R \cap [x]_S = [x]_T \in A/T$.
21. $\mathcal{F} \otimes \mathcal{F} = \{\mathbb{R}^+ \times \mathbb{R}^+, \mathbb{R}^- \times \mathbb{R}^+, \mathbb{R}^- \times \mathbb{R}^-, \mathbb{R}^+ \times \mathbb{R}^-, \mathbb{R}^+ \times \{0\}, \mathbb{R}^- \times \{0\}, \{0\} \times \mathbb{R}^+, \{0\} \times \mathbb{R}^-, \{(0, 0)\}\}$. In geometric terms these are the four

quadrants of the plane, the positive and negative x-axes, the positive and negative y-axes, and the origin.

23. (a) Hint: Let $T = \{(X, Y) \in A/S \times A/S \mid \exists x \in X \exists y \in Y(xRy)\}$.
    (b) Suppose $x, y, x', y' \in A$, xSx', and ySy'. Then $[x]_S = [x']_S$ and $[y]_S = [y']_S$, so xRy iff $[x]_S T [y]_S$ iff $[x']_S T [y']_S$ iff $x'Ry'$.

Chapter 5

Section 5.1

1. (a) Yes.
   (b) No.
   (c) Yes.
3. (a) $f(a) = b$, $f(b) = b$, $f(c) = a$.
   (b) $f(2) = 0$.
   (c) $f(\pi) = 3$ and $f(-\pi) = -4$.
5. $L \circ H : N \to N$, and for every $n \in N$, $(L \circ H)(n) = n$. Thus, $L \circ H = i_N$.
   $H \circ L : C \to C$, and for every $c \in C$, $(H \circ L)(c) =$ the capital of the country in which c is located.
7. (a) Suppose that $c \in C$. We must prove that there is a unique $b \in B$ such that $(c, b) \in f \restriction C$.
   Existence: Let $b = f(c) \in B$. Then $(c, b) \in f$ and $(c, b) \in C \times B$, and therefore $(c, b) \in f \cap (C \times B) = f \restriction C$.
   Uniqueness: Suppose that $(c, b_1) \in f \restriction C$ and $(c, b_2) \in f \restriction C$. Then $(c, b_1) \in f$ and $(c, b_2) \in f$, so since f is a function, $b_1 = b_2$.
   This proves that $f \restriction C$ is a function from C to B. Finally, to derive the formula for $(f \restriction C)(c)$, suppose that $c \in C$, and let $b = f(c)$. We showed in the *existence* half of the proof that $(c, b) \in f \restriction C$. It follows that
   $$f(c) = b = (f \restriction C)(c).$$
   (b) ($\rightarrow$) Suppose $g = f \restriction C$. Then $g = f \cap (C \times B)$, so clearly $g \subseteq f$.
   ($\leftarrow$) Suppose $g \subseteq f$. Suppose $c \in C$, and let $b = g(c)$. Then $(c, b) \in g$, so $(c, b) \in f$, and therefore $f(c) = b$. But then by part (a), $(f \restriction C)(c) = f(c) = b = g(c)$. Since c was arbitrary, it follows by Theorem 5.1.4 that $g = f \restriction C$.
   (c) $h \restriction \mathbb{Z} = h \cap (\mathbb{Z} \times \mathbb{R}) = \{(x, y) \in \mathbb{R} \times \mathbb{R} \mid y = 2x + 3\} \cap (\mathbb{Z} \times \mathbb{R}) = \{(x, y) \in \mathbb{Z} \times \mathbb{R} \mid y = 2x + 3\} = g$.

10. (a) Suppose $b \in B$. Since $\text{Dom}(S) = B$, we know that there is some $c \in C$ such that $(b, c) \in S$. To see that it is unique, suppose that $c' \in C$ and $(b, c') \in S$. Since $\text{Ran}(R) = B$, we can choose some $a \in A$ such that $(a, b) \in R$. But then $(a, c) \in S \circ R$ and $(a, c') \in S \circ R$, and since $S \circ R$ is a function, it follows that $c = c'$.
    (b) $A = \{1\}$, $B = \{2, 3\}$, $C = \{4\}$, $R = \{(1, 2), (1, 3)\}$, $S = \{(2, 4), (3, 4)\}$.
12. (a) No. Example: $A = \{1\}$, $B = \{2, 3\}$, $f = \{(1, 2)\}$, $R = \{(1, 1)\}$.
    (b) Yes. Suppose R is symmetric. Suppose $(x, y) \in S$. Then we can choose some u and v in A such that $f(u) = x$, $f(v) = y$, and $(u, v) \in R$. Since R is symmetric, $(v, u) \in R$, and therefore $(y, x) \in S$.
    (c) No. Example: $A = \{1, 2, 3, 4\}$, $B = \{5, 6, 7\}$, $f = \{(1, 5), (2, 6), (3, 6), (4, 7)\}$, $R = \{(1, 2), (3, 4)\}$.
16. (a) Let $a = 3$ and $c = 8$. Then for any $x > a = 3$,

$$|f(x)| = |7x + 3| = 7x + 3 < 7x + x = 8x < 8x^2 = c|g(x)|.$$

This shows that $f \in O(g)$.

Now suppose that $g \in O(f)$. Then we can choose $a \in \mathbb{Z}^+$ and $c \in \mathbb{R}^+$ such that $\forall x > a(|g(x)| \leq c|f(x)|)$, or in other words, $\forall x > a$ $(x^2 \leq c(7x + 3))$. Let x be any positive integer larger than both a and $10c$. Multiplying both sides of the inequality $x > 10c$ by x, we can conclude that $x^2 > 10cx$. But since $x > a$, we also have $x^2 \leq c(7x + 3) \leq c(7x + 3x) = 10cx$, so we have reached a contradiction. Therefore $g \notin O(f)$.

    (b) Clearly for any function $f \in \mathcal{F}$ we have $\forall x \in \mathbb{Z}^+(|f(x)| \leq 1 \cdot |f(x)|)$, so $f \in O(f)$, and therefore $(f, f) \in S$. Thus, S is reflexive. To see that it is also transitive, suppose $(f, g) \in S$ and $(g, h) \in S$. Then there are positive integers a_1 and a_2 and positive real numbers c_1 and c_2 such that $\forall x > a_1(|f(x)| \leq c_1|g(x)|)$ and $\forall x > a_2(|g(x)| \leq c_2|h(x)|)$. Let a be the maximum of a_1 and a_2, and let $c = c_1c_2$. Then for all $x > a$,

$$|f(x)| \leq c_1|g(x)| \leq c_1c_2|h(x)| = c|h(x)|.$$

Thus, $(f, h) \in S$, so S is transitive. Finally, to see that S is not a partial order, we show that it is not antisymmetric. Let f and g be the functions from $\mathbb{Z}^+$ to $\mathbb{R}$ defined by the formulas $f(x) = x$ and $g(x) = 2x$. Then for all $x \in \mathbb{Z}^+$, $|f(x)| \leq |g(x)|$ and $|g(x)| \leq 2|f(x)|$, so $f \in O(g)$ and also $g \in O(f)$. Therefore $(f, g) \in S$ and $(g, f) \in S$, but $f \neq g$.

    (c) Since $f_1 \in O(g)$, we can choose $a_1 \in \mathbb{Z}^+$ and $c_1 \in \mathbb{R}^+$ such that $\forall x > a_1(|f_1(x)| \leq c_1|g(x)|)$. Similarly, since $f_2 \in O(g)$ we can choose $a_2 \in \mathbb{Z}^+$ and $c_2 \in \mathbb{R}^+$ such that $\forall x > a_2(|f_2(x)| \leq c_2|g(x)|)$. Let a be the maximum of a_1 and a_2, and let $c = |s|c_1 + |t|c_2 + 1$. (We have added

1 here just to make sure that c is positive, as required in the definition of O.) Then for all $x > a$,

$$\begin{aligned}|f(x)| &= |sf_1(x) + tf_2(x)| \leq |s||f_1(x)| + |t||f_2(x)| \\ &\leq |s|c_1|g(x)| + |t|c_2|g(x)| = (|s|c_1 + |t|c_2)|g(x)| \leq c|g(x)|.\end{aligned}$$

Therefore $f \in O(g)$.

18. (a) Hint: Let $h = \{(X, y) \in A/R \times B \mid \exists x \in X(f(x) = y)\}$.
    (b) Hint: Use the fact that if xRy then $[x]_R = [y]_R$.

Section 5.2

2. (a) f is not a function.
   (b) f is not a function. g is a function that is onto, but not one-to-one.
   (c) R is one-to-one and onto.
5. (a) Suppose that $x_1 \in A, x_2 \in A$, and $f(x_1) = f(x_2)$. Then we can perform the following algebraic steps:

$$\begin{aligned}\frac{x_1+1}{x_1-1} &= \frac{x_2+1}{x_2-1}, \\ (x_1+1)(x_2-1) &= (x_2+1)(x_1-1), \\ x_1x_2 - x_1 + x_2 - 1 &= x_1x_2 - x_2 + x_1 - 1, \\ 2x_2 - 2x_1 &= 0, \\ x_1 &= x_2.\end{aligned}$$

This shows that f is one-to-one.

To show that f is onto, suppose that $y \in A$. Let

$$x = \frac{y+1}{y-1}.$$

Notice that this is defined, since $y \neq 1$, and also clearly $x \neq 1$, so $x \in A$. Then

$$f(x) = \frac{x+1}{x-1} = \frac{\frac{y+1}{y-1}+1}{\frac{y+1}{y-1}-1} = \frac{\frac{2y}{y-1}}{\frac{2}{y-1}} = y.$$

   (b) For any $x \in A$,

$$(f \circ f)(x) = \frac{\frac{x+1}{x-1}+1}{\frac{x+1}{x-1}-1} = \frac{\frac{2x}{x-1}}{\frac{2}{x-1}} = x = i_A(x).$$

7. (a) $\{1, 2, 3, 4\}$.
   (b) f is onto, but not one-to-one.

10. (a) Suppose that f is one-to-one. Suppose that $c_1 \in C, c_2 \in C$, and $(f \restriction C)(c_1) = (f \restriction C)(c_2)$. By exercise 7(a) of Section 5.1, it follows that $f(c_1) = f(c_2)$, so since f is one-to-one, $c_1 = c_2$.
    (b) Suppose that $f \restriction C$ is onto. Suppose $b \in B$. Then since $f \restriction C$ is onto, we can choose some $c \in C$ such that $(f \restriction C)(c) = b$. But then $c \in A$, and by exercise 7(a) of Section 5.1, $f(c) = b$.
    (c) Let $A = B = \mathbb{R}$ and $C = \mathbb{R}^+$. For (a), use $f(x) = |x|$, and for (b), use $f(x) = x$.
14. (a) Suppose R is reflexive and f is onto. Let $x \in B$ be arbitrary. Since f is onto, we can choose some $u \in A$ such that $f(u) = x$. Since R is reflexive, $(u, u) \in R$. Therefore $(x, x) \in S$.
    (b) Suppose R is transitive and f is one-to-one. Suppose that $(x, y) \in S$ and $(y, z) \in S$. Since $(x, y) \in S$, we can choose some u and v in A such that $f(u) = x$, $f(v) = y$, and $(u, v) \in R$. Similarly, since $(y, z) \in S$ we can choose p and q in A such that $f(p) = y$, $f(q) = z$, and $(p, q) \in R$. Since $f(v) = y = f(p)$ and f is one-to-one, $v = p$. Therefore $(v, q) = (p, q) \in R$. Since we also have $(u, v) \in R$, by transitivity of R it follows that $(u, q) \in R$, so $(x, z) \in S$.
17. (a) Let $b \in B$ be arbitrary. Since f is onto, we can choose some $a \in A$ such that $f(a) = b$. Therefore $g(b) = (g \circ f)(a) = (h \circ f)(a) = h(b)$. Since b was arbitrary, this shows that $\forall b \in B(g(b) = h(b))$, so $g = h$.
    (b) Let c_1 and c_2 be two distinct elements of C. Suppose $b \in B$. Let g and h be functions from B to C such that $\forall x \in B(g(x) = c_1)$, $\forall x \in B \setminus \{b\}(h(x) = c_1)$, and $h(b) = c_2$. (Formally, $g = B \times \{c_1\}$ and $h = [(B \setminus \{b\}) \times \{c_1\}] \cup \{(b, c_2)\}$.) Then $g \neq h$, so by assumption $g \circ f \neq h \circ f$, and therefore we can choose some $a \in A$ such that $g(f(a)) \neq h(f(a))$. But by the way g and h were defined, the only $x \in B$ for which $g(x) \neq h(x)$ is $x = b$, so it follows that $f(a) = b$. Since b was arbitrary, this shows that f is onto.

Section 5.3

1. $R^{-1}(p)$ = the person sitting immediately to the right of p.
3. Let $g(x) = (3x - 5)/2$. Then for any $x \in \mathbb{R}$,

$$f(g(x)) = \frac{2(3x-5)/2+5}{3} = \frac{3x-5+5}{3} = \frac{3x}{3} = x$$

and

$$g(f(x)) = \frac{3(2x+5)/3-5}{2} = \frac{2x+5-5}{2} = \frac{2x}{2} = x.$$

Therefore $f \circ g = i_{\mathbb{R}}$ and $g \circ f = i_{\mathbb{R}}$, and by Theorems 5.3.4 and 5.3.5 it follows that f is one-to-one and onto and $f^{-1} = g$.

5. $f^{-1}(x) = 2 - \log x$.
9. Suppose that $f : A \to B, g : B \to A$, and $f \circ g = i_B$. Let b be an arbitrary element of B. Let $a = g(b) \in A$. Then $f(a) = f(g(b)) = (f \circ g)(b) = i_B(b) = b$. Since b was arbitrary, this shows that f is onto.
11. (a) Suppose that f is one-to-one and $f \circ g = i_B$. By part 2 of Theorem 5.3.3, f is also onto, so $f^{-1} : B \to A$ and $f^{-1} \circ f = i_A$. This gives us enough information to imitate the reasoning in the proof of Theorem 5.3.5:

$$g = i_A \circ g = (f^{-1} \circ f) \circ g = f^{-1} \circ (f \circ g) = f^{-1} \circ i_B = f^{-1}.$$

(b) Hint: Imitate the solution to part (a).

(c) Hint: Use parts (a) and (b), together with Theorem 5.3.3.

14. (a) Suppose $x \in A' = \text{Ran}(g)$. Then we can choose some $b \in B$ such that $g(b) = x$. Therefore $(g \circ f)(x) = g(f(g(b))) = g((f \circ g)(b)) = g(i_B(b)) = g(b) = x$.

(b) By the given information, $(f \restriction A') \circ g = i_B$, and by part (a), $g \circ (f \restriction A') = i_{A'}$. Therefore by Theorem 5.3.4, $f \restriction A'$ is a one-to-one, onto function from A' to B, and by Theorem 5.3.5, $g = (f \restriction A')^{-1}$.

16. Hint: Suppose $x \in \mathbb{R}$. To determine whether or not $x \in \text{Ran}(f)$, you must see if you can find a real number y such that $f(y) = x$. In other words, you must try to solve the equation $4y - y^2 = x$ for y in terms of x. Notice that this is similar to the method we used in part 1 of Example 5.3.6. However, in this case you will find that for some values of x there is no solution for y, and for some values of x there is more than one solution for y.

Chapter 6

Section 6.1

1. Base case: When $n = 0$, both sides of the equation are 0.

Induction step: Suppose that $n \in \mathbb{N}$ and $0 + 1 + 2 + \cdots + n = n(n+1)/2$. Then

$$\begin{aligned} 0 + 1 + 2 + \cdots + (n+1) &= (0 + 1 + 2 + \cdots + n) + (n+1) \\ &= \frac{n(n+1)}{2} + (n+1) \\ &= (n+1)\left(\frac{n}{2} + 1\right) = \frac{(n+1)(n+2)}{2}, \end{aligned}$$

as required.

3. Base case: When $n = 0$, both sides of the equation are 0.

Induction step: Suppose that $n \in \mathbb{N}$ and $0^3 + 1^3 + 2^3 + \cdots + n^3 = [n(n+1)/2]^2$. Then

$$\begin{aligned} 0^3 + 1^3 + 2^3 + \cdots + (n+1)^3 &= (0^3 + 1^3 + 2^3 + \cdots + n^3) + (n+1)^3 \\ &= \left[\frac{n(n+1)}{2}\right]^2 + (n+1)^3 \\ &= (n+1)^2\left[\frac{n^2}{4} + n + 1\right] \\ &= (n+1)^2 \cdot \frac{n^2+4n+4}{4} \\ &= \left[\frac{(n+1)(n+2)}{2}\right]^2. \end{aligned}$$

7. Hint: The formula is $(3^{n+1} - 1)/2$.

10. Base case: When $n = 0$, $9^n - 8n - 1 = 0 = 64 \cdot 0$, so $64 \mid (9^n - 8n - 1)$.

Induction step: Suppose that $n \in \mathbb{N}$ and $64 \mid (9^n - 8n - 1)$. Then there is some integer k such that $9^n - 8n - 1 = 64k$. Therefore

$$\begin{aligned} 9^{n+1} - 8(n+1) - 1 &= 9^{n+1} - 8n - 9 \\ &= 9^{n+1} - 72n - 9 + 64n \\ &= 9(9^n - 8n - 1) + 64n \\ &= 9(64k) + 64n \\ &= 64(9k + n), \end{aligned}$$

so $64 \mid (9^{n+1} - 8(n+1) - 1)$.

14. Base case: When $n = 10$, $2^n = 1024 > 1000 = n^3$.

Induction step: Suppose $n \geq 10$ and $2^n > n^3$. Then

$$\begin{aligned} 2^{n+1} &= 2 \cdot 2^n \\ &> 2n^3 && \text{(by inductive hypothesis)} \\ &= n^3 + n^3 \\ &\geq n^3 + 10n^2 && \text{(since } n \geq 10) \\ &= n^3 + 3n^2 + 7n^2 \\ &\geq n^3 + 3n^2 + 70n && \text{(since } n \geq 10) \\ &= n^3 + 3n^2 + 3n + 67n \\ &> n^3 + 3n^2 + 3n + 1 = (n+1)^3. \end{aligned}$$

19. (a) Base case: When $n = 1$, the statement to be proven is $0 < a < b$, which was given.

Induction step: Suppose that $n \geq 1$ and $0 < a^n < b^n$. Multiplying this inequality by the positive number a we get $0 < a^{n+1} < ab^n$, and multiplying the inequality $a < b$ by the positive number b^n gives us $ab^n < b^{n+1}$. Combining these inequalities, we can conclude that $0 < a^{n+1} < b^{n+1}$.

(b) Hint: First note that $\sqrt[n]{a}$ and $\sqrt[n]{b}$ are both positive. (For n odd, this follows from exercise 18. For n even, each of a and b has two nth roots, one positive and one negative, but $\sqrt[n]{a}$ and $\sqrt[n]{b}$ are by definition the positive roots.) Now use proof by contradiction, and apply part (a).

(c) Hint: The inequality to be proven can be rearranged to read $a^{n+1} - ab^n - ba^n + b^{n+1} > 0$. Now factor the left side of this inequality.

(d) Hint: Use mathematical induction. For the base case, use the $n = 1$ case of part (c). For the induction step, multiply both sides of the inductive hypothesis by $(a + b)/2$ and then apply part (c).

Section 6.2

1. (a) We must prove that R' is reflexive, transitive, and antisymmetric. For the first, suppose $x \in A'$. Since R is reflexive, $(x, x) \in R$, so $(x, x) \in R \cap (A' \times A') = R'$. This shows that R' is reflexive.

Next, suppose that $(x, y) \in R'$ and $(y, z) \in R'$. Then $(x, y) \in R$, $(y, z) \in R$, and $x, y, z \in A'$. Since R is transitive, $(x, z) \in R$, so $(x, z) \in R \cap (A' \times A') = R'$. Therefore R' is transitive.

Finally, suppose that $(x, y) \in R'$ and $(y, x) \in R'$. Then $(x, y) \in R$ and $(y, x) \in R$, so since R is antisymmetric, $x = y$. Thus R' is antisymmetric.

(b) To see that T is reflexive, suppose $x \in A$. If $x = a$, then $(x, x) = (a, a) \in \{a\} \times A \subseteq T$. If $x \neq a$, then $x \in A'$, so since R' is reflexive, $(x, x) \in R' \subseteq T' \subseteq T$.

For transitivity, suppose that $(x, y) \in T$ and $(y, z) \in T$. If $x = a$ then $(x, z) = (a, z) \in \{a\} \times A \subseteq T$. Now suppose $x \neq a$. Then $(x, y) \notin \{a\} \times A$, so since $(x, y) \in T = T' \cup (\{a\} \times A)$ we must have $(x, y) \in T'$. But $T' \subseteq A' \times A'$, so $y \in A'$ and therefore $y \neq a$. Similar reasoning now shows that $(y, z) \in T'$. Since T' is transitive, it follows that $(x, z) \in T' \subseteq T$.

To show that T is antisymmetric, suppose $(x, y) \in T$ and $(y, x) \in T$. If $x = a$ then $(y, x) \notin T'$, so $(y, x) \in \{a\} \times A$ and therefore $y = a = x$. Similarly, if $y = a$ then $x = y$. Now suppose $x \neq a$ and $y \neq a$. Then as in the proof of transitivity it follows that $(x, y) \in T'$ and $(y, x) \in T'$, so by antisymmetry of T', $x = y$.

We now know that T is a partial order. To see that it is total, suppose $x \in A$ and $y \in A$. If $x = a$ then $(x, y) \in \{a\} \times A \subseteq T$. Similarly, if $y = a$ then $(y, x) \in T$. Now suppose $x \neq a$ and $y \neq a$. Then $x \in A'$ and $y \in A'$, so since T' is a total order, either $(x, y) \in T' \subseteq T$ or $(y, x) \in T' \subseteq T$.

Finally, to see that $R \subseteq T$, suppose that $(x, y) \in R$. If $x = a$ then $(x, y) \in \{a\} \times A \subseteq T$. Now suppose $x \neq a$. If $y = a$ then the fact that $(x, y) \in R$ would contradict the R-minimality of a. Therefore $y \neq a$. But then $(x, y) \in R \cap (A' \times A') = R' \subseteq T' \subseteq T$.

4. (a) We will prove the statement: For every $n \geq 1$, for every $B \subseteq A$, if B has n elements then there is some $x \in B$ such that $\forall y \in B((x, y) \in R \circ R)$. We proceed by induction on n.

Base case: Suppose $n = 1$. If $B \subseteq A$ and B has one element, then for some $x \in B$, $B = \{x\}$. Since R is reflexive, $(x, x) \in R$, and therefore $(x, x) \in R \circ R$. But x is the only element in B, so $\forall y \in B((x, y) \in R \circ R)$, as required.

Induction step: Suppose that $n \geq 1$ and for every $B \subseteq A$, if B has n elements then $\exists x \in B \forall y \in B((x, y) \in R \circ R)$. Now suppose that $B \subseteq A$ and B has $n + 1$ elements. Choose some $b \in B$, and let $B' = B \setminus \{b\}$. Then $B' \subseteq A$ and B' has n elements, so by inductive hypothesis there is some $x \in B'$ such that $\forall y \in B'((x, y) \in R \circ R)$. We now consider two cases.

Case 1: $(x, b) \in R \circ R$. Then $\forall y \in B((x, y) \in R \circ R)$, so we are done.

Case 2: $(x, b) \notin R \circ R$. In this case, we will prove that $\forall y \in B((b, y) \in R \circ R)$. To do this, let $y \in B$ be arbitrary. If $y = b$, then since R is reflexive, $(b, b) \in R$, and therefore $(b, y) = (b, b) \in R \circ R$. Now suppose $y \neq b$. Then $y \in B'$, so by the choice of x we know that $(x, y) \in R \circ R$. This means that for some $z \in A$, $(x, z) \in R$ and $(z, y) \in R$. We have $(x, z) \in R$, so if $(z, b) \in R$ then $(x, b) \in R \circ R$, contrary to the assumption for this case. Therefore $(z, b) \notin R$, so by the hypothesis on R, $(b, z) \in R$. But then since $(b, z) \in R$ and $(z, y) \in R$, we have $(b, y) \in R \circ R$, as required.

(b) Hint: Let $A = B =$ the set of contestants and let $R = \{(x, y) \in A \times A \mid x \text{ beats } y\} \cup i_A$. Now apply part (a).

8. (a) Let $m = (a + b)/2$, the arithmetic mean of a and b, and let $d = (a - b)/2$. Then it is easy to check that $m + d = a$ and $m - d = b$, so

$$\sqrt{ab} = \sqrt{(m + d)(m - d)} = \sqrt{m^2 - d^2} \leq \sqrt{m^2} = m = \frac{a + b}{2}.$$

(b) We use induction on n.

Base case: $n = 1$. This case is taken care of by part (a).

Induction step: Suppose $n \geq 1$, and the arithmetic-geometric mean inequality holds for lists of length 2^n. Now let $a_1, a_2, \ldots, a_{2^{n+1}}$ be a list of 2^{n+1} positive real numbers. Let

$$m_1 = \frac{a_1 + a_2 + \cdots + a_{2^n}}{2^n} \quad \text{and} \quad m_2 = \frac{a_{2^n+1} + a_{2^n+2} + \cdots + a_{2^{n+1}}}{2^n}.$$

Notice that $a_1 + a_2 + \cdots + a_{2^n} = m_1 2^n$, and similarly $a_{2^n+1} + a_{2^n+2} + \cdots + a_{2^{n+1}} = m_2 2^n$. Also, by inductive hypothesis, we know that $m_1 \geq \sqrt[2^n]{a_1 a_2 \cdots a_{2^n}}$ and $m_2 \geq \sqrt[2^n]{a_{2^n+1} a_{2^n+2} \cdots a_{2^{n+1}}}$. Therefore

$$\begin{aligned}
\frac{a_1 + a_2 + \cdots + a_{2^{n+1}}}{2^{n+1}} &= \frac{m_1 2^n + m_2 2^n}{2^{n+1}} = \frac{m_1 + m_2}{2} \geq \sqrt{m_1 m_2} \\
&\geq \sqrt{\sqrt[2^n]{a_1 a_2 \cdots a_{2^n}}\, \sqrt[2^n]{a_{2^n+1} a_{2^n+2} \cdots a_{2^{n+1}}}} \\
&= \sqrt[2^{n+1}]{a_1 a_2 \cdots a_{2^{n+1}}}.
\end{aligned}$$

(c) We use induction on n.

Base case: If $n = n_0$, then by assumption the arithmetic-geometric mean inequality fails for some list of length n.

Induction step: Suppose $n \geq n_0$, and there are positive real numbers $a_1, a_2, \ldots, a_n$ such that

$$\frac{a_1 + a_2 + \cdots + a_n}{n} < \sqrt[n]{a_1 a_2 \cdots a_n}.$$

Let $m = (a_1 + a_2 + \cdots + a_n)/n$, and let $a_{n+1} = m$. Then we have $m < \sqrt[n]{a_1 a_2 \cdots a_n}$, so $m^n < a_1 a_2 \cdots a_n$. Multiplying both sides of this inequality by m gives us $m^{n+1} < a_1 a_2 \cdots a_n m = a_1 a_2 \cdots a_{n+1}$, so $m < \sqrt[n+1]{a_1 a_2 \cdots a_{n+1}}$. But notice that we also have $mn = a_1 + a_2 + \cdots + a_n$, so

$$\frac{a_1 + \cdots + a_{n+1}}{n+1} = \frac{mn + m}{n+1} = \frac{m(n+1)}{n+1} = m < \sqrt[n+1]{a_1 a_2 \cdots a_{n+1}}.$$

Thus, we have a list of length $n + 1$ for which the arithmetic-geometric mean inequality fails.

(d) Suppose that the arithmetic-geometric mean inequality fails for some list of positive real numbers. Let n_0 be the length of this list, and choose an integer $n \geq 1$ such that $n_0 \leq 2^n$. (In fact, we could just let $n = n_0$, as you will show in exercise 12(a) in Section 6.3.) Then by part (b), the arithmetic-geometric mean inequality holds for all lists of length

2^n, but by part (c), it must fail for some list of length 2^n. This is a contradiction, so the inequality must always hold.

10. We proceed by induction on n.

Base case: $n = 0$. If A has 0 elements, then $A = \emptyset$, so $\mathscr{P}(A) = \{\emptyset\}$, which has $1 = 2^0$ elements.

Induction step: Suppose that for every set A with n elements, $\mathscr{P}(A)$ has 2^n elements. Now suppose that A has $n + 1$ elements. Let a be any element of A, and let $A' = A \setminus \{a\}$. Then A' has n elements, so $\mathscr{P}(A')$ has 2^n elements. There are two kinds of subsets of A: those that contain a as an element, and those that don't. The subsets that don't contain a are just the subsets of A', and by inductive hypothesis there are 2^n of these. Those that do contain a are the sets of the form $X \cup \{a\}$, where $X \in \mathscr{P}(A')$, and there are also 2^n of these, since by inductive hypothesis there are 2^n possible choices for X. Thus the total number of elements of $\mathscr{P}(A)$ is $2^n + 2^n = 2^{n+1}$.

13. Base case: $n = 1$. One chord cuts the circle into two regions, and $(n^2 + n + 2)/2 = 2$.

Induction step: Suppose that when n chords are drawn, the circle is cut into $(n^2 + n + 2)/2$ regions. When another chord is drawn, it will intersect each of the first n chords exactly once. Therefore it will pass through $n + 1$ regions, cutting each of those regions in two. (Each time it crosses one of the first n chords, it passes from one region to another.) Therefore the number of regions after the next chord is drawn is

$$\frac{n^2 + n + 2}{2} + (n + 1) = \frac{n^2 + 3n + 4}{2} = \frac{(n + 1)^2 + (n + 1) + 2}{2},$$

as required.

Section 6.3

1. Hint: The formula is

$$\sum_{i=1}^{n} \frac{1}{i(i + 1)} = \frac{n}{n + 1}.$$

6. Base case: $n = 1$. Then

$$\sum_{i=1}^{n} \frac{1}{i^2} = 1 \leq 1 = 2 - \frac{1}{n}.$$

Induction step: Suppose that

$$\sum_{i=1}^{n} \frac{1}{i^2} \leq 2 - \frac{1}{n}.$$

Then

$$\sum_{i=1}^{n+1}\frac{1}{i^2}=\sum_{i=1}^{n}\frac{1}{i^2}+\frac{1}{(n+1)^2}\le 2-\frac{1}{n}+\frac{1}{(n+1)^2}$$
$$=2-\frac{n^2+n+1}{n(n+1)^2}<2-\frac{n^2+n}{n(n+1)^2}=2-\frac{1}{n+1}.$$

8. (a) We let m be arbitrary and then prove by induction that for all $n \ge m$, $H_n - H_m \ge (n-m)/n$.
   Base case: $n = m$. Then $H_n - H_m = 0 \ge 0 = (n-m)/n$.
   Induction step: Suppose that $n \ge m$ and $H_n - H_m \ge (n-m)/n$.

Then

$$H_{n+1}-H_m = H_n+\frac{1}{n+1}-H_m \ge \frac{n-m}{n}+\frac{1}{n+1}$$
$$=\frac{n^2+n-mn-m+n}{n(n+1)}\ge\frac{n^2+n-mn}{n(n+1)}$$
$$=\frac{n+1-m}{n+1}.$$

   (b) Base case: If $n = 0$ then $H_{2^n} = H_1 = 1 \ge 1 = 1 + n/2$.
   Induction step: Suppose $n \ge 0$ and $H_{2^n} \ge 1 + n/2$. By part (a),

$$H_{2^{n+1}} - H_{2^n} \ge \frac{2^{n+1}-2^n}{2^{n+1}} = \frac{1}{2}.$$

   Therefore

$$H_{2^{n+1}} \ge H_{2^n} + \frac{1}{2} \ge 1 + \frac{n}{2} + \frac{1}{2} = 1 + \frac{n+1}{2}.$$

   (c) Since $\lim_{n\to\infty} 1 + n/2 = \infty$, by part (b) $\lim_{n\to\infty} H_{2^n} = \infty$. Clearly the H_n's form an increasing sequence, so $\lim_{n\to\infty} H_n = \infty$.

12. (a) Hint: Try proving that $2^n \ge n + 1$, from which the desired conclusion follows.
   (b) Base case: $n = 9$. Then $n! = 362880 \ge 262144 = (2^n)^2$.
   Induction step: Suppose that $n \ge 9$ and $n! \ge (2^n)^2$. Then

$$(n+1)! = (n+1)\cdot n! \ge (n+1)\cdot(2^n)^2 \ge 10\cdot 2^{2n} \ge 2^2\cdot 2^{2n}$$
$$= 2^{2n+2} = (2^{n+1})^2.$$

   (c) Base case: $n = 0$. Then $n! = 1 \le 1 = 2^{(n^2)}$.
   Induction step: Suppose that $n! \le 2^{(n^2)}$. Then

$$2^{((n+1)^2)} = 2^{n^2+2n+1} = 2^{(n^2)}\cdot 2^{2n+1} \ge 2^{(n^2)}\cdot 2^{n+1} > n!\cdot(n+1)$$
$$= (n+1)!$$

(Notice that the second to last step uses both the inductive hypothesis and part (a).)

15. Base case: $n = 0$. Then $a_n = a_0 = 0 = 2^0 - 0 - 1 = 2^n - n - 1$.

    Induction step: Suppose that $n \in \mathbb{N}$ and $a_n = 2^n - n - 1$. Then

$$a_{n+1} = 2a_n + n = 2(2^n - n - 1) + n$$
$$= 2^{n+1} - 2n - 2 + n = 2^{n+1} - n - 2 = 2^{n+1} - (n+1) - 1.$$

18. (a) $\binom{n}{0} = \frac{n!}{0! \cdot n!} = 1$ and $\binom{n}{n} = \frac{n!}{n! \cdot 0!} = 1$.

    (b)

$$\binom{n}{k} + \binom{n}{k-1} = \frac{n!}{k!(n-k)!} + \frac{n!}{(k-1)!(n-k+1)!}$$
$$= \frac{n!(n-k+1)}{k!(n-k+1)!} + \frac{n!k}{k!(n-k+1)!}$$
$$= \frac{n!(n+1)}{k!(n+1-k)!} = \binom{n+1}{k}.$$

(c) We follow the hint.

Base case: $n = 0$. Suppose A is a set with 0 elements. Then $A = \varnothing$, the only value of k we have to worry about is $k = 0$, $\mathscr{P}_0(A) = \{\varnothing\}$, which has 1 element, and $\binom{0}{0} = 1$.

Induction step: Suppose the desired conclusion holds for sets with n elements, and A is a set with $n + 1$ elements. Let a be an element of A, and let $A' = A \setminus \{a\}$, which is a set with n elements. Now suppose $0 \leq k \leq n + 1$. We consider three cases.

Case 1: $k = 0$. Then $\mathscr{P}_k(A) = \{\varnothing\}$, which has 1 element, and $\binom{n+1}{k} = 1$.

Case 2: $k = n + 1$. Then $\mathscr{P}_k(A) = \{A\}$, which has 1 element, and $\binom{n+1}{k} = 1$.

Case 3. $0 < k \leq n$. There are two kinds of k-element subsets of A: those that contain a as an element, and those that don't. The k-element subsets that don't contain a are just the k-element subsets of A', and by inductive hypothesis there are $\binom{n}{k}$ of these. Those that do contain a are the sets of the form $X \cup \{a\}$, where $X \in \mathscr{P}_{k-1}(A')$, and by inductive hypothesis there are $\binom{n}{k-1}$ of these, since this is the number of possibilities for X. Therefore by part (b), the total number of k-element subsets of A is

$$\binom{n}{k} + \binom{n}{k-1} = \binom{n+1}{k}.$$

(d) We let x and y be arbitrary and then prove the equation by induction on n.

Base case: $n = 0$. Then both sides of the equation are equal to 1.

Induction step: We will make use of parts (a) and (b). Suppose that

$$(x+y)^n = \sum_{k=0}^{n} \binom{n}{k} x^{n-k} y^k.$$

Then

$$\begin{aligned}
(x+y)^{n+1} &= (x+y)(x+y)^n \\
&= (x+y)\sum_{k=0}^{n}\binom{n}{k}x^{n-k}y^k \quad \text{(by inductive hypothesis)} \\
&= (x+y)\left[\binom{n}{0}x^n + \binom{n}{1}x^{n-1}y + \binom{n}{2}x^{n-2}y^2 + \cdots \right. \\
&\quad \left. + \binom{n}{n}y^n\right] \\
&= \binom{n}{0}x^{n+1} + \binom{n}{0}x^n y + \binom{n}{1}x^n y + \binom{n}{1}x^{n-1}y^2 + \cdots \\
&\quad + \binom{n}{n}xy^n + \binom{n}{n}y^{n+1} \\
&= x^{n+1} + \left[\binom{n}{0} + \binom{n}{1}\right]x^n y + \left[\binom{n}{1} + \binom{n}{2}\right]x^{n-1}y^2 \\
&\quad + \cdots + \left[\binom{n}{n-1} + \binom{n}{n}\right]xy^n + y^{n+1} \\
&= \binom{n+1}{0}x^{n+1} + \binom{n+1}{1}x^n y + \binom{n+1}{2}x^{n-1}y^2 \\
&\quad + \cdots + \binom{n+1}{n}xy^n + \binom{n+1}{n+1}y^{n+1} \\
&= \sum_{k=0}^{n+1}\binom{n+1}{k}x^{n+1-k}y^k.
\end{aligned}$$

20. Hint: Surprisingly, it is easier to prove that for all $n \geq 1$, $0 < a_n < 1/2$.

Section 6.4

1. (a) ($\rightarrow$) Suppose that $\forall n\, Q(n)$. Let n be arbitrary. Then $Q(n+1)$ is true, which means $\forall k < n+1(P(k))$. In particular, since $n < n+1$, $P(n)$ is true. Since n was arbitrary, this shows that $\forall n\, P(n)$.

($\leftarrow$) Suppose that $\forall n\, P(n)$. Then for any n, it is clearly true that $\forall k < n P(k)$, which means that $Q(n)$ is true.

(b) Base case: $n = 0$. Then $Q(n)$ is the statement $\forall k < 0 P(k)$, which is vacuously true.

Induction step: Suppose $Q(n)$ is true. This means that $\forall k < n P(k)$ is true, so by assumption, it follows that $P(n)$ is true. Therefore $\forall k < n+1(P(k))$ is true, which means that $Q(n+1)$ is true.

3. (a) Suppose $\sqrt{6}$ is rational. Let $S = \{q \in \mathbb{Z}^+ \mid \exists p \in \mathbb{Z}^+(p/q = \sqrt{6}\}$. Then $S \neq \varnothing$, so we can let q be the smallest element of S, and we can choose a positive integer p such that $p/q = \sqrt{6}$. Therefore $p^2 = 6q^2$, so p^2 is even, and hence p is even. This means that $p = 2\bar{p}$, for some integer $\bar{p}$. Thus $4\bar{p}^2 = 6q^2$, so $2\bar{p}^2 = 3q^2$ and therefore $3q^2$ is even. It is easy to check that if q is odd then $3q^2$ is odd, so q must be even, which means that $q = 2\bar{q}$ for some integer $\bar{q}$. But then $\sqrt{6} = \bar{p}/\bar{q}$ and $\bar{q} < q$, contradicting the fact that q is the smallest element of S.

(b) Suppose that $\sqrt{2} + \sqrt{3} = p/q$. Squaring both sides gives us $5 + 2\sqrt{6} = p^2/q^2$, so $\sqrt{6} = (p^2 - 5q^2)/(2q^2)$, which contradicts part (a).

6. (a) We use ordinary induction on n.

Base case: $n = 0$. Both sides of the equation are equal to 0.

Induction step: Suppose that $\sum_{i=0}^{n} F_i = F_{n+2} - 1$. Then

$$\sum_{i=0}^{n+1} F_i = \sum_{i=0}^{n} F_i + F_{n+1} = (F_{n+2} - 1) + F_{n+1} = F_{n+3} - 1.$$

(b) We use ordinary induction on n.

Base case: $n = 0$. Both sides of the equation are equal to 0.

Induction step. Suppose that $\sum_{i=0}^{n}(F_i)^2 = F_n F_{n+1}$. Then

$$\begin{aligned}\sum_{i=0}^{n+1}(F_i)^2 &= \sum_{i=0}^{n}(F_i)^2 + (F_{n+1})^2 = F_n F_{n+1} + (F_{n+1})^2 \\ &= F_{n+1}(F_n + F_{n+1}) = F_{n+1}F_{n+2}.\end{aligned}$$

(c) We use ordinary induction on n.

Base case: $n = 0$. Both sides of the equation are equal to 1.

Induction step: Suppose that $\sum_{i=0}^{n} F_{2i+1} = F_{2n+2}$. Then

$$\begin{aligned}\sum_{i=0}^{n+1} F_{2i+1} &= \sum_{i=0}^{n} F_{2i+1} + F_{2n+3} = F_{2n+2} + F_{2n+3} \\ &= F_{2n+4} = F_{2(n+1)+2}.\end{aligned}$$

(d) The formula is $\sum_{i=0}^{n} F_{2i} = F_{2n+1} - 1$.

8. (a) ($\rightarrow$) Suppose $a_0, a_1, a_2, \ldots$ is a Gibonacci sequence. Then in particular $a_2 = a_0 + a_1$, which means $c^2 = 1 + c$. Solving this quadratic equation by the quadratic formula leads to the conclusion $c = (1 \pm \sqrt{5})/2$.

($\leftarrow$) Suppose either $c = (1 + \sqrt{5})/2$ or $c = (1 - \sqrt{5})/2$. Then $c^2 = 1 + c$, and therefore for every $n \geq 2$, $a_n = c^n = c^{n-2}c^2 = c^{n-2}(1 + c) = c^{n-2} + c^{n-1} = a_{n-2} + a_{n-1}$.

(b) It will be convenient to introduce the notation $c_1 = (1 + \sqrt{5})/2$ and $c_2 = (1 - \sqrt{5})/2$. Then for any $n \geq 2$, $a_n = sc_1^n + tc_2^n = sc_1^{n-2}c_1^2 + tc_2^{n-2}c_2^2 = sc_1^{n-2}(1 + c_1) + tc_2^{n-2}(1 + c_2) = (sc_1^{n-2} + tc_2^{n-2}) + (sc_1^{n-1} + tc_2^{n-1}) = a_{n-2} + a_{n-1}$.

(c) Hint: Let $s = (5a_0 + (2a_1 - a_0)\sqrt{5})/10$ and $t = (5a_0 - (2a_1 - a_0)\sqrt{5})/10$.

10. Hint: The formula is $a_n = 2 \cdot 3^n - 3 \cdot 2^n$.

14. Let a be the larger of $5k$ and $k(k + 1)$. Now suppose $n > a$, and by the division algorithm choose q and r such that $n = kq + r$ and $0 \leq r < k$. Now note that if $q \leq 4$ then $n = kq + r \leq 4k + r < 5k \leq a$, which is a contradiction. Therefore $q > 4$, so $q \geq 5$, and by Example 6.1.3 it follows that $2^q \geq q^2$. Similar reasoning shows that $q \geq k + 1$, so $q^2 \geq q(k + 1) = kq + q > kq + r = n$. Therefore $2^n \geq 2^{kq} = (2^q)^k \geq (q^2)^k \geq n^k$.

18. We proceed by induction on j.

Base case: $j = 1$. Suppose that $p_1, q_1, q_2, \ldots, q_k$ are prime numbers and $p_1 = q_1 q_2 \cdots q_k$. Since p_1 is prime, we must have $k = 1$ and $p_1 = q_1$.

Induction step: Suppose that $j \geq 1$, and for all $k \geq 1$ and all nondecreasing sequences of primes $p_1, p_2, \ldots, p_j$ and $q_1, q_2, \ldots, q_k$, if $p_1 p_2 \cdots p_j = q_1 q_2 \cdots q_k$ then $j = k$ and $p_i = q_i$ for all i, $1 \leq i \leq j$. Now suppose $p_1, p_2, \ldots, p_{j+1}$ and $q_1, q_2, \ldots, q_k$ are nondecreasing sequences of primes and $p_1 p_2 \cdots p_{j+1} = q_1 q_2 \cdots q_k$. Then $k \geq 2$, since otherwise we have $p_1 p_2 \cdots p_{j+1} = q_1$, contradicting the fact that q_1 is prime. Also, $p_{j+1} \mid (q_1 q_2 \cdots q_k)$, so by exercise 17(b), $p_{j+1} \mid q_i$ for some i. But then $p_{j+1} = q_i \leq q_k$. Similar reasoning shows that $q_k \leq p_{j+1}$, so $p_{j+1} = q_k$. Therefore $p_1 p_2 \cdots p_j = q_1 q_2 \cdots q_{k-1}$, and by inductive hypothesis it follows that $k - 1 = j$ and $p_i = q_i$ for $1 \leq i \leq j$.

20. Hint: The formula is $a_n = F_{n+2}/F_{n+1}$.

Section 6.5

1. (a) To see that $\mathcal{F} \neq \varnothing$, notice that $B \subseteq A \subseteq A$ and A is closed under f, so $A \in \mathcal{F}$. It follows that $\cap\mathcal{F}$ is defined, and in fact by exercise 9 of Section 3.3, $\cap\mathcal{F} \subseteq A$. According to the definition of $\mathcal{F}$, for every $C \in \mathcal{F}$, $B \subseteq C$, so by exercise 10 of Section 3.3, $B \subseteq \cap\mathcal{F}$. Thus, we have $B \subseteq \cap\mathcal{F} \subseteq A$. To see that $\cap\mathcal{F}$ is closed under f, suppose that $x \in \cap\mathcal{F}$. Let $C \in \mathcal{F}$ be arbitrary. Then $x \in C$ and C is closed under f, so $f(x) \in C$. Since C was arbitrary, this shows that $\forall C \in \mathcal{F}(f(x) \in C)$, so $f(x) \in \cap\mathcal{F}$. Finally, to see that $\mathcal{F}$ is *smallest*, suppose that $B \subseteq C \subseteq A$ and C is closed under f. Then $C \in \mathcal{F}$, and therefore, applying exercise 9 of Section 3.3 again, $\cap\mathcal{F} \subseteq C$.

(b) Let $C = \cup_{n \in \mathbb{Z}^+} B_n$. Clearly $B = B_1 \subseteq C \subseteq A$. To see that C is closed under f, suppose $x \in C$. Then for some $n \in \mathbb{Z}^+, x \in B_n$, so $f(x) \in B_{n+1}$ and therefore $f(x) \in C$. Finally, to see that C is smallest, suppose that $B \subseteq D \subseteq A$ and D is closed under f. We prove by induction that $\forall n \in \mathbb{Z}^+, B_n \subseteq D$, from which it follows that $C \subseteq D$.

Base case: $n = 1$. Then $B_n = B \subseteq D$ by assumption.

Induction step: Suppose $n \in \mathbb{Z}^+$ and $B_n \subseteq D$. Now let $y \in B_{n+1}$ be arbitrary. Then by the definition of B_{n+1}, we can choose $x \in B_n$ such that $y = f(x)$. Since $B_n \subseteq D, x \in D$, and since D is closed under f, $y = f(x) \in D$. Thus $B_{n+1} \subseteq D$.

5. $\{n \in \mathbb{Z} \mid n \geq 2\}$.
8. (a) $R \cap S \subseteq R$ and $R \cap S \subseteq S$. Therefore by exercise 7, for every positive integer $n, (R \cap S)^n \subseteq R^n$ and $(R \cap S)^n \subseteq S^n$, so $(R \cap S)^n \subseteq R^n \cap S^n$. However, the two need not be equal. For example, if $A = \{1, 2, 3, 4\}$, $R = \{(1, 2), (2, 4)\}$, and $S = \{(1, 3), (3, 4)\}$, then $(R \cap S)^2 = \varnothing$ but $R^2 \cap S^2 = \{(1, 4)\}$.

   (b) $R^n \cup S^n \subseteq (R \cup S)^n$, but they need not be equal. (You should be able to prove the first statement, and find a counterexample to justify the second.)
10. (a) We use induction.

   Base case: $n = 1$. Suppose $(a, b) \in R^1 = R$. Let $f = \{(0, a), (1, b)\}$. Then f is an R-path from a to b of length 1. For the other direction, suppose f is an R-path from a to b of length 1. By the definition of R-path, this means that $f(0) = a$, $f(1) = b$, and $(f(0), f(1)) \in R$. Therefore $(a, b) \in R = R^1$.

   Induction step: Suppose n is a positive integer and $R^n = \{(a, b) \in A \times A \mid$ there is an R-path from a to b of length $n\}$. Now suppose $(a, b) \in R^{n+1} = R^1 \circ R^n$. Then there is some c such that $(a, c) \in R^n$ and $(c, b) \in R$. By inductive hypothesis, there is an R-path f from a to c of length n. Then $f \cup \{(n + 1, b)\}$ is an R-path from a to b of length $n + 1$. For the other direction, suppose f is an R-path from a to b of length $n + 1$. Let $c = f(n)$. Then $f \setminus \{(n + 1, b)\}$ is an R-path from a to c of length n, so by inductive hypothesis $(a, c) \in R^n$. But also $(c, b) = (f(n), f(n + 1)) \in R$, so $(a, b) \in R^1 \circ R^n = R^{n+1}$.

   (b) This follows from part (a) and Theorem 6.5.2.
14. We use induction on n.

   Base case: $n = 1$. Then $x = 2! + 2 = 4$. The only value of i we have to worry about is $i = 0$, and for this value of i we have $i + 2 = 2$ and $x + i = 4$. Since $2 \mid 4$, we have $(i + 2) \mid (x + i)$, as required.

   Induction step: Suppose that n is a positive integer, and for every integer i, if $0 \leq i \leq n - 1$ then $(i + 2) \mid ((n + 1)! + 2 + i)$. Now let

$x = (n+2)! + 2$, and suppose that $0 \leq i \leq n$. If $i = n$ then we have

$$x + i = (n+2)! + 2 + i = (i+2)! + (i+2) = (i+2)((i+1)! + 1),$$

so $(i+2) \mid (x+i)$. Now suppose $0 \leq i \leq n-1$. By inductive hypothesis, we know that $(i+2) \mid ((n+1)! + 2 + i)$, so we can choose some integer k such that $(n+1)! + 2 + i = k(i+2)$, and therefore $(n+1)! = (k-1)(i+2)$. Therefore

$$\begin{aligned} x + i &= (n+2)! + 2 + i = (n+2)(n+1)! + (i+2) \\ &= (n+2)(k-1)(i+2) + (i+2) = (i+2)((n+2)(k-1) + 1), \end{aligned}$$

so $(i+2) \mid (x+i)$.

Chapter 7

Section 7.1

1. (a) Define $f : \mathbb{Z}^+ \to \mathbb{N}$ by the formula $f(n) = n - 1$. It is easy to check that f is one-to-one and onto.
   (b) Let $E = \{n \in \mathbb{Z} \mid n \text{ is even}\}$, and define $f : \mathbb{Z} \to E$ by the formula $f(n) = 2n$. It is easy to check that f is one-to-one and onto, so $\mathbb{Z} \sim E$. But we already know that $\mathbb{Z}^+ \sim \mathbb{Z}$, so by Theorem 7.1.3, $\mathbb{Z}^+ \sim E$, and therefore E is denumerable.
4. (a) No. Counterexample: Let $A = B = C = \mathbb{Z}^+$ and $D = \{1\}$.
   (b) No. Counterexample: Let $A = B = \mathbb{N}$, $C = \mathbb{Z}^-$, and $D = \varnothing$.
6. (a) We proceed by induction on n.

   Base case: $n = 0$. Suppose that $m \in \mathbb{N}$ and there is a one-to-one, onto function $f : I_n \to I_m$. Since $n = 0$, $I_n = \varnothing$. But then since f is onto, we must also have $I_m = \varnothing$, so $m = 0 = n$.

   Induction step: Suppose that $n \in \mathbb{N}$, and for all $m \in \mathbb{N}$, if $I_n \sim I_m$ then $n = m$. Now suppose that $m \in \mathbb{N}$ and $I_{n+1} \sim I_m$. Let $f : I_{n+1} \to I_m$ be a one-to-one, onto function. Let $k = f(n+1)$, and notice that $1 \leq k \leq m$, so m is positive. Using the fact that f is onto, choose some $j \leq n+1$ such that $f(j) = m$.

   We now define $g : I_n \to I_{m-1}$ as follows:

$$g(i) = \begin{cases} f(i) & \text{if } i \neq j, \\ k & \text{if } i = j. \end{cases}$$

   We leave it to the reader to verify that g is one-to-one and onto. By inductive hypothesis, it follows that $n = m - 1$, so $n + 1 = m$.

(b) Suppose A is finite. Then by definition of "finite," we know that there is at least one $n \in \mathbb{N}$ such that $I_n \sim A$. To see that it is unique, suppose that n and m are natural numbers, $I_n \sim A$, and $I_m \sim A$. Then by Theorem 7.1.3, $I_n \sim I_m$, so by part (a), $n = m$.

8. (a) We use induction on n.

Base case: $n = 0$. Suppose $A \subseteq I_n = \varnothing$. Then $A = \varnothing$, so $|A| = 0$.

Induction step: Suppose that $n \in \mathbb{N}$, and for all $A \subseteq I_n$, A is finite, $|A| \leq n$, and if $A \neq I_n$ then $|A| < n$. Now suppose that $A \subseteq I_{n+1}$. If $A = I_{n+1}$ then clearly $A \sim I_{n+1}$, so A is finite and $|A| = n + 1$. Now suppose that $A \neq I_{n+1}$. If $n + 1 \notin A$, then $A \subseteq I_n$, so by inductive hypothesis, A is finite and $|A| \leq n$. If $n + 1 \in A$, then there must be some $k \in I_n$ such that $k \notin A$. Let $A' = (A \cup \{k\}) \setminus \{n + 1\}$. Then by matching up k with $n + 1$ it is not hard to show that $A' \sim A$. Also, $A' \subseteq I_n$, so by inductive hypothesis, A' is finite and $|A'| \leq n$. Therefore by exercise 7, A is finite and $|A| \leq n$.

(b) Suppose A is finite and $B \subseteq A$. Let $n = |A|$, and let $f : A \to I_n$ be one-to-one and onto. Then $f(B) \subseteq I_n$, so by part (a), $f(B)$ is finite, $|f(B)| \leq n$, and if $B \neq A$ then $f(B) \neq I_n$, so $|f(B)| < n$. Since $B \sim f(B)$, the desired conclusion follows.

12. It will be helpful first to verify two facts about the function f. Both of the facts below can be checked by straightforward algebra:

(1) For all $j \in \mathbb{Z}^+$, $f(1, j + 1) - f(1, j) = j$.

(2) For all $i, j \in \mathbb{Z}^+$, $f(1, i + j - 1) \leq f(i, j) < f(1, i + j)$. It follows that $i + j$ is the smallest $k \in \mathbb{Z}^+$ such that $f(i, j) < f(1, k)$.

To see that f is one-to-one, suppose that $f(i_1, j_1) = f(i_2, j_2)$. Then by fact (2) above,

$$\begin{aligned} i_1 + j_1 &= \text{the smallest } k \in \mathbb{Z}^+ \text{such that } f(i_1, j_1) < f(1, k) \\ &= \text{the smallest } k \in \mathbb{Z}^+ \text{such that } f(i_2, j_2) < f(1, k) \\ &= i_2 + j_2. \end{aligned}$$

Using the definition of f, it follows that

$$\begin{aligned} i_1 &= f(i_1, j_1) - \frac{(i_1 + j_1 - 2)(i_1 + j_1 - 1)}{2} \\ &= f(i_2, j_2) - \frac{(i_2 + j_2 - 2)(i_2 + j_2 - 1)}{2} \\ &= i_2. \end{aligned}$$

But then since $i_1 = i_2$ and $i_1 + j_1 = i_2 + j_2$, we must also have $j_1 = j_2$, so $(i_1, j_1) = (i_2, j_2)$. This shows that f is one-to-one.

To see that f is onto, suppose $n \in \mathbb{Z}^+$. Let k be the smallest positive integer such that $f(1,k) > n$, and notice that $f(1,1) = 1 \leq n$, so $k \geq 2$. Since k is smallest, $f(1,k-1) \leq n$, and therefore by fact (1),

$$0 \leq n - f(1,k-1) < f(1,k) - f(1,k-1) = k - 1.$$

Adding 1 to all terms, we get

$$1 \leq n - f(1,k-1) + 1 < k.$$

Thus, if we let $i = n - f(1,k-1) + 1$ then $1 \leq i < k$. Let $j = k - i$, and notice that $i \in \mathbb{Z}^+$ and $j \in \mathbb{Z}^+$. With this choice for i and j we have

$$\begin{aligned} f(i,j) &= \frac{(i+j-2)(i+j-1)}{2} + i \\ &= \frac{(k-2)(k-1)}{2} + n - f(1,k-1) + 1 \\ &= \frac{(k-2)(k-1)}{2} + n - \left[\frac{(k-2)(k-1)}{2} + 1\right] + 1 = n. \end{aligned}$$

14. (a) If $B \setminus \{f(m) \mid m \in \mathbb{Z}^+, m < n\} = \varnothing$ then $B = \{f(m) \mid m \in \mathbb{Z}^+, m < n\}$, so by exercise 10, B is finite. But we assumed that B was infinite, so this is impossible.

(b) We use strong induction. Suppose that $\forall m < n, f(m) \geq m$. Now suppose that $f(n) < n$. Let $m = f(n)$. Then by inductive hypothesis, $f(m) \geq m$. Also, by the definition of $f(n)$, $m = f(n) \in B \setminus \{f(k) \mid k \in \mathbb{Z}^+, k < n\} \subseteq B \setminus \{f(k) \mid k \in \mathbb{Z}^+, k < m\}$. But since $f(m)$ is the *smallest* element of this last set, it follows that $f(m) \leq m$. Since we have $f(m) \geq m$ and $f(m) \leq m$, we can conclude that $f(m) = m$. But then $m \notin B \setminus \{f(k) \mid k \in \mathbb{Z}^+, k < n\}$, so we have a contradiction.

(c) Suppose that $i \in \mathbb{Z}^+, j \in \mathbb{Z}^+$, and $i \neq j$. Then either $i < j$ or $j < i$. Suppose first that $i < j$. Then according to the definition of $f(j)$, $f(j) \in B \setminus \{f(m) \mid m \in \mathbb{Z}^+, m < j\}$, and clearly $f(i) \in \{f(m) \mid m \in \mathbb{Z}^+, m < j\}$. It follows that $f(i) \neq f(j)$. A similar argument shows that if $j < i$ then $f(i) \neq f(j)$. This shows that f is one-to-one.

To see that f is onto, suppose that $n \in B$. By part (b), $f(n+1) \geq n + 1 > n$. But according to the definition of f, $f(n+1)$ is the smallest element of $B \setminus \{f(m) \mid m \in \mathbb{Z}^+, m < n+1\}$. It follows that $n \notin B \setminus \{f(m) \mid m \in \mathbb{Z}^+, m < n+1\}$. But $n \in B$, so it must be the case that also $n \in \{f(m) \mid m \in \mathbb{Z}^+, m < n+1\}$. In other words, for some positive integer $m < n+1$, $f(m) = n$.

17. Following the hint, we recursively define partial orders R_n, for $n \in \mathbb{N}$, so that $R = R_0 \subseteq R_1 \subseteq R_2 \subseteq \ldots$ and

$$\forall i \in I_n \forall j \in \mathbb{Z}^+((a_i, a_j) \in R_n \vee (a_j, a_i) \in R_n). \qquad (*)$$

Let $R_0 = R$. Given R_n, to define R_{n+1} we apply exercise 2 of Section 6.2, with $B = \{a_i \mid i \in I_n\}$. Finally, let $T = \cup_{n\in\mathbb{N}} R_n$. Clearly T is reflexive, because every R_n is. To see that T is transitive, suppose that $(a, b) \in T$ and $(b, c) \in T$. Then for some natural numbers m and n, $(a, b) \in R_m$ and $(b, c) \in R_n$. If $m \leq n$ then $R_m \subseteq R_n$, and therefore $(a, b) \in R_n$ and $(b, c) \in R_n$. Since R_n is transitive, it follows that $(a, c) \in R_n \subseteq T$. A similar argument shows that if $n < m$ then $(a, c) \in T$, so T is transitive. The proof that T is antisymmetric is similar. Finally, to see that T is a total order, suppose $x \in A$ and $y \in A$. Since we have numbered the elements of A, we know that for some positive integers m and n, $x = a_m$ and $y = a_n$. But then by (*) we know that either (a_m, a_n) or (a_n, a_m) is an element of R_n, and therefore also an element of T.

20. (a) We follow the hint.

Base case: $n = 0$. Suppose A and B are finite sets and $|B| = 0$. Then $B = \varnothing$, so $A \times B = \varnothing$ and $|A \times B| = 0 = |A| \cdot 0$.

Induction step: Let n be an arbitrary natural number, and suppose that for all finite sets A and B, if $|B| = n$ then $A \times B$ is finite and $|A \times B| = |A| \cdot n$. Now suppose A and B are finite sets and $|B| = n + 1$. Choose an element $b \in B$, and let $B' = B \setminus \{b\}$, a set with n elements. Then $A \times B = A \times (B' \cup \{b\}) = (A \times B') \cup (A \times \{b\})$, and since $b \notin B'$, $A \times B'$ and $A \times \{b\}$ are disjoint. By inductive hypothesis, $A \times B'$ is finite and $|A \times B'| = |A| \cdot n$. Also, it is not hard to see that $A \sim A \times \{b\}$–just match up each $x \in A$ with $(x, b) \in A \times \{b\}$–so $A \times \{b\}$ is finite and $|A \times \{b\}| = |A|$. By Theorem 7.1.7, it follows that $A \times B$ is finite and $|A \times B| = |A \times B'| + |A \times \{b\}| = |A| \cdot n + |A| = |A| \cdot (n + 1)$.

(b) To order a meal, you name an element of $A \times B$, where $A =$ {steak, chicken, pork chops, shrimp, spaghetti} and $B =$ {ice cream, cake, pie}. So the number of meals is $|A \times B| = |A| \cdot |B| = 5 \cdot 3 = 15$.

22. (a) Base case: $n = 0$. If $|A| = 0$ then $A = \varnothing$, so $F = \{\varnothing\}$, and $|F| = 1 = 0!$.

Induction step: Suppose n is a natural number, and the desired conclusion holds for n. Now let A be a set with $n + 1$ elements, and let $F = \{f \mid f$ is a one-to-one, onto function from I_{n+1} to $A\}$. Let $g : I_{n+1} \to A$ be a one-to-one, onto function. For each $i \in I_{n+1}$, let $A_i = A \setminus \{g(i)\}$, a set with n elements, and let $F_i = \{f \mid f$ is a one-to-one, onto function from I_n to $A_i\}$. By inductive hypothesis, F_i is finite and $|F_i| = n!$. Now let $F_i' = \{f \in F \mid f(n+1) = g(i)\}$. Define a function $h : F_i \to F_i'$ by the formula $h(f) = f \cup \{(n+1, g(i))\}$. It is not hard to check that h is one-to-one and onto, so F_i' is finite and $|F_i'| = |F_i| = n!$. Finally, notice that $F = \cup_{i \in I_{n+1}} F_i'$ and $\forall i \in I_{n+1} \forall j \in I_{n+1}(i \neq j \to F_i' \cap F_j' = \varnothing)$. It follows, by exercise 19, that F is finite and $|F| = \sum_{i=1}^{n+1} |F_i'| = (n+1) \cdot n! = (n+1)!$.

(b) Hint: Define $h : F \to L$ by the formula $h(f) = \{(a, b) \in A \times A \mid f^{-1}(a) \leq f^{-1}(b)\}$. (You should check that this set is a total order on A.) To see that h is one-to-one, suppose that $f \in F$, $g \in F$, and $f \neq g$. Let i be the smallest element of I_n for which $f(i) \neq g(i)$. Now show that $(f(i), g(i)) \in h(f)$ but $(f(i), g(i)) \notin h(g)$, so $h(f) \neq h(g)$. To see that h is onto, suppose R is a total order on A. Define $g : A \to I_n$ by the formula $g(a) = |\{x \in A \mid x R a\}|$. Show that $\forall a \in A \forall b \in A(aRb \leftrightarrow g(a) \leq g(b))$, and use this fact to show that $g^{-1} \in F$ and $h(g^{-1}) = R$.

(c) $5! = 120$.

25. Base case: $n = 1$. Then $I_n = \{1\}$, $P = \{\{1\}\}$, and $A_{\{1\}} = A_1$. Therefore $|\cup_{i \in I_n} A_i| = |A_1|$ and $\sum_{S \in P}(-1)^{|S|+1}|A_S| = (-1)^2|A_{\{1\}}| = |A_1|$.

Induction step: Suppose the Inclusion–Exclusion Principle holds for n sets, and suppose $A_1, A_2, \ldots, A_{n+1}$ are finite sets. Let $P_n = \mathscr{P}(I_n) \setminus \{\varnothing\}$ and $P_{n+1} = \mathscr{P}(I_{n+1}) \setminus \{\varnothing\}$. By exercise 24(a), exercise 22(a) of Section 3.4, and the inductive hypothesis,

$$\begin{aligned} |\cup_{i \in I_{n+1}} A_i| &= |(\cup_{i \in I_n} A_i) \cup A_{n+1}| \\ &= |\cup_{i \in I_n} A_i| + |A_{n+1}| - |(\cup_{i \in I_n} A_i) \cap A_{n+1}| \\ &= \sum_{S \in P_n} (-1)^{|S|+1}|A_S| + |A_{n+1}| - |\cup_{i \in I_n} (A_i \cap A_{n+1})|. \end{aligned}$$

Now notice that for every $S \in P_n$, $\cap_{i \in S}(A_i \cap A_{n+1}) = (\cap_{i \in S} A_i) \cap A_{n+1} = A_{S \cup \{n+1\}}$. Therefore, by another application of the inductive hypothesis, $|\cup_{i \in I_n} (A_i \cap A_{n+1})| = \sum_{S \in P_n}(-1)^{|S|+1}|A_{S \cup \{n+1\}}|$. Thus

$$\begin{aligned} |\cup_{i \in I_{n+1}} A_i| &= \sum_{S \in P_n} (-1)^{|S|+1}|A_S| + |A_{n+1}| - \sum_{S \in P_n} (-1)^{|S|+1}|A_{S \cup \{n+1\}}| \\ &= \sum_{S \in P_n} (-1)^{|S|+1}|A_S| + (-1)^2|A_{\{n+1\}}| \\ &\quad + \sum_{S \in P_n} (-1)^{|S \cup \{n+1\}|+1}|A_{S \cup \{n+1\}}|. \end{aligned}$$

Finally, notice that there are three kinds of elements of P_{n+1}: those that are elements of P_n, the set $\{n+1\}$, and sets of the form $S \cup \{n+1\}$, where $S \in P_n$. It follows that the last formula above is just $\sum_{S \in P_{n+1}} (-1)^{|S|+1} |A_S|$, as required.

Section 7.2

1. (a) By Theorem 7.1.6, $\mathbb{Q}$ is countable. If $\mathbb{R} \setminus \mathbb{Q}$ were countable then, by Theorem 7.2.1, $\mathbb{Q} \cup (\mathbb{R} \setminus \mathbb{Q}) = \mathbb{R}$ would be countable, contradicting Theorem 7.2.6. Thus, $\mathbb{R} \setminus \mathbb{Q}$ must be uncountable.
   (b) Let $A = \{\sqrt{2} + n \mid n \in \mathbb{Z}^+\}$. It is not hard to see that A and $\mathbb{Q}$ are disjoint, since $\sqrt{2}$ is irrational, and A is denumerable. Now apply Theorems 7.1.6 and 7.2.1 to conclude that $A \cup \mathbb{Q}$ is denumerable, and therefore $A \cup \mathbb{Q} \sim A$. Finally, observe that $\mathbb{R} = (\mathbb{R} \setminus (A \cup \mathbb{Q})) \cup (A \cup \mathbb{Q})$ and $\mathbb{R} \setminus \mathbb{Q} = (\mathbb{R} \setminus (A \cup \mathbb{Q})) \cup A$, and apply part 2 of Theorem 7.1.2.
4. Suppose that $A \sim \mathscr{P}(A)$. Then there is a one-to-one, onto function $f : A \to \mathscr{P}(A)$. Let $X = \{a \in A \mid a \notin f(a)\} \in \mathscr{P}(A)$. Since f is onto, there must be some $a \in A$ such that $f(a) = X$. But then according to the definition of X, $a \in X$ iff $a \notin f(a)$, so $X \neq f(a)$, which is a contradiction.
7. Hint: Define $f : \mathscr{P}(A) \times \mathscr{P}(B) \to \mathscr{P}(A \cup B)$ by the formula $f(X, Y) = X \cup Y$, and prove that f is one-to-one and onto.
9. Hint: First note that if $\mathcal{F} = \varnothing$ then g can be any function. If $\mathcal{F} \neq \varnothing$, then since $\mathcal{F}$ is countable, we can write its elements in a list: $\mathcal{F} = \{f_1, f_2, \ldots\}$. Now define $g : \mathbb{Z}^+ \to \mathbb{R}$ by the formula $g(n) = \max\{|f_1(n)|, |f_2(n)|, \ldots, |f_n(n)|\}$.

Section 7.3

1. (a) The function $i_A : A \to A$ is one-to-one.
   (b) Suppose $A \precsim B$ and $B \precsim C$. Then there are one-to-one functions $f : A \to B$ and $g : B \to C$. By part 1 of Theorem 5.2.5, $g \circ f : A \to C$ is one-to-one, so $A \precsim C$.
5. Let $g : A \to B$ and $h : C \to D$ be one-to-one functions.
   (a) Since $A \neq \varnothing$, we can choose some $a_0 \in A$. Notice that $g^{-1} : \mathrm{Ran}(g) \to A$. Now define $j : B \to A$ as follows:

$$j(b) = \begin{cases} g^{-1}(b) & \text{if } b \in \mathrm{Ran}\ g, \\ a_0 & \text{otherwise.} \end{cases}$$

   We let the reader verify that j is onto.

Now define $F : {}^{A}C \to {}^{B}D$ by the formula $F(f) = h \circ f \circ j$. To see that F is one-to-one, suppose that $f_1 \in {}^{A}C$, $f_2 \in {}^{A}C$, and $F(f_1) = F(f_2)$, which means $h \circ f_1 \circ j = h \circ f_2 \circ j$. Let $a \in A$ be arbitrary. Since j is onto, there is some $b \in B$ such that $j(b) = a$. Therefore $h(f_1(a)) = (h \circ f_1 \circ j)(b) = (h \circ f_2 \circ j)(b) = h(f_2(a))$, and since h is one-to-one, it follows that $f_1(a) = f_2(a)$. Since a was arbitrary, this shows that $f_1 = f_2$.

(b) Yes. (You should be able to justify this answer with a counterexample.)

8. (a) Let n be arbitrary, and then proceed by induction on m. The base case is $m = n + 1$, and it is taken care of by exercise 7. For the induction step, apply exercise 2(b).

(b) $\bigcup_{n \in \mathbb{Z}^+} A_n$ is an infinite set that is not equinumerous with A_n for any $n \in \mathbb{Z}^+$. In fact, for every positive integer n, $A_n \prec \bigcup_{n \in \mathbb{Z}^+} A_n$. Can you find even larger infinite sets?

10. (a) Note that $\mathcal{E} \subseteq \mathscr{P}(\mathbb{Z}^+ \times \mathbb{Z}^+)$. It follows, using exercise 5 of Section 7.1, that $\mathcal{E} \precsim \mathscr{P}(\mathbb{Z}^+ \times \mathbb{Z}^+) \sim \mathscr{P}(\mathbb{Z}^+)$.

(b) Suppose $f(X) = f(Y)$. Then $X \cup \{1\} \in f(X) = f(Y) = \{Y \cup \{1\}, (A \setminus Y) \cup \{2\}\}$, so either $X \cup \{1\} = Y \cup \{1\}$ or $X \cup \{1\} = (A \setminus Y) \cup \{2\}$. But clearly $2 \notin X \cup \{1\}$, so the second possibility can be ruled out. Therefore $X \cup \{1\} = Y \cup \{1\}$. Since neither X nor Y contains 1, it follows that $X = Y$.

(c) Clearly A is denumerable, and we showed at the end of Section 5.3 that $\mathcal{P} \sim \mathcal{E}$. It follows that $\mathscr{P}(\mathbb{Z}^+) \sim \mathscr{P}(A) \precsim \mathcal{P} \sim \mathcal{E}$. Combining this with part (a) and applying the Cantor–Schröder–Bernstein Theorem gives the desired conclusion.

14. (a) According to the definition of function, ${}^{\mathbb{R}}\mathbb{R} \subseteq \mathscr{P}(\mathbb{R} \times \mathbb{R})$, and therefore by exercise 12(b) and exercise 5 of Section 7.1, ${}^{\mathbb{R}}\mathbb{R} \precsim \mathscr{P}(\mathbb{R} \times \mathbb{R}) \sim \mathscr{P}(\mathbb{R})$.

Clearly $\{\text{yes}, \text{no}\} \precsim \mathbb{R}$, so by exercise 5(c) of Section 7.2 and exercise 5, $\mathscr{P}(\mathbb{R}) \sim {}^{\mathbb{R}}\{\text{yes}, \text{no}\} \precsim {}^{\mathbb{R}}\mathbb{R}$. Since we have both ${}^{\mathbb{R}}\mathbb{R} \precsim \mathscr{P}(\mathbb{R})$ and $\mathscr{P}(\mathbb{R}) \precsim {}^{\mathbb{R}}\mathbb{R}$, by the Cantor–Schröder–Bernstein theorem, ${}^{\mathbb{R}}\mathbb{R} \sim \mathscr{P}(\mathbb{R})$.

(b) By Theorems 7.1.6 and 7.3.3, exercise 21(a) of Section 7.1, and exercise 5(d) of Section 7.2, ${}^{\mathbb{Q}}\mathbb{R} \sim {}^{\mathbb{Z}^+}\mathscr{P}(\mathbb{Z}^+) \sim \mathscr{P}(\mathbb{Z}^+) \sim \mathbb{R}$.

(c) Define $F : \mathcal{C} \to {}^{\mathbb{Q}}\mathbb{R}$ by the formula $F(f) = f \restriction \mathbb{Q}$. (See exercise 7 of Section 5.1 for the meaning of the notation used here.) Suppose $f \in \mathcal{C}$, $g \in \mathcal{C}$, and $F(f) = F(g)$. Then $f \restriction \mathbb{Q} = g \restriction \mathbb{Q}$, which means that for all $x \in \mathbb{Q}$, $f(x) = g(x)$. Now let x be an arbitrary real number. Use Lemma 7.3.4 to construct a sequence $x_1, x_2, \ldots$ of rational numbers such that $\lim_{n\to\infty} x_n = x$. Then since f and g are continuous,

$f(x) = \lim_{n\to\infty} f(x_n) = \lim_{n\to\infty} g(x_n) = g(x)$. Since x was arbitrary, this shows that $f = g$. Therefore F is one-to-one, so $\mathcal{C} \precsim {}^{\mathbb{Q}}\mathbb{R}$. Combining this with part (b), we can conclude that $\mathcal{C} \precsim \mathbb{R}$.

Now define $G : \mathbb{R} \to \mathcal{C}$ by the formula $G(x) = \mathbb{R} \times \{x\}$. In other words, $G(x)$ is the constant function whose value at every real number is x. Clearly G is one-to-one, so $\mathbb{R} \precsim \mathcal{C}$. By the Cantor–Schröder–Bernstein Theorem, it follows that $\mathcal{C} \sim \mathbb{R}$.

Appendix 2

Proof Designer

Proof Designer is a Java applet that writes outlines of proofs in elementary set theory, under the guidance of the user. It should work on any computer with a sufficiently up-to-date version of Java, including both PC and Macintosh computers. To use Proof Designer, open your web browser and go to the Proof Designer website:

`http://www.cs.amherst.edu/~djv/pd/pd.html`

Complete instructions for using Proof Designer can be found at the website. Here we will provide an outline of these instructions.

When you open the Proof Designer home page in your browser you should see a button at the bottom of the page that says "Write a Proof." (If you don't, you may need to follow the instructions at the website for setting up your computer to use Proof Designer.) To start writing a proof, click on the Write a Proof button.

A dialog box will open, asking you to enter the hypotheses and conclusion of the theorem you want to prove. The hypotheses and conclusion are entered using ordinary set-theoretic and logical notation.[1] When you have entered the hypotheses and conclusion, click OK.

The dialog box will close, and a new window will open. The window will contain three menus, called "Edit," "Strategy," and "Infer." Below the menus will be the statement of the theorem you are proving, and then a place where

[1] There is one significant difference between the way set theory notation is used in Proof Designer and the way we have used it in this book. The difference concerns intersections of families of sets. In this book, we have used the notation $\cap\mathcal{F}$ only in contexts in which we could be sure that $\mathcal{F} \neq \varnothing$; the reason for this rule is discussed in exercise 14 of Section 2.3. Proof Designer enforces this rule by restricting the contexts in which the notation $\cap\mathcal{F}$ can be used. For further details about this, and a more complete explanation of how to type statements into Proof Designer, see the website.

the proof will go. As you give commands in Proof Designer, an outline of the proof will gradually take shape in this window. While you are working on the proof, there will usually be one or more *gaps* in the proof where additional steps need to be filled in. Each gap will be indented and enclosed in a box, and it will have a button labeled "?" in the upper left corner. The gap will say what needs to be filled in at that point in the proof, and then it will give a list of *givens* – statements that are known to be true at that point in the proof – and the *goal* of the gap. Usually the goal is a statement that needs to be proven, but occasionally the goal indicates that you need to assign a value to a variable, and you can also have gaps that have no goal at all. Initially, the entire proof consists of a gap whose givens are the hypotheses of the theorem, and whose goal is the conclusion of the theorem.

To add a step to the proof, you click on a given or goal to select it and then give a command from one of the menus at the top of the window. Sometimes you will need to select several items. To do this, select the first item, and then hold down the shift key on the keyboard and click on additional items to add them to the selection. You can also select an entire gap by clicking either on the sentence that introduces the gap or in the margin to the left of the gap.

As you give commands, steps will be added to the proof, and the givens and goal lists will be updated. Sometimes a step will be justified by a *subproof* – a sequence of steps that, together, justify an assertion. Each subproof is indented and enclosed in a box, and has a button labeled "∵" in the upper left corner. Subproofs can be nested inside each other, and a subproof may also contain a gap. You can select a step in the proof by clicking on it. If the step is justified by one or more subproofs, the subproofs get selected as well.

Some commands will add *variant* forms to givens or goals. A variant of a statement is another statement that is equivalent to the original statement. A variant of a given or goal is listed below the original and indented. You use a variant just like the original given or goal. In particular, you can select a variant by clicking on it.

You can change the order of the givens in a givens list by pointing to a given, pressing and holding down the mouse button to "grab" it, and then dragging it to a new location in the list. Any variants of the given get moved with it.

If the structure of the proof you are creating gets complicated, you can hide some of the details by clicking on a "∵" or "?" button in the upper left corner of a subproof or gap. When you click on the button, the details of the subproof or gap are hidden. Click again to show the details again.

Applets are not allowed to print. However, you may be able to print your proof by using one of the Export HTML commands. See the website for details on this.

Suggestions for Further Reading

Barwise, J. and Echemendy, J. *Language, Proof and Logic*. Stanford: CSLI Publications, 2002.

Burton, D. *Elementary Number Theory*, 5th edition. Boston: McGraw-Hill, 2002.

Eccles, P. *An Introduction to Mathematical Reasoning: Numbers, Sets and Functions*. Cambridge: Cambridge University Press, 1997.

Enderton, H. *A Mathematical Introduction to Logic*, 2nd edition. San Diego: Harcourt/Academic Press, 2001.

Enderton, H. *Elements of Set Theory*. San Diego: Academic Press, 1977.

Epp, S. *Discrete Mathematics with Applications*, 3rd edition. Belmont, Calif.: Brooks Cole, 2004.

Halmos, P. *Naive Set Theory*. New York: Springer-Verlag, 1974.

Hamilton, A. *Logic for Mathematicians*, revised edition. Cambridge: Cambridge University Press, 1988.

Hamilton, A. *Numbers, Sets and Axioms: The Apparatus of Mathematics*. Cambridge: Cambridge University Press, 1982.

Mendelson, E. *Introduction to Mathematical Logic*, 4th edition. London: Chapman & Hall, 1997.

Polya, G. *How to Solve It: A New Aspect of Mathematical Method*, 2nd edition. Princeton: Princeton University Press, 1957.

Rosen, K. *Discrete Mathematics and Its Applications*, 5th edition. Boston: McGraw-Hill, 2003.

Ross, K. and Wright, C. *Discrete Mathematics*, 5th edition. Upper Saddle River, N.J.: Prentice Hall, 2003.

Stark, H. *An Introduction to Number Theory*. Cambridge, Mass.: MIT Press, 1978.

van Dalen, D., Doets, H., and deSwart, H. *Sets: Naive, Axiomatic, and Applied*. Oxford: Pergamon Press, 1978.

Summary of Proof Techniques

Note: Paragraphs marked with the symbol PD explain how to use each technique in Proof Designer.

To prove a goal of the form:

1. $\neg P$:
   (a) Reexpress as a positive statement.
   PD: Select the goal, give the Reexpress command in the Strategy menu, and use the Reexpress Negative button in the Reexpress dialog box.
   (b) Use proof by contradiction; that is, assume that P is true and try to reach a contradiction.
   PD: Select the goal and give the Contradiction command in the Strategy menu. If you already know which given you are planning to contradict, you can select it too before giving the Contradiction command, and Proof Designer will indicate what you have to prove to achieve the desired contradiction.
2. $P \rightarrow Q$:
   (a) Assume P is true and prove Q.
   PD: Select the goal and give the Direct command in the Strategy menu.
   (b) Prove the contrapositive; that is, assume that Q is false and prove that P is false.
   PD: Select the goal and give the Contrapositive command in the Strategy menu.
3. $P \wedge Q$:
   Prove P and Q separately. In other words, treat this as two separate goals: P, and Q.
   PD: Select the goal and give the Conjunction command in the Infer menu. If you already have either P or Q as a given, you can select it too,

and Proof Designer will only ask you to prove the statement that you don't already know.

4. $P \vee Q$:
   (a) Assume P is false and prove Q, or assume Q is false and prove P.
   PD: Select the goal and give the Disjunction command in the Strategy menu. Proof Designer will ask which statement you are planning to prove.
   (b) Use proof by cases. In each case, either prove P or prove Q.
   PD: Use the Cases command in the Strategy menu to break your proof into cases. Your goal in each case will be $P \vee Q$. In each case, select this goal and give the Disjunction command in the Strategy menu, and Proof Designer will ask you which statement you plan to prove in that case. If you don't want to assume the negation of the other statement, remove the check mark from the "Assume negations of others" check box by clicking on it.
5. $P \leftrightarrow Q$:
   Prove $P \rightarrow Q$ and $Q \rightarrow P$, using the methods listed under part 2.
   PD: Select the goal and give the Biconditional command in the Strategy menu.
6. $\forall x P(x)$:
   Let x stand for an arbitrary object, and prove $P(x)$. (If the letter x already stands for something in the proof, you will have to use a different letter for the arbitrary object.)
   PD: Select the goal and give the Arbitrary Object command in the Strategy menu.
7. $\exists x P(x)$:
   Find a value of x that makes $P(x)$ true. Prove $P(x)$ for this value of x.
   PD: Select the goal and give the Existence command in the Strategy menu. Proof Designer will ask you what value you want to use for x. If you're not sure what to use for x, you can choose a variable to stand for this value, and fill in the choice of a value for that variable later.
8. $\exists! x P(x)$:
   (a) Prove $\exists x P(x)$ (existence) and $\forall y \forall z((P(y) \wedge P(z)) \rightarrow y = z)$ (uniqueness).
   PD: Select the goal and give the Existence & Uniqueness command in the Strategy menu.
   (b) Prove the equivalent statement $\exists x(P(x) \wedge \forall y(P(y) \rightarrow y = x))$.
   PD: Select the goal, give the Reexpress command in the Strategy menu, and click on the Apply Definition button in the Reexpress dialog box.

9. $\forall n \in \mathbb{N} P(n)$:
   (a) Mathematical Induction: Prove $P(0)$ (base case) and $\forall n \in \mathbb{N}(P(n) \to P(n+1))$ (induction step).
   (b) Strong Induction: Prove $\forall n \in \mathbb{N}[(\forall k < n P(k)) \to P(n)]$.

To use a given of the form:

1. $\neg P$:
   (a) Reexpress as a positive statement.
      PD: Select the given, give the Reexpress command in the Strategy menu, and use the Reexpress Negative button in the Reexpress dialog box.
   (b) In a proof by contradiction, you can reach a contradiction by proving P.
      PD: Select the given and give the Contradiction command in the Strategy menu.
2. $P \to Q$:
   (a) If you are also given P, or you can prove that P is true, then you can conclude that Q is true.
      PD: Select the givens P and $P \to Q$ and give the Modus Ponens command in the Infer menu, and Proof Designer will infer Q. (If you don't already have P as a given but you think you can prove it, you can use the Insert command in the Edit menu to insert a proof of P.)
   (b) Use the contrapositive: If you are given or can prove that Q is false, then you can conclude that P is false.
      PD: Select the givens $\neg Q$ and $P \to Q$ and give the Modus Tollens command in the Infer menu, and Proof Designer will infer $\neg P$.
3. $P \land Q$:
   Treat this as two givens: P, and Q.
      PD: Select the given and give the Split Up command in the Infer menu.
4. $P \lor Q$:
   (a) Use proof by cases. In case 1 assume that P is true, and in case 2 assume that Q is true.
      PD: Select the given and give the Cases command in the Strategy menu.
   (b) If you are also given that P is false, or you can prove that P is false, then you can conclude that Q is true. Similarly, if you know that Q is false then you can conclude that P is true.
      PD: Select the givens $\neg P$ (or $\neg Q$) and $P \lor Q$ and give the Disjunctive Syllogism command in the Infer menu.
5. $P \leftrightarrow Q$:
   Treat this as two givens: $P \to Q$, and $Q \to P$.
      PD: Select the given and give the Split Up command in the Infer menu.

6. $\forall x P(x)$:

   You can plug in any value, say a, for x, and conclude that $P(a)$ is true.

   ᴾᴰ: Select the given and give the Universal Instantiation command in the Infer menu. Proof Designer will ask you what you want to plug in for x. As with proofs of goals of the form $\exists x P(x)$, if you're not sure what to plug in for x, you can choose a variable to stand for the object to be plugged in, and fill in the choice of a value for that variable later.

7. $\exists x P(x)$:

   Introduce a new variable, say x_0, into the proof, to stand for a particular object for which $P(x_0)$ is true.

   ᴾᴰ: Select the given and give the Existential Instantiation command in the Infer menu.

8. $\exists! x P(x)$:

   Introduce a new variable, say x_0, into the proof, to stand for a particular object for which $P(x_0)$ is true. You may also assume that $\forall y(P(y) \to y = x_0)$.

   ᴾᴰ: Select the given and give the Existential Instantiation command in the Infer menu.

Techniques that can be used in any proof:

1. Proof by contradiction: Assume the goal is false and derive a contradiction.

   ᴾᴰ: Select the goal and give the Contradiction command in the Strategy menu. If you already know which given you are planning to contradict, you can select it too before giving the Contradiction command.

2. Proof by cases: Consider several cases that are *exhaustive*, that is, that include all the possibilities. Prove the goal in each case.

   ᴾᴰ: If you select a given of the form $P \vee Q$ and give the Cases command in the Strategy menu, then Proof Designer will break the proof into the cases determined by this given. If you select a goal and give the Cases command, then Proof Designer will ask you to type in some statement P that will be used to distinguish the cases. In case 1, Proof Designer will assume that P is true, and in case 2 it will assume that P is false.

Index